ANNALS *of* THE NEW YORK ACADEMY OF SCIENCES

DIRECTOR AND EXECUTIVE EDITOR
Douglas Braaten

ASSISTANT EDITOR
Joseph Abrajano

PROJECT MANAGER
Steven E. Bohall

PROJECT COORDINATOR
Ralph W. Brown

CREATIVE DIRECTOR
Ash Ayman Shairzay

The New York Academy of Sciences
7 World Trade Center
250 Greenwich Street, 40th Floor
New York, NY 10007-2157

annals@nyas.org
www.nyas.org/annals

D1523058

THE NEW YORK ACADEMY OF SCIENCES BOARD OF GOVERNORS
SEPTEMBER 2009 - SEPTEMBER 2010

CHAIR
John E. Sexton

VICE-CHAIR
Bruce McEwen

TREASURER
Jay Furman

PRESIDENT
Ellis Rubinstein [ex officio]

SECRETARY
Larry Smith [ex officio]

CHAIRMAN EMERITUS
Torsten N. Wiesel

HONORARY LIFE GOVERNORS
Karen E. Burke
Herbert J. Kayden
John F. Niblack

GOVERNORS
Seth F. Berkley
Len Blavatnik
Nancy Cantor
Robert Catell
Virgina W. Cornish
Kenneth L. Davis
Robin L. Davisson
Brian Ferguson
Brian Greene
William A. Haseltine
Steven Hochberg
Toni Hoover
Morton P. Hyman
Mehmood Khan
Abraham Lackman
Russell Read
Jeffrey D. Sachs
David J. Skorton
George E. Thibault
Iris Weinshall
Anthony Welters
Frank Wilczek
Deborah Wiley
Michael Zigman
Nancy Zimpher

INTERNATIONAL GOVERNORS
Manuel Camacho Solis
Gerald Chan
Rajendra K. Pachauri
Paul Stoffels

Published by Blackwell Publishing
On behalf of the New York Academy of Sciences

Boston, Massachusetts
2010

ANNALS *of* THE NEW YORK ACADEMY OF SCIENCES

VOLUME 1200

ISSUE

Phylogenetic Aspects of Neuropeptides

From Invertebrates to Humans

ISSUE EDITORS

Hubert Vaudry and Seiji Shioda

This volume presents manuscripts stemming from an international symposium that was held on October 2–3, 2009 in Yakushima, Japan.

TABLE OF CONTENTS

Become a Member Today of the New York Academy of Sciences

The New York Academy of Sciences is dedicated to identifying the next frontiers in science and catalyzing key breakthroughs. As has been the case for 200 years, many of the leading scientific minds of our time rely on the Academy for key meetings and publications that serve as the crucial forum for a global community dedicated to scientific innovation.

Select one FREE *Annals* volume and up to five volumes for only $40 each.

Network and exchange ideas with the leaders of academia and industry.

Broaden your knowledge across many disciplines.

Gain access to exclusive online content.

Join Online at **www.nyas.org**

Or by phone at **800.344.6902** (516.576.2270 if outside the U.S.).

TABLE OF CONTENTS, CONTINUED

The New York Academy of Sciences believes it has a responsibility to provide an open forum for discussion of scientific questions. The positions taken by the authors and issue editors of the *Annals of the New York Academy of Sciences* are their own and not necessarily those of the Academy unless specifically stated. The Academy has no intent to influence legislation by providing such forums.

Ann. N.Y. Acad. Sci. ISSN 0077-8923

ANNALS OF THE NEW YORK ACADEMY OF SCIENCES
Issue: *Phylogenetic Aspects of Neuropeptides*

Coevolution of neuropeptidergic signaling systems: from worm to man

Tom Janssen, Marleen Lindemans, Ellen Meelkop, Liesbet Temmerman, and Liliane Schoofs
Functional Genomics and Proteomics Unit, Department of Biology, KULeuven, Leuven, Belgium
Address for correspondence: Tom Janssen, Naamsestraat 59, 3000 Leuven, Belgium. Tom.Janssen@bio.kuleuven.be

Despite the general knowledge and repeated predictions of peptide G protein–coupled receptors following the elucidation of the *Caenorhabditis elegans* genome in 1998, only a few have been deorphanized so far. This was attributed to the apparent lack of coevolution between (neuro)peptides and their cognate receptors. To resolve this issue, we have used an *in silico* genomic data mining tool to identify the real putative peptide GPCRs in the *C. elegans* genome and then made a well-considered selection of orphan peptide GPCRs. To maximize our chances of a successful deorphanization, we adopted a combined reverse pharmacology approach. At this moment, we have successfully uncovered four *C. elegans* neuropeptide signaling systems that support the theory of receptor–ligand coevolution. All four systems are extremely well conserved within nematodes and show a high degree of similarity with their vertebrate and arthropod counterparts. Our data indicate that these four neuropeptide signaling systems have been well conserved during the course of evolution and that they were already well established prior to the divergence of protostomes and deuterostomes.

Keywords: coevolution; GPCR; neuropeptide; *C. elegans*

Introduction

Caenorhabditis elegans is a small, free-living bacteriovorous nematode that has become one of the most important animal model organisms in the postgenomic era. The apparent simplicity and uniformity of its nervous system belies a rich diversity of putative signaling molecules. Like other metazoan organisms, *C. elegans* uses small molecule neurotransmitters such as acetylcholine (ACh), γ-aminobutyric acid (GABA), and nitric oxide (NO); excitatory amino acids such as glutamate; and biogenic amines such as octopamine, tyramine, serotonin (5-HT), and dopamine, which are packaged into synaptic vesicles and are released by exocytosis.[1–6] In addition to these classical neurotransmitters, the *C. elegans* genome encodes a wide variety of bioactive peptides (~265), which can be subdivided into three major families according to their conserved motifs. The best studied (neuro)peptide group in nematodes is the FMRFamide-like peptide (*flp*) gene family, consisting of 70 different peptides encoded by 33 precursor genes.[7–10] A second family encloses the *ins* genes (40), which encode 40 different insulin-like peptides,[11,12] and finally, peptides without sequence resemblance with FMRFamide or insulin are derived from the so-called neuropeptide-like protein (*nlp*) genes. The *nlp* family comprises the largest group with 155 different peptides encoded by 46 precursor genes.[10,13]

Neuropeptides are a diverse family of signaling molecules, which play an important role in neurotransmission and neuromodulation in both vertebrates and invertebrates. Expression studies show that the NLP-type neuropeptides are found almost exclusively in neurons.[13] Thus, neuropeptides represent far and away the most abundant signaling molecules in the *C. elegans* nervous system.[14] Most of these neuropeptides are thought to function through the activation of G protein–coupled receptors.[14–16]

The *C. elegans* genome is predicted to encode approximately 1300 GPCR genes that encompass more than 6.5% of all predicted genes.[17–21] The majority (~1000) are putative chemoreceptors and on average 50–60 of these have been predicted

doi: 10.1111/j.1749-6632.2010.05506.x

to encode peptide GPCRs.[17,18,22–24] Many of the predicted peptide GPCRs share a significant degree of homology to some of the known insect and mammalian peptide GPCR families, including the somatostatin/galanin/allatostatin-like, the NPY/NPF-like, the CCK/gastrin-like, the tachykinin-like, the neurotensin- and TRH-like, and the vasopressin-like receptors.[17,23] In contrast, no clear mammalian or arthropod peptide orthologue could be identified for most of the 265 predicted *C. elegans* peptides, except for some of the FLPs.[7,10,13]

Despite all the *in silico* knowledge and intense efforts in many laboratories over the last decade, the ligands for most of the predicted peptide GPCRs therefore still remained elusive. Between 2003 and 2007, a few FLP receptors were deorphaned by testing synthetic peptides,[25–31] but NLP-type ligands had not found their cognate receptor by this. This was attributed to the apparent lack of coevolution between NLP neuropeptides and their cognate receptors, which has impeded the deorphanization progress in *C. elegans* for many years now. The concept of coevolution is mostly known in terms of the evolution of species in scenarios such as host–parasite and predator–prey interactions, symbiosis, and mutualism. This concept can be extrapolated, however, to proteins and their interacting partners and in this case more specifically to receptors and their ligands. The functions of receptors in biological systems are determined by the physical interactions they have with their ligands. Proteins (e.g., receptors) and their interaction partners (i.e., ligands) must coevolve so that any divergent changes in one partner's binding surface are complemented at the interface by their interaction partner. Otherwise, the interaction between the proteins is lost, along with its function. Based on sequence alignments alone, however, coevolution of receptor–ligand pairs in evolutionary distant clades is sometimes hard to identify, and this might explain the apparent lack of coevolution between NLP neuropeptides and their cognate receptors in *C. elegans*.

To resolve this, we first used an *in silico* genomic data mining tool to identify the real putative peptide GPCRs in the *C. elegans* genome and then made a well-considered orphan GPCR selection. To maximize our chances of a successful deorphanization, we adopted a combined reverse pharmacology approach. This strategy has now proven its worth as it has resulted in the recent discovery of four *C. elegans* NLP neuropeptide signaling systems that do support the theory receptor–ligand coevolution.

Results and discussion

The identification of potential peptide GPCRs in C. elegans

The estimated number of peptide GPCRs in *C. elegans* varies a lot between different publications, ranging from 31 to 120.[17,18,22,23] Because of this discrepancy, we opted to perform an in depth *in silico* genomic data mining using the available protein sequences of the already deorphanized GPCRs as a template. To identify *C. elegans* GPCRs as potential peptide GPCRs, we used a bioinformatics tool available on the web named Multiple Expectation Maximization for Motif Elicitation (MEME[32]). It is designed to discover motifs in a group of related DNA or protein sequences. A motif is defined as a sequence pattern that occurs repeatedly in a group of related protein or DNA sequences. MEME uses the statistical algorithm expectation–maximization to fit a statistical model to its input sequences and automatically chooses the best width, number of occurrences, and description for each motif. To find new sequence patterns typical to peptide GPCRs, we subjected the already deorphanized *C. elegans* peptide receptors ("the training set," Table 1) to MEME analysis (http://meme.sdsc.edu/meme/meme.html). MEME was run using default parameters, except that the distribution of the motifs was set to "any number of repetitions," the maximum width was set to "50" and the maximum number of motifs to find was set to "15." This resulted in the identification of 10 typical sequence motifs (*E*-value $<1.0e^{-20}$), which are listed in Table 2. These motifs were subsequently used in a MAST

Table 1. List of the deorphanized *C. elegans* peptide receptors used as the training set for MEME analysis

GPCR gene	Protein	Reference
C39E6.6	CE06941	26–27
C10C6.2	CE08056	25
C26F1.6	CE06880	28
T19F4.1a	CE29348	29
T19F4.1b	CE29349	29
Y59H11AL.1	CE31260	30

Table 2. Result of the MEME analysis

Motif no.	Length	Motif sequence	*E*-value
1	32	KYYGTMVMVLQFVVPQAVMAYCYGHIGQKMWK	9.7e−50
2	21	FCSAWTLVAISLDRWMAICRP	2.6e−33
3	21	AMSHSMWNPIIYFWFNEKFRQ	5.1e−28
4	21	NVFLCNLAFSDLCMCVFCIPI	8.8e−27
5	50	ASEIMMRNEPTTTENPAVQEMNHIYHLTPSMKMLCILFYSILCVCCVYGN	3.9e−26
6	15	EDQYCGHFCDENWWQ	8.0e−24
7	21	YYDDKDWAFGSMMCHLVPFLQ	1.1e−21
8	23	QNPGMCQGATQKQRVDRHERKKK	5.2e−20
9	29	QIAENIAPQGYPDINVWGYIRYIWWFAHG	1.3e−21
10	50	WCTPRKSRNAILVIIVSAFLYNFVRFFEYRFVVTESGALYEKWLRDPGKH	2.7e−20

Note: Protein motifs with an *E*-value below $1.0e^{-20}$ were considered significant.

analysis (Motif Alignment and Search Tool[32]) (http://meme.sdsc.edu/meme/mast.html) to find gene products with similar motifs in the entire *C. elegans* genome (WormBase WS187). This resulted in 91 potential peptide receptor genes, encoding 125 different receptors (Table 3), all belonging to the rhodopsin class (class A) of G protein–coupled receptors.

This list is pretty accurate as almost all of the previously predicted peptide GPCRs[17,23] are present. Recently, Fredriksson *et al.* predicted only 31 peptide GPCR genes in *C. elegans* and all of these (except one) are present in the MAST list.[18,22]

GPCR selection and reverse pharmacological deorphanization

From the MAST list, four putative peptide receptors were selected based on homology to known insect and mammalian receptors,[17,23] ontology, number of predicted transmembrane domains, Wormbase knowledge, etc. Three additional receptors (C13B9.4a,b,c) were selected purely based on homology to the newly deorphanized *Drosophila* PDF receptor and the mammalian calcitonin receptor.[33] These receptors were, however, not present in the MAST list. This is likely due to the fact that they belong to the secretin class of GPCRs (class B) and therefore contain protein motifs that could not be identified through a rhodopsin-type motif-based MEME/MAST search.

To identify the corresponding peptide ligands, we adopted a combined reverse pharmacology approach. Sequence-specific primers were designed and used to amplify, clone, and verify the open reading frame of each selected GPCR. We used a combination of three reverse pharmacological approaches, the tissue extract-based approach, the library-based approach, and the information-based approach[34] (Fig. 1). For the tissue extract-based approach, a peptide extract from approximately 30,000,000 whole body mix stage worms was made and then subjected to reversed-phase HPLC fractionation on a Delta-pack C_{18} (2× [25 × 100 mm]) column (Waters) (Fig. S1). See Fig. S2, for an overview of the peptide extraction and purification strategy.

In addition to their direct use in the cellular pharmacological assays, the resulting 80 peptide fractions were also studied in detail by mass spectrometric analysis (MALDI-TOF and ESI-Qq-TOF MS/MS Mass Spectrometry) to identify potential new peptides of interest. Based on bioinformatic predictions ("library-based approach") and the mass spectrometric analysis of the reversed-phase HPLC fractions ("information-based approach"), a collection of 156 peptides, belonging to the established FLP and NLP families of peptides, was selected and custom synthesized (Sigma-Genosys). The peptides were selected in a way that maximizes the number of different potential activating ligands in the combined library of synthetic and natural (endogenous) peptides (Fig. 1). This combined approach enabled us to test approximately 215 putative peptide ligands of *C. elegans* for activation of the orphan GPCRs. This was done in heterologues cellular assays (CHO and HEK293) using aequorin or

Table 3. List of potential *C. elegans* peptide GPCRs as predicted from the MAST analysis of the entire *C. elegans* genome

Sequence name	Description	*E*-value	Length
Y59H11AL.1b	CE38456 WBGene00022004	1.4e−200	434
Y59H11AL.1a	CE31260 WBGene00022004	1.8e−200	427
Y58G8A.4a	CE33345 WBGene00021983	2.7e−138	397
Y58G8A.4b	CE36962 WBGene00021983	5.9e−138	433
T19F4.1a	CE29348 WBGene00020586	6.4e−119	402
T19F4.1b	CE29349 WBGene00020586	4.5e−118	432
C39E6.6	CE06941 WBGene00003807	1.1e−76	457
C10C6.2	CE08056 WBGene00007006	1.7e−58	376
T05A1.1b	CE32925 WBGene00003808	3.1e−39	387
T05A1.1a	CE32924 WBGene00003808	8.4e−39	430
ZC412.1	CE35920 WBGene00013883	2.5e−37	450
C16D6.2	CE37317 WBGene00007635	2.9e−28	468
C53C7.1b	CE36989 WBGene00008278	8.8e−27	362
C53C7.1a	CE19767 WBGene00008278	9.4e−27	365
F41E7.3	CE31509 WBGene00009619	1.7e−24	402
C26F1.6	CE06880 WBGene00016149	1.9e−21	360
ZK455.3	CE03814 WBGene00013974	9.1e−21	444
C54A12.2	CE36809 WBGene00016909	2.4e−20	381
AC7.1a	CE38261 WBGene00006428	4.4e−20	395
AC7.1b	CE38262 WBGene00006428	4.4e−20	386
Y54E2A.1	CE31259 WBGene00013187	5.8e−19	613
C25G6.5	CE04086 WBGene00016110	1.2e−18	455
T22D1.12	CE17256 WBGene00020689	4.4e−18	391
Y39A3B.5b	CE37727 WBGene00021439	7.2e−18	552
Y39A3B.5a	CE21646 WBGene00021439	1.8e−17	582
T23B3.4	CE31369 WBGene00020712	3.1e−17	395
C50F7.1b	CE04239 WBGene00016842	2.4e−16	381
F21C10.12	CE32868 WBGene00017661	9.8e−16	349
F35G8.1	CE39498 WBGene00018067	2.6e−14	380
F53A9.5	CE40067 WBGene00018728	3.0e−14	305
K06C4.8	CE11816 WBGene00019444	3.6e−14	355
C49A9.7	CE16937 WBGene00016761	6.1e−14	391
R12C12.3	CE02848 WBGene00020023	1.3e−13	340
F01E11.5a	CE38313 WBGene00017157	2.3e−13	462
T27D1.3	CE14245 WBGene00012084	4.1e−13	356
K09G1.4b	CE35994 WBGene00001053	4.6e−13	840
M03F4.3c	CE39602 WBGene00006475	5.0e−13	485
M03F4.3a	CE31203 WBGene00006475	5.5e−13	564
M03F4.3b	CE12360 WBGene00006475	5.7e−13	592
K09G1.4a	CE35993 WBGene00001053	6.6e−13	705
T07D4.1	CE36450 WBGene00011578	9.2e−13	393
C52B11.3	CE39139 WBGene00016872	1.3e−12	517
Y54G2A.35	CE31653 WBGene00021897	1.7e−12	401
W05B5.2	CE38879 WBGene00012275	1.7e−12	362

Continued.

Table 3. *Continued*

Sequence name	Description	*E*-value	Length
K06C4.9	CE29506 WBGene00019445	1.6e−10	492
F15A8.5c	CE33617 WBGene00001052	2.2e−10	399
F15A8.5b	CE33616 WBGene00001052	2.3e−10	402
F16D3.7	CE19797 WBGene00008890	4.5e−10	516
F15A8.5a	CE33615 WBGene00001052	4.8e−10	460
F14D12.6a	CE35478 WBGene00006411	5.6e−10	406
F14D12.6b	CE35479 WBGene00006411	5.8e−10	408
F59C12.2b	CE38526 WBGene00004776	8.2e−10	635
C38C10.1	CE00104 WBGene00006576	8.9e−10	374
F57H12.4	CE17125 WBGene00019019	9.7e−10	454
T23C6.5	CE35783 WBGene00020727	1.2e−09	362
F59C12.2a	CE28481 WBGene00004776	1.3e−09	683
T07D10.2	CE13377 WBGene00011582	1.7e−09	379
F31B9.1	CE17727 WBGene00009278	1.7e−09	399
T02E9.1	CE13062 WBGene00011381	1.8e−09	376
Y40H4A.1a	CE27783 WBGene00001519	2.3e−09	585
T14E8.3b	CE39179 WBGene00020506	3.3e−09	607
T14E8.3c	CE39180 WBGene00020506	5.9e−09	590
C56G3.1a	CE04283 WBGene00016984	6.4e−09	490
T14E8.3a	CE39178 WBGene00020506	7.0e−09	605
Y40H4A.1b	CE35800 WBGene00001519	7.7e−09	611
C56G3.1b	CE30923 WBGene00016984	8.5e−09	510
C56A3.3	CE09031 WBGene00008342	9.5e−09	426
C06G4.5	CE38997 WBGene00015559	2.5e−08	436
K10B4.4	CE38395 WBGene00019616	4.5e−08	403
R106.2	CE39612 WBGene00020086	1.1e−07	373
C30F12.6	CE16886 WBGene00016265	1.2e−07	395
F14F4.1	CE17670 WBGene00008808	1.8e−07	398
F47D12.1c	CE24987 WBGene00001518	1.9e−07	667
Y116A8B.5	CE39640 WBGene00013782	2.2e−07	435
K10C8.2	CE06168 WBGene00010735	3.00E−07	425
K07E8.5	CE35992 WBGene00019496	3.5e−07	297
C09B7.1c	CE17391 WBGene00004780	4.4e−07	410
T14E8.3d	CE39181 WBGene00020506	4.7e−07	245
C09B7.1b	CE30332 WBGene00004780	5.4e−07	422
T02D1.6	CE30684 WBGene00011372	6.00E−07	433
C09B7.1a	CE28532 WBGene00004780	6.6e−07	435
F47D12.1b	CE30135 WBGene00001518	8.6e−07	614
F47D12.1a	CE30134 WBGene00001518	1.00E−06	627
C30B5.5	CE02524 WBGene00016246	1.2e−06	790
F02E8.2b	CE07017 WBGene00017176	1.3e−06	425
C02D4.2f	CE32299 WBGene00004777	1.3e−06	421
C02D4.2b	CE36089 WBGene00004777	1.5e−06	429
C02D4.2a	CE36088 WBGene00004777	1.5e−06	432
C02D4.2e	CE36092 WBGene00004777	2.1e−06	455

Continued.

Table 3. *Continued*

Sequence name	Description	*E*-value	Length
Y22D7AR.13	CE29370 WBGene00004779	4.8e−06	445
C43C3.2	CE01524 WBGene00008065	5.9e−06	498
F59D12.1	CE11490 WBGene00010329	8.3e−06	318
C48C5.1	CE04224 WBGene00016747	1.1e−05	378
F56B6.5b	CE39375 WBGene00006864	2.3e−05	420
F56B6.5a	CE31186 WBGene00006864	2.6e−05	426
T21B4.4b	CE40127 WBGene00011878	3.4e−05	360
T21B4.4a	CE40126 WBGene00011878	3.5e−05	362
F02E8.2a	CE33990 WBGene00017176	4.6e−05	445
F42C5.2	CE04555 WBGene00018344	6.1e−05	426
F59B2.13	CE37246 WBGene00010315	0.00019	515
C04E6.9	CE25755 WBGene00005094	0.0002	340
T02D1.4	CE40469 WBGene00011371	0.0002	446
F54D7.3	CE17102 WBGene00018798	0.00026	401
R13H7.2a	CE39027 WBGene00020069	0.00027	440
K12D9.4	CE34344 WBGene00005862	0.0003	367
C44C3.5	CE08711 WBGene00005865	0.00031	378
R13H7.2c	CE39029 WBGene00020069	0.00039	402
R13H7.2b	CE39028 WBGene00020069	0.00042	408
F47D12.1d	CE01946 WBGene00001518	0.00067	245
F53B7.2	CE05921 WBGene00009965	0.00084	572
T11F9.1	CE06413 WBGene00011709	0.00085	444
C02D4.2c	CE36423 WBGene00004777	0.0013	199
C02D4.2d	CE36424 WBGene00004777	0.0014	196
F16C3.1	CE09432 WBGene00008885	0.0015	399
Y41D4A.8	CE21846 WBGene00021510	0.0019	347
F57A8.4	CE05985 WBGene00010179	0.002	396
C15B12.5c	CE29668 WBGene00001517	0.0023	622
C15B12.5b	CE29667 WBGene00001517	0.0025	713
C44C3.11	CE34754 WBGene00005870	0.0037	365
C24A8.1	CE39225 WBGene00016037	0.0038	582
C15B12.5a	CE29666 WBGene00001517	0.004	682
H27D07.5	CE34574 WBGene00005869	0.0042	372
C48C5.3	CE04226 WBGene00016748	0.0057	326
Y62E10A.4	CE24541 WBGene00013375	0.0087	310
E04D5.2	CE03123 WBGene00008481	0.0091	404
F01E11.5b	CE35711 WBGene00017157	0.0092	342

Note: The peptide receptors deorphaned and published/patented are indicated in light gray (Lowery *et al.*, 2003, Pharmacia & Upjohn Company, United States. G protein–coupled receptor-like receptors and modulators thereof, Patent 6,632[25–31]) and the receptors selected for this study are indicated in dark gray. The amino acid sequence length of each receptor is also presented.

Fluo-4 as calcium reporters (Refs. 35–37 for details). Following deorphanization, each signaling system was studied further at a structural and functional level, the results of which are summarized later.

Cholecystokinin/sulfakinin-like signaling

Although gene Y39A3B.5 was predicted to encode four different GPCR isoforms, we found that the actual expressed gene products are two

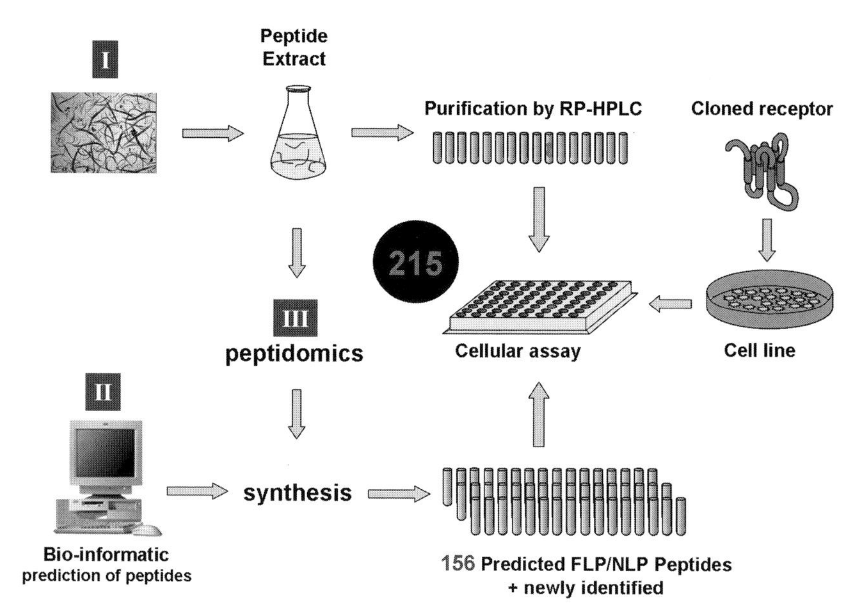

Figure 1. Overview of the composition of the *C. elegans* peptide library used for screening of the orphan GPCRs. (I) Tissue extract-based strategy: endogenous peptides were extracted from whole body mixed stage animals and subjected to reversed phase HPLC (RP-HPLC) for further fractionation. (II) Library-based strategy and (III) information-based strategy: based on bioinformatic predictions and the peptidomic analysis of the RP-HPLC fractions, a collection of 156 peptides (FLP and NLP) was selected and custom synthesized (Sigma-Genosys). Both the RP-HPLC fractions and the synthetic peptides, encompassing approximately 215 different *C. elegans* peptides in total, were tested for activation of each orphan receptor.

functional isoforms drawn up of a combination of the four predicted ones. Both receptors share more than 60% sequence similarity with the vertebrate cholecystokinin (CCK) receptors CCK1R (>63%) and CCK2R (>67%) and the insect sulfakinin receptors (e.g., Drosophila SK-R1 [>64%]) (Fig. 2A).[36] The Y39A3B.5 gene is also very well conserved in other nematodes, including the free-living species *C. briggsae* and *C. remanei* and the parasitic nematodes Brugia malayi (human filarial worm) and Ancylostoma ceylanicum (human hookworm).[36,38] No clear CCK or sulfakinin orthologues, however, could be found through *in silico* searches. By using the potential CCK receptors as bait, we were able to isolate and identify two NLP-type neuropeptides, known as NLP-12a and b as the ligands of both these nematode receptors. Also the two HPLC peptide fractions containing the endogenous NLP-12 peptides (no other) were able to activate both receptors indicating that these are indeed the natural ligands of these receptors (Supporting Fig. S1, nos. 36 and 37).[36] Both NLP-12 peptides share a high degree of sequence similarity to the vertebrate CCK/gastrin hormones and the arthropod sulfakinins (Fig. 2B).[36,38] In addition, we could demonstrate that both *C. elegans* NLP-12 peptides are recognized by an antibody specifically directed against human CCK-8, confirming the structural similarities with vertebrate CCK/gastrin.[36] The NLP-12 peptides are widely conserved within nematodes and almost identical in at least 11 species belonging to three different clades. Members of the CCK/gastrin peptide family, including the arthropod sulfakinins and their cognate receptors, play an important role in the regulation of feeding behavior and energy homeostasis. *Nlp-12* is only expressed in a single tail-neuron and a functional analysis of *C. elegans* lacking *nlp-12* or its receptors revealed that this neuropeptide signaling system is involved in the control of digestive enzyme secretion and fat storage, similar to their counterparts in arthropods and vertebrates. The *C. elegans* receptors both signal through a $G_{\alpha q}$ type of G protein, which also is in agreement with the Drosophila sulfakinin receptor and mammalian CCK receptors where CCKR

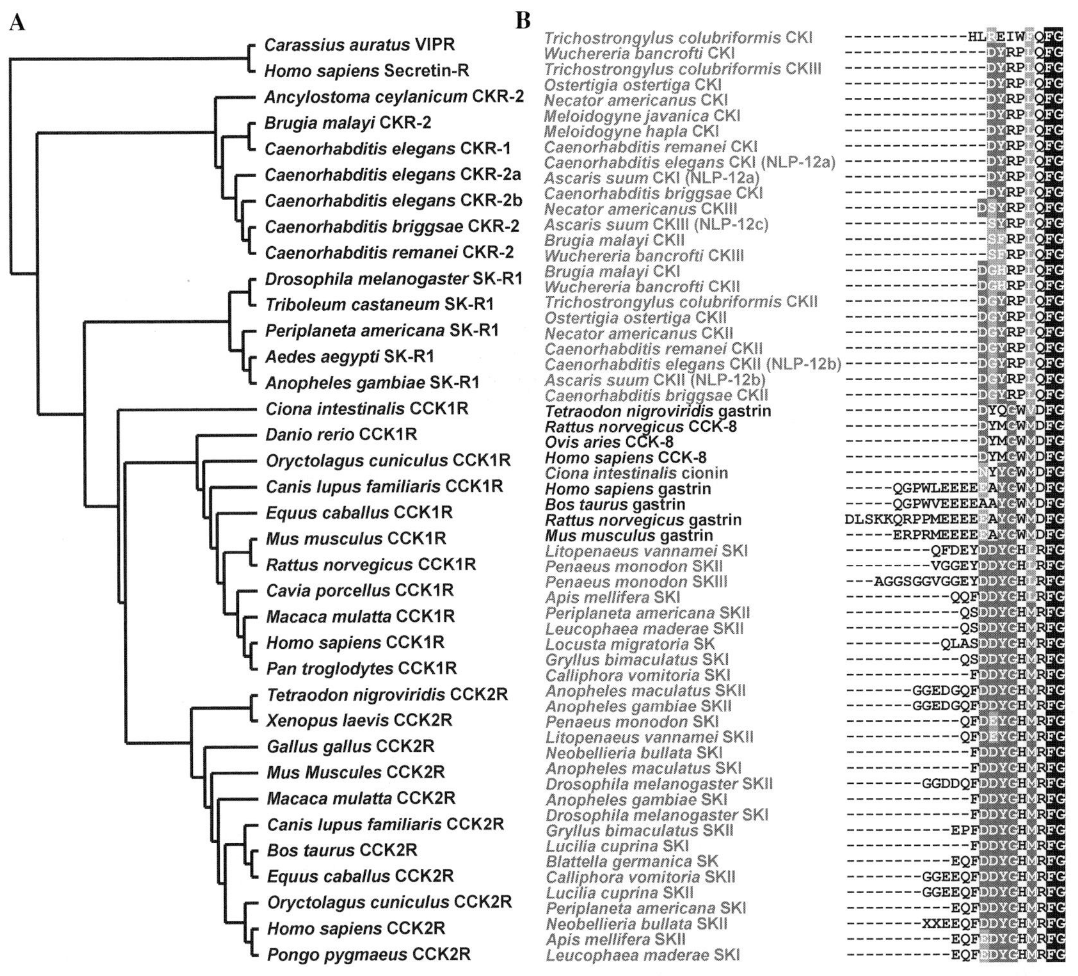

Figure 2. (**A**) Dendrogram showing the phylogenetic relationship between the nematode CCK receptor-like receptors and the most closely related invertebrate and vertebrate orthologues (AlignX software, Vector NTI Advance 10, Invitrogen), selected based on BLASTP and tBLASTn analysis. The *Carassius auratus* VIPR and the *Homo sapiens* secretin receptor were used as an outgroup. (**B**) Alignment and phylogenetic analysis of the nematode NLP-12 peptides with the known arthropod sulfakinins and some vertebrate cholecystokinin and gastrins (AlignX software, Vector NTI Advance 10, Invitrogen). The nematode CCK-like peptides are indicated in blue, the arthropod sulfakinins in dark green, the tunicate cionin in purple, and the vertebrate CCK/gastrins in black. Identical amino acids are highlighted in black, conserved amino acids in dark grey, and similar amino acids in light grey. The C-terminal glycine is amidated *in vivo.*

activates phospholipase C, triggering an increase in inositol trisphosphate production.[36,39,40] Together, our results indicate that this newly identified neuropeptide signaling system in *C. elegans* constitutes not only the structural, but also the functional homologue of the CCK/gastrin signaling system of vertebrates and the sulfakinin signaling system of arthropods. In terms of evolution, the *C. elegans* CCK-like peptides would represent the oldest members of the CCK/gastrin family of peptides known to date.

Vasoactive intestinal peptide/pigment dispersing factor-like signaling system

The three secretin class receptors were selected, based on a BLAST analysis using the amino acid sequence of the recently characterized *D. melanogaster* pigment dispersing factor (PDF) receptor[33,41,42] as a query. This revealed three potential *C. elegans* orthologues with 36/54%, 35/53%, and 36/54% sequence identity and % similarity, respectively. These GPCRs, hereafter named PDFR-1a, b, and c, represent differentially spliced isoforms of the same gene, called C13B9.4 (a, b, and c). Each cloned splice isoform was confirmed to be identical to the predicted cDNA sequence (Wormbase, WS187).[35] A phylogenetic analysis of the *C. elegans* PDF receptors show a clear clustering of nematode PDF receptor-like receptors (*C. elegans, C. remanei, C. briggsae, B. malayi*), the insect PDF receptors (*D. melanogaster, D. pseudoobscurae, A. melifera, A. gambiae*), and the vertebrate calcitonin receptors (*H. sapiens, C. porcellus, D. rerio, O. gorbuscha*) (Fig. 3A).[35] In general, the receptors are highly conserved in nematodes and very closely related to the PDF receptors from insects. They are also related to the vertebrate calcitonin GPCRs and, to a lesser extent, to the vertebrate vasoactive intestinal peptide (VIP) receptors (*H. sapiens, M. musculus*) (Fig. 3A).

In this case, a combinatorial approach of biochemistry and peptidomics has lead us to the biochemical isolation, identification, and characterization of three peptides, which we called PDF-1a, b, and PDF-2, and has enabled us to couple the natural *C. elegans* PDF peptides to the three PDF receptors.[35,43,44] Two HPLC-separated peptide fractions (Supporting Fig. S1, nos. 53 and 55) containing all three endogenous PDF-like peptides were able to activate all three receptors. All the *C. elegans* PDF peptides display a moderate-to-strong sequence homology to the insect PDF neuropeptides and the crustacean pigment-dispersing hormones (PDH) (Fig. 3B).[43,44] tBLASTn analysis of the peptide precursors also reveals a very strong conservation within the phylum of nematodes.[43,44] Fifteen species of nematodes, 12 of which are parasitic, have a strong *pdf-1* orthologue, encoding both PDF-1a- and PDF-1b-like peptides and 18 nematode species, 13 of which are parasitic, display a clear *pdf-2* orthologue (10 of them contain both *pdf-1-* and *pdf-2*-like precursor genes). These species are representatives of four different nematode clades (I, III, IV, and V) including free-living species and parasites of plants, insects, and mammals.[43–45] Together, our data indicate that the PDF signaling system has been well conserved in arthropods and nematodes. *In vivo* localization studies revealed that many of the PDF- and PDFR-expressing cells in *C. elegans* play a role

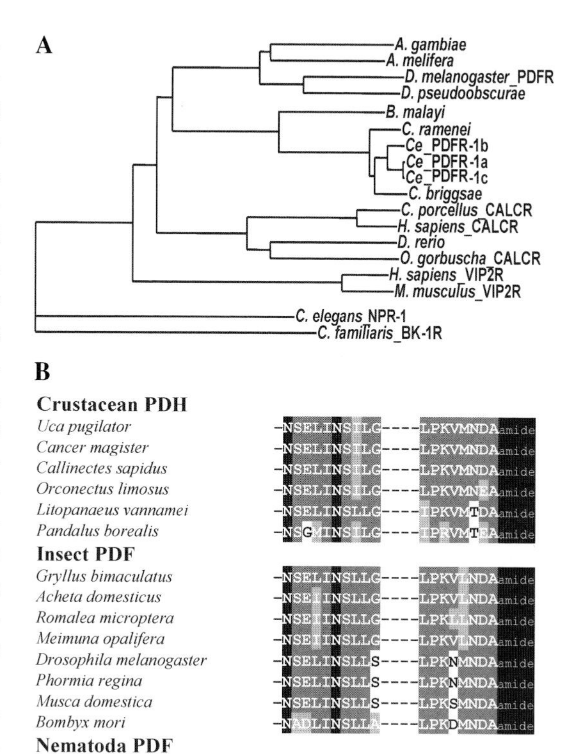

Figure 3. (A) Phylogenetic relationship of the *C. elegans* PDF receptors (*Ce*_PDFR-1a, b and c; C13B9.4a,b,c) and related receptors (AlignX software, Vector NTI Advance 10, Invitrogen). *C. elegans* NPR-1 and the *Canis familiaris* bradykinin receptor B1 were used as an outgroup. (B) Amino acid sequence alignment of the *C. elegans* PDF-like peptides with the crustacean pigment dispersing hormones (PDH) and the insect PDF peptides. Identical amino acids are highlighted in *black*, conserved amino acids in *dark grey*, and similar amino acids in *light grey*.

in the integration of environmental stimuli and the control of locomotion (i.e., body wall muscle cells and sensory-, command-, and interneurons). We also demonstrated that this neuropeptide system is involved in the regulation of locomotor behavior, similar to their counterparts in insects.[35,43,44] The transcription of both PDF genes seems to be regulated by *atf-2* and *ces-2*, which encode bZIP transcription factors homologous to the well-known *Drosophila* circadian clock components *vrille* and *par domain protein 1* (*Pdp1e*), respectively.[43] In insects, PDF has become known as a key player in the circadian clock system. Although the sequence conservation of PDF peptides appears to be restricted to the invertebrate lineages, their receptors are clearly related to certain mammalian receptors (i.e., calcitonin receptors and to a lesser extent to either PACAP or VIP receptors[35,46]) which all exert influences in the mammalian circadian system. Although Class II peptide GPCRs, to which all of these receptors belong, typically couple to Gs, many Class II receptors also signal via calcium, as has been shown for the *Drosophila* PDFR and mammalian VIP receptors[47,48], but not the *C. elegans* PDFRs. All these observations suggest that there is conservation of the PDF signaling pathway between nematodes, arthropods, and mammals. Some recently discovered functions for PDF in *Drosophila* show striking similarities with that of the neuropeptide VIP in the mammalian circadian system, that is, mediating circadian rhythmicity and synchrony in the clock neurons.[49–51] This has led some researchers to believe that PDF might be the functional counterpart of the mammalian VIP with respect to the circadian clock.[51] Although the structural conservation between insects, nematodes, and mammals is only apparent for the PDF receptors and not their ligands, these neuropeptide signaling systems seem to share some distinct parallels at a functional level.[35,43,44]

Gonadotropin-releasing hormone/adipokinetic hormone-like signaling system

Gene F54D7.3 (*gnrr-1*) encodes a GPCR that was previously identified as one of the *C. elegans* orthologues of the vertebrate gonadotropin-releasing hormone receptor (GnRHR) (Fig. 4A).[52] In insects, the natural ligands for GnRH-like receptors belong to the adipokinetic hormone (AKH) peptide family.[53] Although AKHs do not show very obvious sequence similarities to GnRHs, there is a surprising resemblance in their precursor organization, which may refer to an evolutionary ancient, common origin. Based on this knowledge, we recently used a bioinformatics approach to identify a *C. elegans* NLP-type peptide precursor, designated NLP-47. The peptide and its precursor are reminiscent of both the insect AKH precursors and the GnRH-preprohormone precursors from tunicates and higher vertebrates (Fig. 4B).[37,54,55] It encodes a single peptide that was able to activate GNRR-1

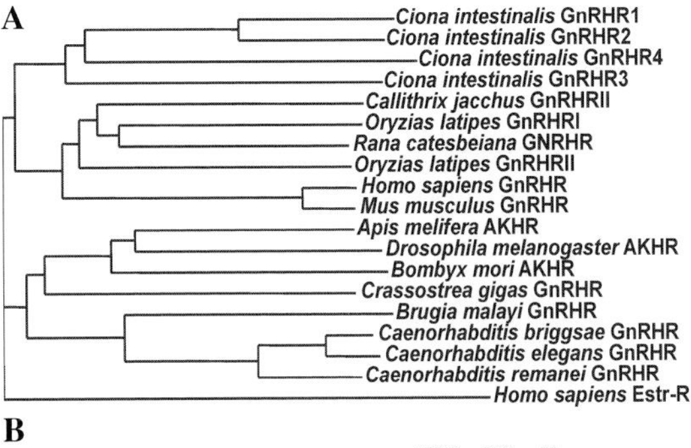

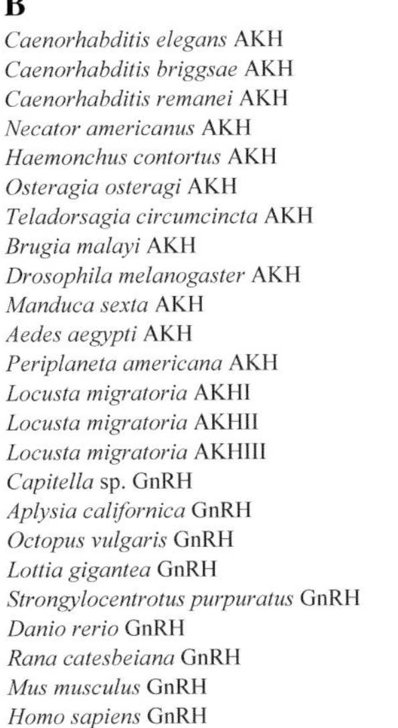

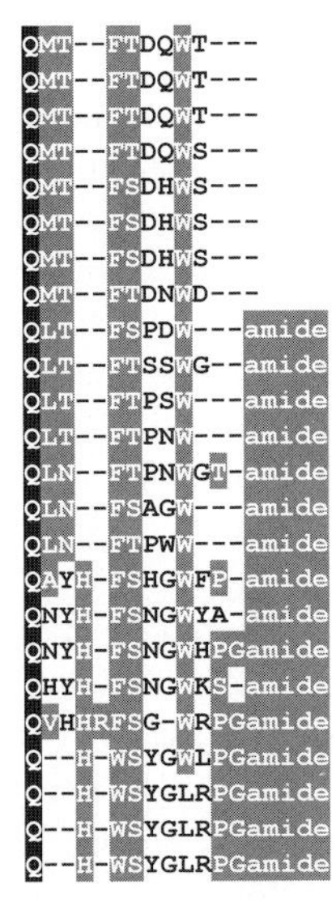

Figure 4. (A) Phylogenetic representation of the relationship between the *Ce*-GnRHR and insect AKH receptors, a mollusk GnRH receptor and tunicate and vertebrate GnRH receptors (AlignX software, Vector NTI Advance 10, Invitrogen). The human estrogen receptor was used as outgroup. (B) Amino acid sequence alignment of AKH and GnRH peptides of different protostomian and deuterostomian representatives. Identical amino acids are highlighted in *black*, conserved amino acids in *dark grey*, and similar amino acids in *light grey*.

in a dose-dependent manner through a $G_{\alpha q}$ type of G protein, analogous to both insect AKH receptor and vertebrate GnRH receptor signaling. GNRR-1 could also be activated by *Drosophila* AKH and other nematode AKH–GnRHs that we found in EST databases.[37]

In mammals, GnRH is a neuropeptide that stimulates the release of gonadotropins from the anterior pituitary. Insect AKHs, on the other hand, are mainly synthesized in the glandular part of the corpora cardiaca and play an important role in the mobilization of energy substrates from the fat body during energy-requiring activities, such as flight and locomotion. Based on the close relationship between metabolism and reproduction, a possible (indirect) role of AKH-like peptides in insect reproductive physiology cannot readily be excluded. The nematode GnRH/AKH receptor ortholog is expressed in germline cells, the pharynx, and the intestine, suggesting a possible role in the coupling of food intake with reproduction.[52] Gene silencing of the GnRH/AKH receptor and/or its ligand in *C. elegans* results in a delay in the egg-laying process, comparable to a delay in puberty in mammals with a mutated GnRH or GnRHR gene.[37,55] These data support the view that the AKH–GnRH signaling system probably arose very early in metazoan evolution and that its role in reproduction might have been developed before the divergence of protostomians and deuterostomians.

Neuromedin/pyrokinin-like signaling system

K10B4.4 was selected because of its close homology to the human neuromedin U receptor (NMU-R1; *E*-value $9e^{-43}$) and the fruitfly pyrokinin receptor (PK-1-R; *E*-value $3e^{-48}$) (Fig. 5A). Neuromedin U is a structurally highly conserved vertebrate neuropeptide with pleiotropic activities, including the regulation of smooth–muscle contraction, gastric acid secretion and feeding behavior.[56] In *Drosophila*, two neuropeptide genes encoding pyrokinins (PKs), capability (capa), and hugin, are known as putative homologues of the vertebrate NMU and also exhibit a very broad range of physiological functions.[57] We recently identified the *C. elegans* pyrokinin-like peptide precursor through a PSI-BLAST search using the known periviscerokinins (PVK) and pyrokinin peptides from the CAPA precursor of *D. melanogaster*, *A. gambiae*, and *M. sexta* as a query. The precursor, encoded by *nlp-44*, contains three NLP-type peptides that are expressed in two pairs of neuronal cells, which is reminiscent to the CAPA peptides in insects and NMUs in vertebrates (Fig. 5B).[58] This neuropeptide precursor is homologous to the insect CAPA precursors because it encodes a PK-like peptide and two periviscerokinin-like peptides (PVKs). Very similar sequences could be found in *C. briggsae* and *C. remanei*. When expressed in HEK293 cells, only the

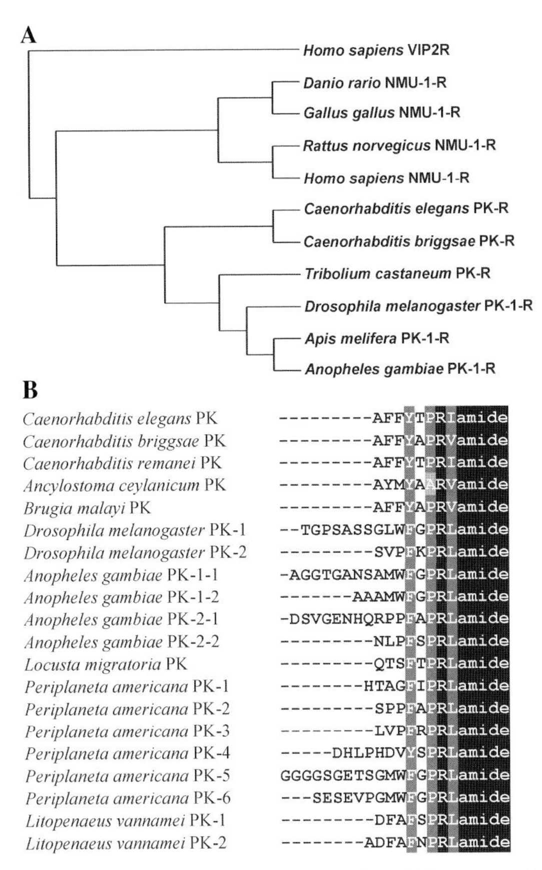

Figure 5. **(A) Phylogenetic relationship of the nematode and insect (protostomian) pyrokinin receptors and vertebrate (deuterostomian) NMU receptors. The human VIP2R was used as an outgroup. (B) Amino acid sequence alignment of nematode and insect pyrokinin peptides. Identical amino acids are highlighted in *black*, conserved amino acids in *dark grey*, and similar amino acids in *light grey*.**

synthetic PK neuropeptide was able to activate the orphan *C. elegans* neuromedin U-like receptor in a dose-dependent manner.[58] This is consistent with the observations in *Drosophila* and *Anopheles*, where the CAPA–PVKs and CAPA–PK activate separate receptors.[59–61] Whether the nematode pyrokinin signaling system is also conserved at a functional level, remains to be determined.

Conclusions

Although the resulting MAST list of putative *C. elegans* peptide GPCRs is rather extensive compared to previous predictions, the successful deorphanization of these seven peptide GPCRs confirms the value of the chosen bioinformatics-based selection procedure. As more and more peptide GPCRs become deorphaned, the training set for MEME analysis will grow larger, and therefore the resulting MAST list will become, without a doubt, more accurate in the future. Together with a combined reverse pharmacology strategy, as was adopted here, this will certainly speed up the GPCR deorphanization process in *C. elegans* and nematodes in general. The identification and functional characterization of new receptor–ligand interactions is a critical step in the discovery of potentially valuable novel agro-pharmaceutical or therapeutic receptor targets. Moreover, as *C. elegans* has rapidly gained the supermodel status, not only in the study of parasitic nematodes, but also as a model for human diseases, it is imperative that we learn more about the wealth of neuropeptidergic signaling systems that this tiny worm has to offer.

Our data indicate that at least these four neuropeptide signaling systems seem to be conserved, not only at a structural level, but also functionally, during the course of evolution and were already well established prior to the divergence of protostomes and deuterostomes. This emphasizes the functional importance of these systems in both vertebrates and invertebrates. All of the neuropeptide signaling systems described here thus represent fine examples of receptor–ligand coevolution.

Supporting Information

Additional Supporting Information may be found in the online version of this article:

Figure S1. HPLC purification of the 0–60% CH_3CN/0.1% TFA peptide fraction from approximately 30,000,000 worms on a Waters Delta-pack C_{18} (2× [25×100 mm]) column.

Figure S2. Overview of the peptide extraction and purification strategy.

Please note: Wiley-Blackwell are not responsible for the content or functionality of any supporting materials supplied by the authors. Any queries (other than missing material) should be directed to the corresponding author for the article.

Conflicts of interest

The authors declare no conflict of interest.

References

1. Brownlee, D.J. & I. Fairweather. 1999. Exploring the neurotransmitter labyrinth in nematodes. *Trends Neurosci.* **22:** 16–24.
2. Weimer, R.M. & E.M. Jorgensen. 2003. Controversies in synaptic vesicle exocytosis. *J. Cell Sci.* **116:** 3661–3666.
3. Alkema, M.J. *et al.* 2005. Tyramine functions independently of octopamine in the *Caenorhabditis elegans* nervous system. *Neuron* **46:** 247–260.
4. Scalettar, B.A. 2006. How neurosecretory vesicles release their cargo. *Neuroscientist* **12:** 164–176.
5. Richmond, J. 2007. Synaptic function. In: WormBook, doi/10.1895/wormbook.1.69.1, http://www.wormbook.org.
6. Chase, D.L. & M.R. Koelle. 2007. Biogenic amine neurotransmitters in *C. elegans*. In: WormBook, doi/10.1895/wormbook.1.132.1, http://www.wormbook.org.
7. Li, C. 2005. The ever-expanding neuropeptide gene families in the nematode *Caenorhabditis elegans*. *Parasitology* **131:** S109–S127.
8. McVeigh, P. *et al.* 2006. The FLP-side of nematodes. *Trends Parasitol.* **22:** 385–396.
9. McVeigh, P. *et al.* 2006. Gene expression and pharmacology of nematode NLP-12 neuropeptides. *Int. J. Parasitol.* **36:** 633–640.
10. Husson, S.J. *et al.* 2007. Neuropeptidergic signaling in the nematode *Caenorhabditis elegans*. *Prog. Neurobiol.* **82:** 33–55.
11. Pierce, S.B. *et al.* 2001. Regulation of DAF-2 receptor signaling by human insulin and ins-1, a member of the unusually large and diverse *C. elegans* insulin gene family. *Genes. Dev.* **15:** 672–686.
12. Li, W. *et al.* 2003. *daf-28* encodes a *C. elegans* insulin superfamily member that is regulated by environmental cues and acts in the DAF-2 signaling pathway. *Genes. Dev.* **17:** 844–858.
13. Nathoo, A.N. *et al.* 2001. Identification of neuropeptide-like protein gene families in *Caenorhabditis elegans* and other species. *Proc. Natl. Acad. Sci. USA* **98:** 14000–14005.
14. Richmond, J.E. & K.S. Broadie. 2002. The synaptic vesicle cycle: exocytosis and endocytosis in *Drosophila* and *C. elegans*. *Curr. Opin. Neurobiol.* **12:** 499–507.
15. Zupanc, G.K. 1996. Peptidergic transmission: from morphological correlates to functional implications. *Micron* **27:** 35–91.
16. Vanden Broeck, J. 2001. Insect G protein-coupled receptors and signal transduction. *Arch. Insect. Biochem. Physiol.* **48:** 1–12.

17. Bargmann, C.I. 1998 Neurobiology of the *Caenorhabditis elegans* genome. *Science* **282:** 2028–2033.
18. Fredriksson, R. & H.B. Schiöth. 2005. The repertoire of G-protein-coupled receptors in fully sequenced genomes. *Mol. Pharmacol.* **67:** 1414–1425.
19. Thomas, J.H. *et al.* 2005. Adaptive evolution in the SRZ chemoreceptor families of *Caenorhabditis elegans* and *Caenorhabditis briggsae*. *Proc. Natl. Acad. Sci. USA* **102:** 4476–4481.
20. Robertson, H.M. & J.H. Thomas. 2006. The putative chemoreceptor families of *C. elegans*. In: WormBook, doi/10.1895/wormbook.1.66.1, http://www.wormbook.org.
21. Wistrand, M. *et al.* 2006. A general model of G protein-coupled receptor sequences and its application to detect remote homologs. *Protein Sci.* **15:** 509–521.
22. Fredriksson, R. *et al.* 2003. The G-protein-coupled receptors in the human genome form five main families. Phylogenetic analysis, paralogon groups, and fingerprints. *Mol. Pharmacol.* **63:** 1256–1272.
23. Keating, C.D. *et al.* 2003. Whole-genome analysis of 60 G protein-coupled receptors in *Caenorhabditis elegans* by gene knockout with RNAi. *Curr. Biol.* **13:** 1715–1720.
24. Geary, T.G. & T.M. Kubiak. 2005. Neuropeptide G-protein-coupled receptors, their cognate ligands and behavior in *Caenorhabditis elegans*. *Trends Pharmacol. Sci.* **26:** 56–58.
25. Kubiak, T.M. *et al.* 2003. Functional annotation of the putative orphan *Caenorhabditis elegans* G-protein-coupled receptor C10C6.2 as a FLP15 peptide receptor. *J. Biol. Chem.* **278:** 42115–42120.
26. Kubiak, T.M. *et al.* 2003. Differential activation of "social" and "solitary" variants of the *Caenorhabditis elegans* G protein-coupled receptor NPR-1 by its cognate ligand AF9. *J. Biol. Chem.* **278:** 33724–33729.
27. Rogers, C. *et al.* 2003. Inhibition of *Caenorhabditis elegans* social feeding by FMRFamide-related peptide activation of NPR-1. *Nat. Neurosci.* **6:** 1178–1185.
28. Mertens, I. *et al.* 2004. Functional characterization of the putative orphan neuropeptide G-protein coupled receptor C26F1.6 in *Caenorhabditis elegans*. *FEBS Lett.* **573:** 55–60.
29. Mertens, I. *et al.* 2005. Molecular characterization of two G protein-coupled receptor splice variants as FLP2 receptors in *Caenorhabditis elegans*. *Biochem. Biophys. Res. Commun.* **330:** 967–974.
30. Mertens, I. *et al.* 2006. FMRFamide related peptide ligands activate the *Caenorhabditis elegans* orphan GPCR Y59H11AL.1. *Peptides* **27:** 1291–1296.
31. Kubiak, T.M. *et al.* 2007. FMRFamide-like peptides (FLPs) encoded on the flp-18 precursor gene activate two isoforms of the orphan *Caenorhabditis elegans* G-protein-coupled receptor Y58G8A.4 heterologously expressed in mammalian cells. *Biopolymers* **90:** 339–348.
32. Bailey, T.L. & M. Gribskov. 1998. Combining evidence using p-values: application to sequence homology searches. *Bioinformatics* **14:** 48–54.
33. Mertens, I. *et al.* 2005. PDF receptor signaling in *Drosophila* contributes to both circadian and geotactic behaviors. *Neuron* **48:** 213–219.
34. Kotarsky, K. & E.N. Niclas. 2004. Reverse pharmacology and the de-orphanization of 7TM receptors. *Drug. Discov. Today: Technol.* **1:** 99–104.
35. Janssen, T. *et al.* 2008. Functional characterization of three G protein-coupled receptors for pigment dispersing factors in *Caenorhabditis elegans*. *J. Biol. Chem.* **283:** 15241–15249.
36. Janssen, T. *et al.* 2008. Discovery of a cholecystokinin-gastrin like signaling system in nematodes. *Endocrinology* **149:** 2826–2839.
37. Lindemans, M. *et al.* 2009. Adipokinetic hormone signaling through the gonadotropin-releasing hormone receptor modulates egg-laying behavior in *C. elegans*. *Proc. Natl. Acad. Sci.* **106:** 1642–1647.
38. Janssen, T. *et al.* 2009. Evolutionary conservation of the cholecystokinin/gastrin signaling system in nematodes. *Trends Comp. Endocrinol. Neurobiol.: Ann NY Acad Sci* **1163:** 428–432.
39. Dufresne, M. *et al.* 2006. Cholecystokinin and gastrin receptors. *Physiol Rev* **86:** 805–847.
40. Kubiak, T.M. *et al.* 2002. Cloning and functional expression of the first *Drosophila melanogaster* sulfakinin receptor DSK-R1. *Biochem. Biophys. Res. Commun.* **291:** 313–320.
41. Hyun, S. *et al.* 2005. *Drosophila* GPCR Han is a receptor for the circadian clock neuropeptide PDF. *Neuron* **48:** 267–278.
42. Lear, B.C. *et al.* 2005. A G protein-coupled receptor, groom-of-PDF, is required for PDF neuron action in circadian behavior. *Neuron* **48:** 221–227.
43. Janssen, T. *et al.* 2009. Discovery and characterization of a conserved pigment dispersing factor-like neuropeptide pathway in *Caenorhabditis elegans*. *J. Neurochem.* **111:** 228–241.
44. Meelkop, E., L. Temmerman *et al.* 2009. Invertebrate signaling trhough pigment dispersing-like peptides. *Prog. Neurobiol.*, Submitted for publication.
45. Parkinson, J. *et al.* 2004. A transcriptomic analysis of the phylum Nematoda. *Nat. Genet.* **36:** 1259–1267.

46. Hewes, R.S. & P.H. Taghert. 2001. Neuropeptides and neuropeptide receptors in the *Drosophila melanogaster* genome. *Genome Res.* **11:** 1126–1142.
47. Van Rampelbergh, J. *et al.* 1997. The pituitary adenylate cyclase activating polypeptide (PACAP I) and VIP (PACAP II VIP1) receptors stimulate inositol phosphate synthesis in transfected CHO cells through interaction with different G proteins. *Biochim. Biophys. Acta.* **1357:** 249–255.
48. DeHaven, W.I. & J. Cuevas. 2004. VPAC receptor modulation of neuroexcitability in intracardiac neurons: dependence on intracellular calcium mobilization and synergistic enhancement by PAC1 receptor activation. *J. Biol. Chem.* **279:** 40609–40621.
49. Harmar, A.J. *et al.* 2002. The VPAC(2) receptor is essential for circadian function in the mouse suprachiasmatic nuclei. *Cell* **109:** 497–508.
50. Aton, S.J. *et al.* 2005. Vasoactive intestinal polypeptide mediates circadian rhythmicity and synchrony in mammalian clock neurons. *Nat. Neurosci.* **8:** 476–483.
51. Vosko, A.M. *et al.* 2007. Vasoactive intestinal peptide and the mammalian circadian system. *Gen. Comp. Endocrinol.* **152:** 165–175.
52. Vadakkadath, M.S. *et al.* 2006. Identification of a gonadotropin-releasing hormone receptor orthologue in *Caenorhabditis elegans. BMC Evol. Biol.* **6:** 103–120.
53. Staubli, F. *et al.* 2002. Molecular identification of the insect adipokinetic hormone receptors. *Proc. Natl. Acad. Sci. USA* **99:** 3446–3451.
54. Merzendorfer, H. *et al.* 2009. GnRH evolutionary conserved in reproduction. *J. Exp. Biol.* **212:** doi: 10.1242/jeb.021717.
55. Lindemans, M. *et al.* In press. Gonadotropin-releasing hormone and adipokinetic hormone signaling systems share a common evolutionary origin. *J. Neuroendocrinol.*
56. Brighton, P.G. *et al.* 2004. Neuromedin U and its receptors: structure, function, and physiological roles. *Pharmacol. Rev.* **56:** 231–248.
57. Predel, R. & C. Wegener. 2006. Biology of the CAPA peptides in insects. *Cell Mol. Life Sci.* **63:** 2477–2490.
58. Lindemans, M. *et al.* 2009. Discovery and characterization of a nematode pyrokinin receptor in *Caenorhabditis elegans. Biochem. Biophys. Res. Commun.* **379:** 760–764.
59. Iversen, I. *et al.* 2002. Molecular cloning and functional expression of a *Drosophila* receptor for the neuropeptides capa-1 and -2. *Biochem. Biophys. Res. Commun.* **299:** 628–633.
60. Cazzamali, G. *et al.* 2005. The *Drosophila* gene CG9918 codes for a pyrokinin-1 receptor. *Biochem. Biophys. Res. Commun.* **335:** 14–19.
61. Olsen, S.S. *et al.* 2007. Identification of one capa and two pyrokinin receptors from the malaria mosquito *Anopheles gambiae. Biochem. Biophys. Res. Commun.* **362:** 245–251.

Ann. N.Y. Acad. Sci. ISSN 0077-8923

ANNALS OF THE NEW YORK ACADEMY OF SCIENCES
Issue: *Phylogenetic Aspects of Neuropeptides*

Insights into the evolution of proglucagon-derived peptides and receptors in fish and amphibians

Stephanie Y. L. Ng, Leo T. O. Lee, and Billy K. C. Chow

School of Biological Sciences, The University of Hong Kong, Pokfulam Road, Hong Kong SAR, China

Address for correspondence: Professor Billy K.C. Chow, School of Biological Sciences, The University of Hong Kong, Pokfulam Road, Hong Kong SAR, China. bkcc@hkusua.hku.hk

Glucagon and the glucagon-like peptides (GLP-1 and GLP-2) share a common evolutionary origin and are triplication products of an ancestral glucagon exon. In mammals, a standard scenario is found where only a single proglucagon-derived peptide set exists. However, fish and amphibians have either multiple proglucagon genes or exons that are likely resultant of duplication events. Through phylogenetic analysis and examination of their respective functions, the proglucagon ligand-receptor pairs are believed to have evolved independently before acquiring specificity for one another. This review will provide a comprehensive overview of current knowledge of proglucagon-derived peptides and receptors, with particular focus on fish and amphibian species.

Keywords: gene duplication; glucagon; glucagon-like peptides; receptor; molecular evolution

Introduction

Glucagon and the glucagon-like peptides 1 and 2 (GLP-1 and GLP-2) are pleiotropic hormones encoded by the vertebrate proglucagon gene, whose expression is both species and tissue dependent. Together, they are suggested to have common evolutionary origin, evolving from a primordial exon that arose millions of years ago.[1] Within the proglucagon peptide trio, glucagon was the first to be discovered. It was initially mistaken as an impurity in insulin extracts and described as a pancreatic hyperglycemic glycogenolytic factor.[2] Since the initial isolation of proglucagon cDNAs from the anglerfish (*Lophus americanus*)[3] and chicken (*Gallus gallus*),[4] an explosion of information has followed providing a glimpse into the molecular evolution of the proglucagon gene. Most vertebrates are found to have only a single proglucagon gene, whereas teleosts and amphibians exhibit a distinctive case, possessing either multiple proglucagon genes or derived sequences.

To elicit their biological functions, all vertebrate glucagon and GLPs interact with corresponding receptors: GCG-R, GLP-1R, and GLP-2R. Collectively, these receptors belong to a superfamily of G protein–coupled receptors, also commonly known as the class II B receptor family, representing one of the largest hormone and neuropeptide receptor gene families. They are characterized by a long N-terminal extracellular domain containing seven cysteine residues, N-glycosylation sites for receptor conformation and ligand binding, seven hydrophobic transmembrane (TM) domains within which the third endoloop between TM5 and TM6 is responsible for receptor-specific functions, and together with the intracellular C-terminal domain, are important for G-protein interaction. Like proglucagon peptides, the glucagon receptor family has a suggested origin prior to divergence of fish and tetrapods ~450 million years ago (mya).[5] Although a range of proglucagon ligand and receptor sequences are known to date, most research has been focused on mammals and the information regarding fish and frogs is relatively sparse. As a result, the earlier evolutionary events leading to present-day proglucagon peptide and receptor sequences still remain unclear. To better understand the phylogenetic background of proglucagon derived peptides, a comprehensive overview of current research will be provided, with particular focus on piscine and amphibian species.

doi: 10.1111/j.1749-6632.2010.05505.x

Triplication of an ancestral gene

Duplications of entire or partial gene and genome are important driving forces in evolution, facilitating the increase in complexity[6] by acting as a source of raw materials from which new genes and functions may evolve.[7] As proposed by Ohno in 1970, two or three rounds of genome duplication have occurred throughout vertebrate evolution.[6] Although these "1-2-4" and "1-2-4-8" rules are substantially supported by a vast number of studies, determining these earlier events is complicated by secondary losses of genes or whole chromosomes, chromosomal rearrangements, independent gene duplications, and differences in evolutionary rates for various genes. On a smaller scale, exon duplications also act alongside gene duplications, contributing as sources of novelty, further perplexing our view into evolutionary events.[8] With respect to the incretin hormones, a whole genome duplication is believed to have created independent genes for proglucagon and the closely related glucose-dependent insulinotropic peptide (GIP) early in vertebrate evolution prior to class diversification.[9,10] In particular, proglucagon-derived sequences are believed to be triplication products of an ancestral GLP encoding exon,[10–12] creating the vertebrate proglucagon gene that is known today to encode glucagon, GLP-1, and GLP-2 (Fig. 1). Based on estimated rates of evolution, glucagon is suggested to have the most ancient origin, appearing ~1 billion years ago[11] from initial duplication of an ancestral gene.[10] Exon duplication of the ancestral glucagon presumably followed, creating GLPs that subsequently diverged into GLP-1 and GLP-2 via exon duplication ~700 mya.[10,11] These proglucagon sequences are speculated to be subject to genome and/or gene duplication events, which subsequently have given rise to the variety of vertebrate sequences observed today. Receptors for these proglucagon peptides have also evolved in a similar fashion and will be elaborated in detail later in this review.

Phylogeny of proglucagon-derived peptides

As paramount regulators of carbohydrate, fat, and protein metabolism, interest toward glucagon, and the GLPs has inevitably grown. This is shown in the assortment of known proglucagon hormone sequences, ranging from the extant ancient lamprey fish to humans. Although full-length proglucagon

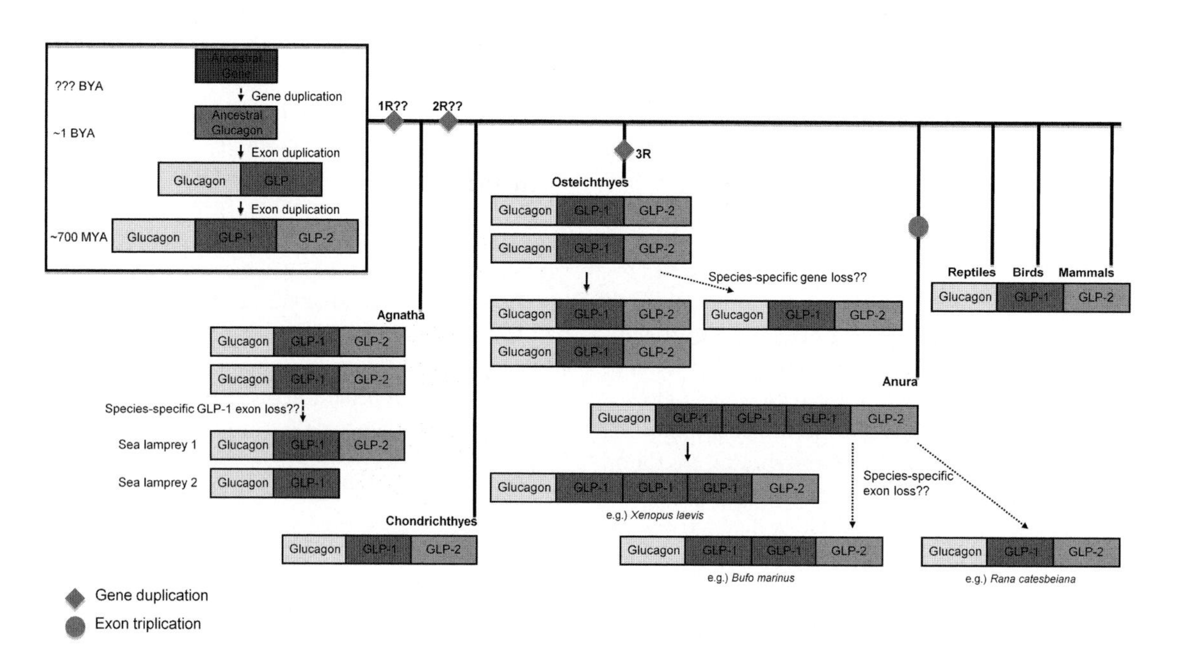

Figure 1. An evolutionary scheme of glucagon, GLP-1, and GLP-2 in vertebrates. Exons are shown as boxes. Unknown or unclear events are denoted by dotted lines or question marks. The phylogenetic timeline for the events are not to scale.

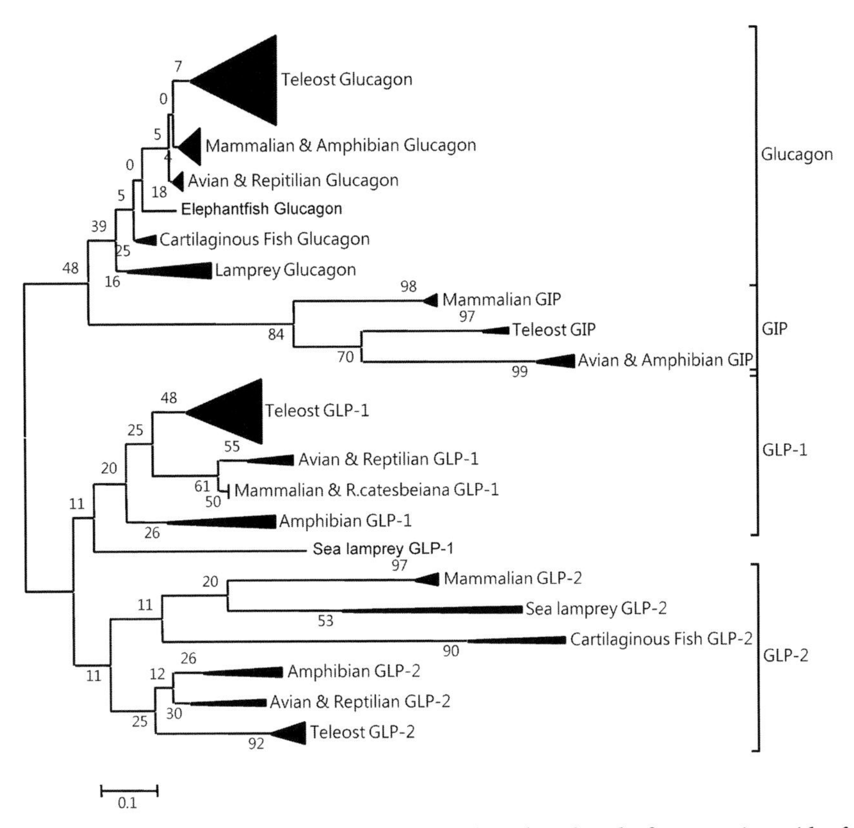

Figure 2. **Phylogenetic analysis of glucagon, GLP-1, GLP-2, and GIP based on the first 29 amino acids of each peptide. The numbers above each branch indicate the percentage of bootstrap replications in which that branch was found based on 500 replications.**

cDNA sequences are known for some species, majority of the sequences obtained are based on sequencing individual peptides and some from data mining of available genomes. To better understand their evolutionary background, the vertebrate proglucagon-derived peptides will first be considered.

Through phylogenetic analyses of both known and predicted sequences, the evolutionary events leading to formation of present-day peptide forms can be seen (Fig. 2). Two main groups are formed, one including glucagon and GIP, and the other of GLPs. The former is reflective of the whole genome duplication creating separate proglucagon and GIP genes during the beginning of proglucagon evolution; supporting the theory that glucagon is the first hormone to have evolved among the peptide trio. The existence of individual genes is further supported by the chromosomal synteny analysis of preproglucagon and GIP (Fig. 3). For the preproglucagon gene, the chromosomal locations were highly conserved in human, mouse, and chicken where the SLC4A10, DPP4, FAP, IFIH1, GCA, and KCNH7 genes were found in all of the tested species with the exception of GCA and KCNH7 in chicken. For the GIP gene, the chromosomal loci are highly conserved in tetrapods and to a lesser extent in amphibians and teleosts. The latter phylogenetic group indicates GLPs to be more closely related and agrees with the duplication event creating individual GLP-1 and GLP-2 encoding exons. After formation of the initial forms, the peptides have not evolved in a uniform manner but instead in an episodic fashion creating variation among species.[1] Both glucagon and GIP are suggested to have experienced two episodes of accelerated evolution, resulting in alterations in primary structures and/or biological function. For glucagon, this has occurred in the Osteichthyes lineage and another in hystricomorphic rodents whose glucagons are known to have had a change in biological activity.[13–15] For GIP, an episode between fish and tetrapods has resulted in lengthening of the peptide whereas another in early mammals has lead to change in function.[15,16] The shorter 31 amino acids length of fish GIP may therefore have greater resemblance to the ancestral form[15] (Table 1).

Unlike glucagon that is very well conserved with a large proportion of characterized sequences highly identical among species, GLP sequences have greater variability and are of more recent origin. Among GLP-1s, mammalian sequences show a conserved pattern suggesting the high sequence constraint whereas frog and fish sequences are less conserved. In frogs, more than one GLP-1 can be found suggesting possibility of independent duplication and point mutation events. Interestingly, mammalian GLP-2s have accumulated a large amount of change, coinciding with a previous hypothesis where mammalian sequences have undergone a period of rapid evolution that is often associated with changes in biological properties, as observed with hystricomorphic glucagons.[15] Similar to GIP, such a change may be directed by the selection pressures in acquiring of new biological function of GLP-2 early in mammalian evolution. However, this hypothesis can only be confirmed with characterization of the biological roles of GLP-2 in other vertebrates.

An exceptional case: fish and amphibians

So far, reptiles, birds, and mammals are reflective of the standard scenario whereby only a single set of proglucagon-derived peptides are known. However, fish (Agnathans and Osteichthyes) and amphibians present a case of multiple proglucagon-derived sequences, either by presence of multiple proglucagon genes or multiple peptide sequences within the proglucagon gene.[1] Although the precise evolutionary events remain unclear, a collaboration of proglucagon sequences and/or genome duplication will be considered in this review in attempt to

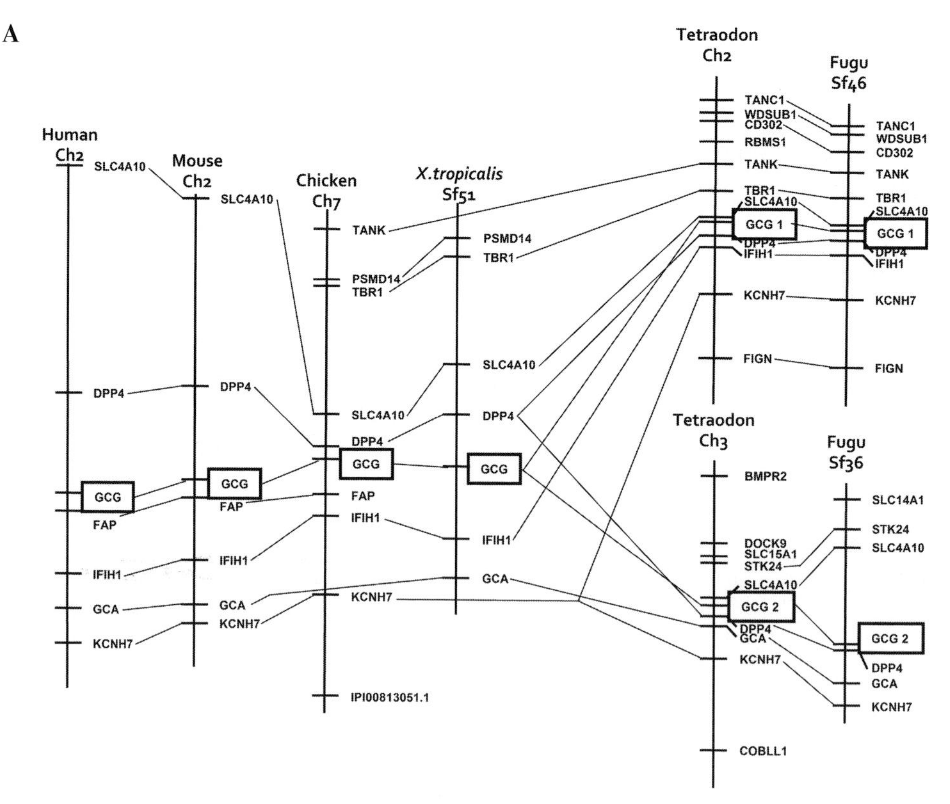

Figure 3. Chromosomal locations of (A) preproglucagon (GCG) and (B) GIP in various vertebrate species. Genes adjacent to GCG and GIP in different genomes are shown. The genes are named according to their annotation in the human genome. GCG and GIP genes are boxed. *Continued.*

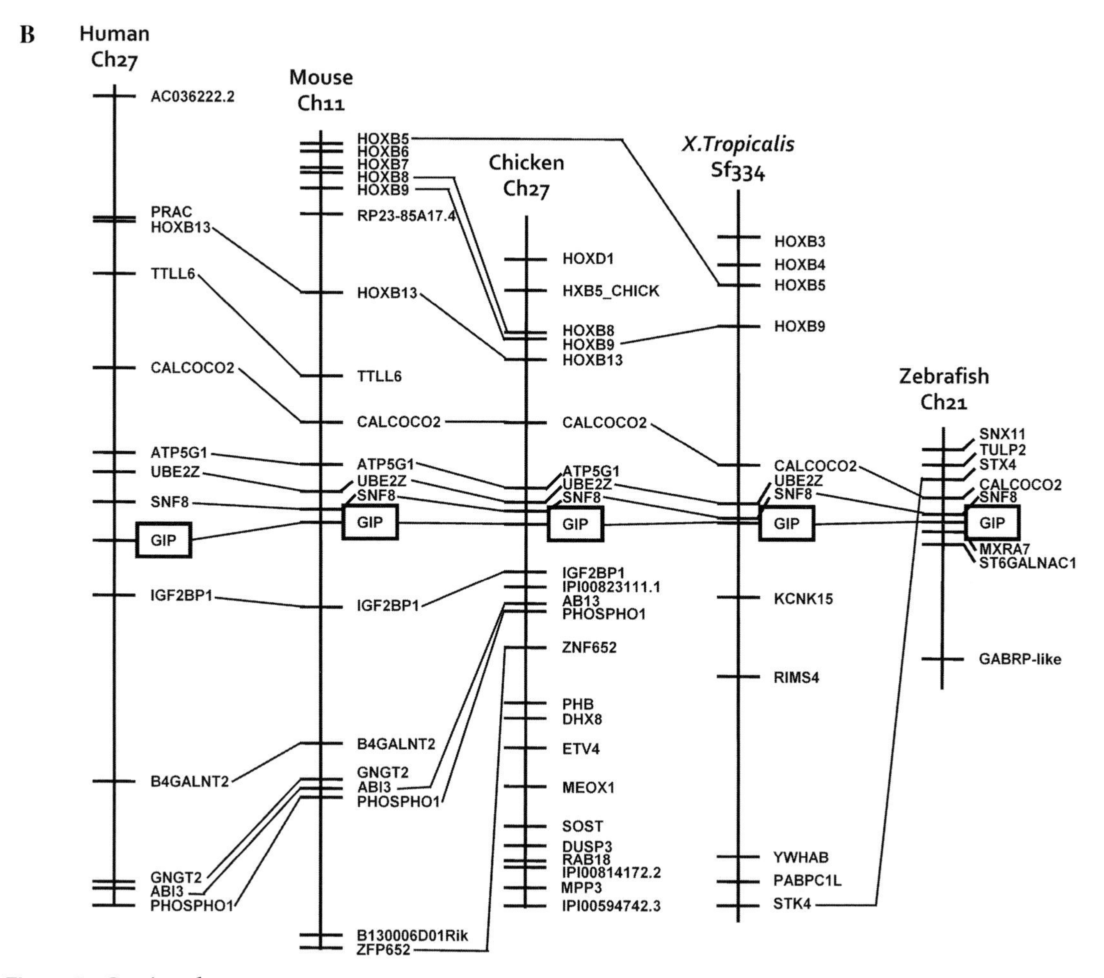

Figure 3. *Continued.*

reveal the earlier events in the fish and amphibian lineages.

Fish

Agnatha. Agnathans (jawless fish) are considered to be the initial pioneers of the ancestral fish line, marking a turning point in evolution, giving rise to the first gnathostomes over 550 mya (Cambrian period).[17] As highly modified survivors of the ancient lineage, lampreys alongside hagfishes are the only modern day representatives of the Agnathan class, and are classified into three families (Petromyzontidae, Mordaciidae, and Geotriidae).[18] In particular, research till now regarding the proglucagon-derived peptides in the Agnatha has been focused solely on lampreys, with the sea lamprey (*Petromyzon marinus*)[19] being the target species in which initial characterization of these islet hormones was restricted to. Interestingly, there are two proglucagon gene copies in the sea lamprey; the first (proglucagon-I) encodes glucagon, GLP-1, and GLP-2, whereas the second (proglucagon-II) only encodes glucagon and GLP-2. As all three proglucagon-derived peptides are encoded for, an origin prior to divergence of jawed and jawless vertebrates is suggested.[11] Of the other Agnathans studied, only glucagon peptide sequences have been determined for the Southern Hemisphere (*Geotria australis*)[20] and river (*Lampetra fluviatilis*)[17] lampreys. Like sea lamprey, two glucagon copies are also present in the Southern Hemisphere lamprey. Although exhibiting relatively high homology, the sea and Southern Hemisphere lamprey glucagon copies differ from one other by a few amino acid substitutions (Table 1), which may be reflective of the split between Petromyzontidae and Geotriidae ~220–280 mya in

Table 1. **The amino acid alignment of glucagon, GLP-1, GLP-2, and GIP from various vertebrates. Identical residues to that of the human homologue are boldfaced. Accession and reference numbers indicate the source of the corresponding sequence. Note the high conservation of primary structure throughout the species**

	10 20 30	
Glucagon		
Human/Hamster/Rat/Mouse/Camel/Dog/Pig	HSQGTFTSDYSKYLDSRRAQDFVQWLMNT	AAH05278,AAA37074,NP_036839,NP_032126,P68273,NP_001003044,NP_999489
Opossum	HSQGTFTSDYSKYLDSRRAQDFVQWLMST	P18108
Degu	HSQGTFTSDYSKFLDSRRAQDFLDWLKNT	AAA40588
Chicken/Turkey	HSQGTFTSDYSKYLDSRRAQDFVQWLMST	CAA68827,ABA77561,ABA77562
Duck/Ostrich/Turtle	HSQGTFTSDYSKYLDTRRAQDFVQWLMST	P68955
Gila monster lizard	HSQGTFTSDYSKYLDTRRAQDFVQWLMNT	O12956
African clawed frog	HSQGTFTSDYSKYLDSRRAQDFVQWLMNT	O42143
Bullfrog	HSQGTFTSDYSKYLDSRRAQDFVQWLMNS	P15438
Western clawed frog	HSQGTFTSDYSKYLDSRRAQDFIQWLMNT	NP_001006913
Salamander	HSQGTFTSDYSKYLDNRRAQDFIQWLMST	[43]
Alligator gar	HSQGTFTNDYSKYLDTRRAQDFVQWLMST	P09566
Eel 1	HSQGTFTNDYSKYLETRRAQDFVQWLMNS	C60840
Eel 2	HSQGTFTNDYSKYQEMKRAQDLVQWLMNS	D60840
Paddlefish 1	HSQGMFTNDYSKYLEEKRAKEFVEWLKNGKS	[30]
Paddlefish 2	HSQGMFTNDYSKYLEEKSAKEFVEWLKNGKS	[30]
Bichir	HSQGTFTNDYTKYQDSRRAQDFVQWLMSN	P0C235
Bowfin	HSQGTFTNDYSKYMDTRRAQDFVQWLMST	AAB28788
Anglerfish 1	HSEGTFSNDYSKYLEDRKAQEFVRWLMNN	P01278
Anglerfish 2/Flounder/Sculpin	HSEGTFSNDYSKYLETRRAQDFVQWLKNS	P04092,P68956,P09686
Catshark	HSEGTFTSDYSKYMDNRRAKDFVQWLMST	AAB31092
Catfish	HSEGTFSNDYSKYLETRRAQDFVQWLMNS	P04093
Elephantfish	HSEGTFSSDYSKYLDSRRAKDFVQWLMST	P13189
Tilapia 1	HSEGTFSNDYSKYLEDRKAQDFVRWLMNN	P81026
Tilapia 2	HAGTYTSDVSSYLQDQAAKEFVSWLKTG	P81027
Pacu	HSEGTFSNDYSKYLETQRAQDFVQWLMNS	P81880
Rainbow trout 1/Salmon	HSEGTFSNDYSKYQEERMAQDFVQWLMNS	NP_001118170,P07499
Rainbow trout 2	QSEGTFSNYYSKYQEERMARDFLHWLMNS	NP_001118172
Goldfish	HSEGTFSNDYSKYLETRRAQDFVEWLMNS	AAB39563
Electric ray	HSEGTFTSDYSKYLDNRRAKDFVQWLMNT	P09567
Sea lamprey 1	HSEGTFTSDYSKYLENKQAKDFVRWLMNA	Q9PUR1
Sea lamprey 2	HSQGSFTSDYSKHLDCKWAKDFVTWLLNT	Q9PUR0
Southern Hemisphere lamprey 1	HSQGSFTSDYSKYLEAKQAKDFVTWLMNT	[20]
Southern Hemisphere lamprey 2	HSEGTFTSDYSKYMENKQAKDFVRWLMNS	[20]
River lamprey	HSQGSFTSDYSKYLDSKQAKDFVIWLMNT	AAB36060

Continued.

Table 1. *Continued*

	10 20 30	
GLP-1		
Human/Hamster/Rat/Mouse/Guinea Pig/Degu	HDEFERHAEGTFTSDVSSYLEGQAAKEFIAWLVKGR	AH05278, AAA37074, NP_036839,AAH12975,BAA00010,AAA40588
Chicken	HSEFERHAEGTYTSDITSYLEGQAAKEFIAWLVNGR	CAA68827
Gila monster lizard	HAEYERHADGTYTSDISSYLEGQAAKEFIAWLVNGR	O12956
Bullfrog	HADGTFTSDMSSYLEEKAAKEFVDWLIKGRPK	P15438_1
African clawed frog A	HAEGTFTSDVTQQLDEKAAKEFIDWLINGGPSKEIIS	O42413
African clawed frog B	HAEGTYTNDVTEYLEEKAAKEFIEWLIKGKPK	O42413
African clawed frog C	HAEGTFTNDMTNYLEEKAAKEFVGWLIKGRPK	O42413
Salamander	HADGTLTSDISSFLEKQATKEFIAWLVSGRGRRQ	[43]
Alligator gar	HADGTYTSDVSSYLQDQAAKKFVTWLKQGQDRRE	P09566
Eel	HAEGTYTSDVSSYLQDQAAKEFVSWLKTGR	P63294
Paddlefish	HADGTYTSDASSFLQEQAARDFISWLKKGQ	[30]
Bowfin	YADAPYISDVYSYLQDQVAKKWLKSGQDRRE	AAB28787
Anglerfish 1	HADGTFTSDVSSYLKDQAIKDFVDRLKAGQV	[3]
Anglerfish 2	HADGTYTSDVSSYLQDQAAKDFVSWLKAGRGRRE	[3]
Tilapia	HADGTYTSDVSSYLQDQAAKEFVSWLKTGRGRRD	2118237E
Sculpin	HADGTFTSDVSSYLNDQAIKDFVAKLKSGKV	P09686
Rainbow trout 1	HADGTYTSDVSTYLQKQAAKDFVSWLKSGRA	Q91971
Rainbow trout 2	HADGTYTSDVSTYLQDQAAKDFVSWLKSGPA	Q91189
Catfish	HADGTYTSDVSSYLQDQAAKDFITWLKSGQPKPE	P04093
Ratfish	HADGIYTSDVASLTDYLKSKRFVESLSNYNRKQND	P23063
Sculpin	HADGTFTSDVSSYLNDQAIKDFVAKLKSGKV	P09686
Sea lamprey	HADGTFTNSMTSYLDAKAARDFVSWLARSDKS	AAB29398
GLP-2		
Human	HADGSFSDEMNTILDNLAARDFINWLIQTKITD	AAH05278
Hamster/Rat/Guinea pig	HADGSFSDEMNTILDSLATRDFINWLIQTKITD	AAA37074,NP_036839,BAA00010
Degu	HADGSFSDEMNTILDHLATKDFINWLIQTKITD	AAA40588
Mouse	HADGSFSDEMSTILDNLAARDFINWLIQTKITD	NP_032126
Pig	HADGSFSKEMNTVLDNLATRDFINWLLHTKITD	NP_999489
Chicken	HADGTFTSDINKILDDMAAKEFLKWLINTKVTQ	CAA68827
Gila monster lizard	HADGTFTSDYNQLLDDIATQEFLKWLINQKVTQ	O12956
Bullfrog	HADGSFTSDFNKALDIKAAQEFLDWIINTPVKE	P15438_1
African clawed frog	HADGSFTNDINKVLDIIAAQEFLDWVINTQETE	O42143
Salamander	HADGSFTSDINKVLDTIAAKEFLNWLISTKVTE	[43]
Rainbow trout 1 & 2	HVDGSFTSDVNKVLDSLAAKEYLLWVMTSKTSG	Q91971,Q91189
Sea lamprey 1	HAEDVMALILRTMAKTDFENWEKQNSNTQTD	Q9PUR1
Sea lamprey 2	HSDGSFTNDMNVMLDRMSAKNFLEWLKQQGRG	Q9PUR0

Continued.

Table 1. *Continued*

	10 20 30	
GIP		
Human	**YAEGTFISDYSIAMDKIHQQDFVNWLLAQKGKKNDWKHNITQ**	P09681
Pig	**YAEGTFISDYSIAMDKI**R**QQDFVNWLLA**W**KG**G**K**S**DWKHNITQ**	P01281
Rat	**YAEGTFISDYSIAMDKI**R**QQDFVNWLLAQKGKKNDWKHN**L**TQ**	NP_062604
Mouse	**YAEGTFISDYSIAMDKI**R**QQDFVNWLLAQ**R**GKK**S**DWKHNITQ**	NP_032145
African clawed frog 1/2	YSEAILA**SDYS**RSV**D**NMLKKN**FV**D**WLLA**RRE**KK**SENTSKAT	ABL10368,ABL10369
Western clawed frog	YSEAILA**SDYS**RSV**D**NMLKKN**FV**D**WLLA**RRE**KK**SENTSEAT	NP_001090751
Salmon	YAESTIA**SD**MSKI**MD**SMV**Q**KN**FVN**F**LL**N**Q**K**E**	ABW77503
Zebrafish	YAESTIA**SDI**SKIV**D**SMV**Q**KN**FVN**F**LL**N**Q**RE	NP_001093614

the Permian–Triassic period.[18] Because hagfish and lampreys evolved before gnathostome radiation and duplicate proglucagon genes can be found in lampreys, a genome duplication event is likely to have occurred early in the lamprey lineage but prior to divergence of modern lampreys.[18,20] The presence of large number of chromosomes in lampreys can further support the possibility of this duplication.[1,11] However the exact timing of this duplication with respect to proglucagon evolution remains to be confirmed as hagfish sequences are yet to be discovered.

Chondrichthyes. Following the Agnathans, Chondrichthyes (cartilaginous fish) are the next fish class to emerge (late Silurian period). To date, only a scarce number of Chondrichthyes single copy proglucagon peptide sequences are known including the electric ray (*Torpedo marmorata*),[21] dogfish (*Scyliorhinus canicula*),[22,23] Pacific ratfish (*Hydrolagus colliei*),[24] and elephantfish (*Callorhynchus milii*).[25] Although only single proglucagon sequences have been identified in Chondrichthyes, it may be speculated that the duplicate proglucagon gene is either lost or silenced in the cartilaginous lineage.

Osteichthyes. With regard to the information known for fish proglucagon-derived peptides, the Osteichthyes (bony fish) are the most extensively studied. Bony fish comprise the lobe-finned and ray-finned fishes, with the latter considered as the most heterogeneous and affluent vertebrate group based on species diversity and morphological complexity.[26] Multiple proglucagon gene copies are also found in several bony fish: Rainbow trout (*Oncorhynchus mykiss*),[27] anglerfish,[3] sculpin (*Cottus scorpius*),[28] eel (*Anguilla anguilla*),[29] and paddlefish (*Polyodon spathula*).[30] By data mining of publically available genomes, two proglucagon copies were also predicted in the fugu (*Takifugu rubripes*), pufferfish (*Tetraodon nigroviridis*), medaka (*Orzyzias latipes*), and stickleback (*Gasterosteus aculeatus*) where one proglucagon encodes for all three derived peptides and the other encodes for glucagon and GLP-1.

Comparative genomics reveals the duplicate nature in teleosts, whereas only one cognate copy is found in tetrapods (Fig. 3). Tetrapod preproglucagon and neighboring genes are localized on single chromosomes although being found on two different linkage groups in teleosts. The teleost preproglucagon and neighboring genes such as SLC4A10, DPP4, and KCNH7 show synteny among their duplicated chromosomes and also with tetrapod forms. Comparative analyses of completed teleost fish genomes with the human genome[31,32] suggests that there is a third genome duplication event (3R) ~350 mya early in the ray-finned lineage prior to teleost diversification.[33] This can be supported by findings of varying number of *Hox* gene clusters found in teleosts such as the zebrafish (7 *Hox* clusters)[34] and pufferfish (6 *Hox* clusters).[35] This 3R genome duplication is likely responsible for the presence of these duplicate preproglucagon genes in teleosts. Interestingly, only one zebrafish (*Danio rerio*) proglucagon gene was predicted suggesting the possibility of either duplicate gene loss or independent gene duplications within individual teleost lineages.[36]

Amphibians

Amphibians represent the first terrestrial vertebrates with ~4800 species organized into three orders: Anura (frogs and toads), Urodela (salamanders), and Gymnophiona (caecilians). Like Agnathans and Osteichthyes, amphibians present yet another unique case of multiple proglucagon sequences. More specifically, proglucagon transcripts from the African clawed frog (*Xenopus laevis*),[37] Northern leopard frog (*Rana pipiens*),[38] tiger frog (*Rana tigrina rugulosa*),[39] and cane toad (*Bufo marinus*)[40] are found to encode up to three multiple in-series GLP-1s. Although phylogenetic analysis reveals the origin of these GLP-1s to be traced back to a common frog ancestor (~150 mya) prior to the initial divergence and are likely triplication products of the GLP-1 exon early in the frog lineage,[41] the scarcity in known sequences complicates understanding of their evolution. Not all frogs studied are recognized to have multiple GLP-1s, and this may be resultant of alternative splicing of pancreatic or intestinal proglucagon mRNA, or inefficient proteolytic processing.[41] Alternatively, this may be resultant of exon triplication followed by species-specific exon losses, hence accounting for the different number of GLP-1s in amphibian species studied.[42,43]

Tissue distribution and physiological functions

Today, glucagon is a well-known regulator in vertebrate blood glucose levels opposite insulin and the major product of proglucagon processing in the pancreatic A-cells. In all reported vertebrate sequences, glucagon exists as a 29 amino acid form with exception of the paddlefish,[30] where it is 31 amino acids. Like glucagon, the 37 amino acids long GLP-1 also plays a role in metazoan blood glucose homeostasis. In mammals, GLP-1 is secreted from intestinal L-cells and acts as an incretin, stimulating insulin's release from pancreatic B-cells. Whereas in fish, GLP-1 is secreted from the intestines and alongside glucagon in the pancreas, with functions resembling glucagon. Although cosecreted with GLP-1 from the intestinal L-cells, GLP-2 on the other hand has no known regulatory properties in blood glucose. To date, it has only been studied in mammals where it is found to be an intestinal growth factor[44] and a neurotransmitter in feeding behavior.[45–47] Tissue specific production of glucagon and GLPs is tightly regulated. In mammals, this process is determined by alternative proteolytic processing of the single copy proglucagon gene.[47–49] In other vertebrates, alternative mRNA splicing allows tissue specific expression to be achieved. More recently, differential gene expression has also been suggested as a third mechanism based on sea lamprey proglucagon cDNAs.[11]

Fish

Like mammals, fish proglucagon transcripts have also been identified in both the pancreas and intestines. However, evidence from histochemical and immunohistochemical studies in hagfish and lampreys reveal the intestinal mucosa to be predominantly where glucagon-producing cells are restricted to.[17] As pancreatic endocrine cells are hypothesized to have evolved from gut endocrine cells, jawless fish are likely intermediate representatives where only partial migration of endocrine cells has occurred.[50] This unique intestinal locality of glucagon is also found true in the phylogenetically ancient dogfish.[22] Despite this, the pancreas remains to be the key location where majority of fish glucagon and GLP-1 are synthesized, whereas the intestines may play a role in producing all three proglucagon-derived peptides.[51] These synthetic roles can be reflected in the dissimilar 3′ ends of the pancreatic and intestinal proglucagon genes. Compared to the intestinal proglucagon gene, the pancreatic copy is shorter and without a GLP-2 sequence.

As with other vertebrate counterparts, fish glucagon is essential in glucose metabolism, opposing insulin's actions through promoting gluconeogenesis. Aside from this traditional role, fish glucagon is able to stimulate lipid mobilization and lipolysis, as demonstrated in rainbow trout[52,53] and sea bream (*Sparus aurata*).[54] In mammals GLP-1 is insulinotropic, however fish GLP-1 is unique having fish-specific glycogenic and gluconeogenic functions. In fish, GLP-1 can be perceived as a glucagon enhancer, indirectly mimicking its actions through potentiating insulin's release. In channel catfish (*Ictalhurus punctatus*), GLP-1 has also been reported to be a potent inhibitor of food intake.[55]

Amphibians

Based on reverse transcriptase-PCR techniques in tiger frog, proglucagon gene expression in frogs

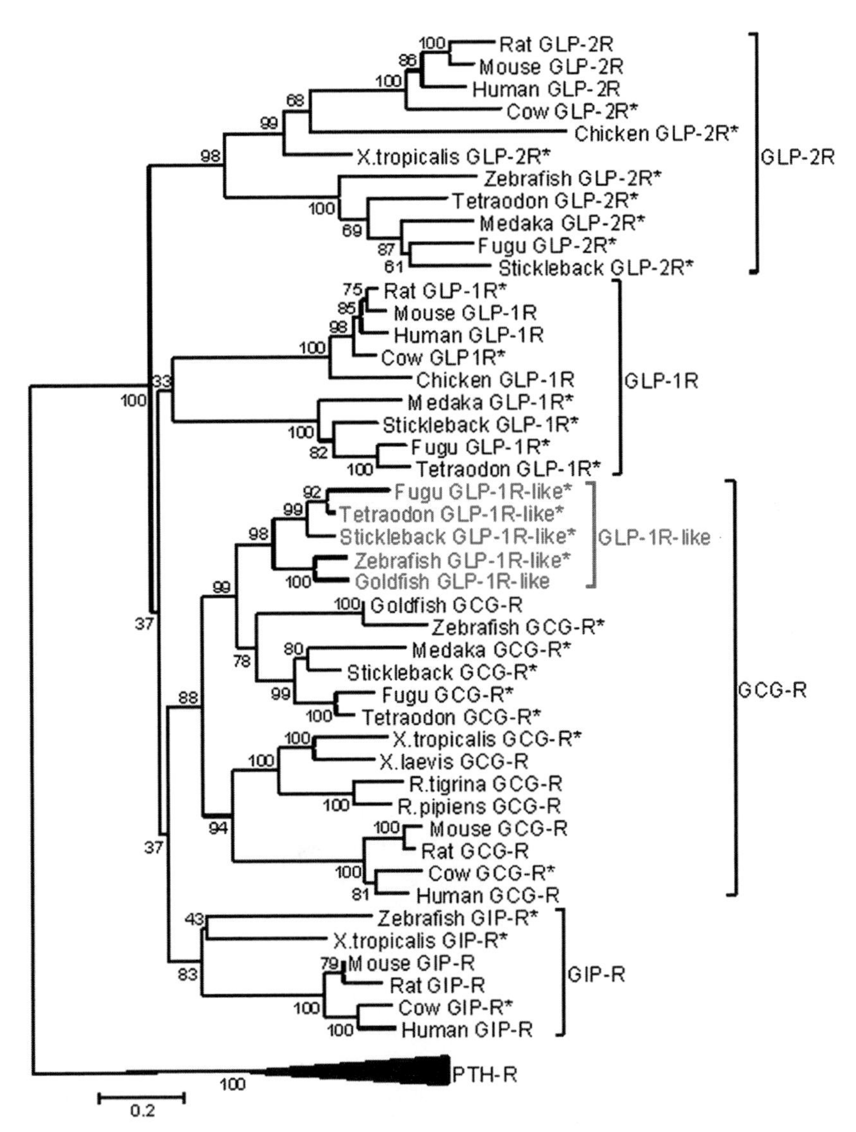

Figure 4. Phylogenetic tree showing vertebrate glucagon, GLP-1, and GLP-2 receptors. The tree was constructed based on the PAM Matrix (Dayhoff) model by Neighbor-Joining method, MEGA 4. The monophyletic groups are indicated on the right. Predicted sequences from the Ensembl genome database are denoted by "*." The numbers above each branch indicate the percentage of bootstrap replications in which that branch was found based on 500 replications. Parathyroid hormone receptor (PTH-R) sequences were used as the out-group.

is present in a range of tissues including the brain, colon, small intestines, liver, lungs, and pancreas suggesting functional diversity.[39] Vertebrate glucagon has a crucial role in blood glucose homeostasis, and this is no exception in amphibians, for example cane toad glucagon stimulates dose-dependent insulin release in rat cells[40] Amphibian GLP-1 also assumes a similar insulinotropic and glycogenolytic role as demonstrated in cane toad[40,56] and African clawed frog.[37,56] Aside from this traditional function, frog GLP-1 may also have paracrine roles in both the lungs and liver.[39]

Glucagon superfamily receptors

Similar to proglucagon sequences, the corresponding receptors (GCG-R, GLP-1R, and GLP-2R) are also products of gene duplication events.[57] These

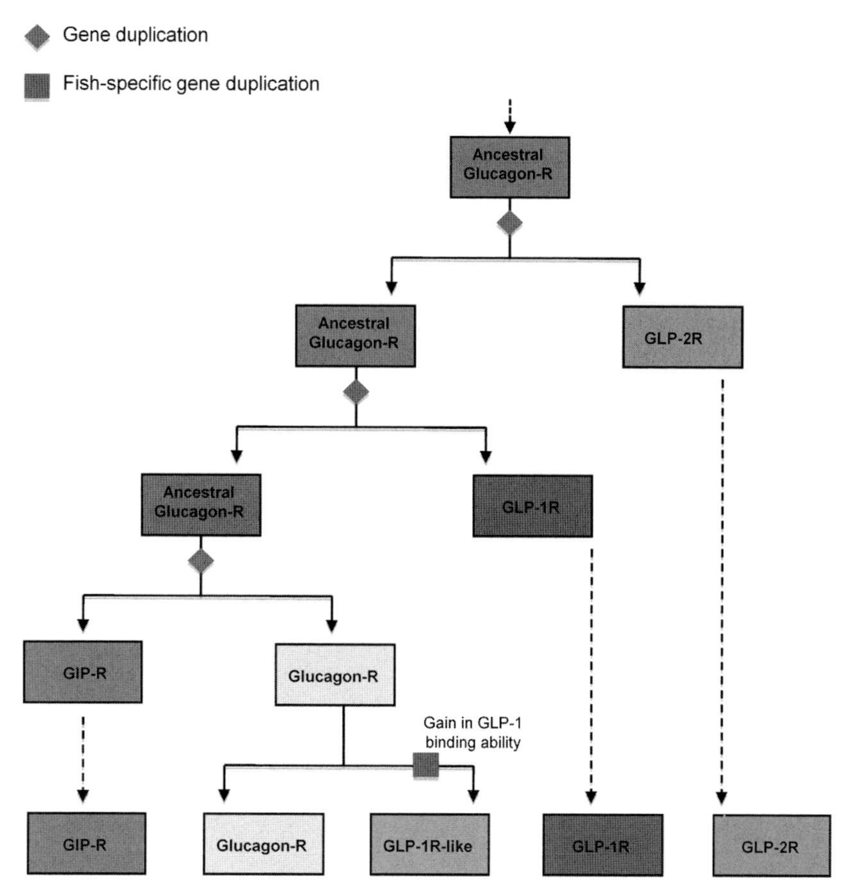

Figure 5. An evolutionary scheme of glucagon, GLP-1, and GLP-2 receptors. Each box represents a single gene. Unknown or unclear events are denoted by dotted lines or question marks. The phylogenetic timeline for the events are not to scale.

receptors have an origin prior to the divergence of fish and mammals and form a monophyletic group within the class II B receptor family. However unlike their ligand counterparts, the glucagon family receptors have not been as extensively studied, with known piscine and amphibian sequences limited to few species. In fish, the goldfish[58,59] and zebrafish[60] are the only species in which molecular cloning of glucagon and/or GLP-1 receptors has been successfully achieved, whereas fugu, pufferfish, medaka, and stickleback constitute the remaining species where receptor sequences have been predicted by data mining of their respective genomes. Amphibian receptors remain understudied, and our knowledge is restricted to GCG-R cloned from the African clawed, Northern leopard,[42] and tiger frogs,[61] and predicted sequences from the Western clawed frog (*Xenopus tropicalis*) genome.

Phylogeny based on structural analysis

To allow study of the genetic relationships within the glucagon receptor family, both known and predicted sequences have been considered, with the parathyroid hormone receptor (PTH-R) as an out-group (Fig. 4). Through phylogeny examination, GLP-2R is indicated to be most ancient in origin, agreeing with previous work.[42] Based on this, GLP-2Rs can be speculated as one of the preliminary products of a gene duplication of the proglucagon receptor gene early in vertebrate evolution (Fig. 5), thus giving rise to preliminary glucagon and GLP-2 receptor forms. A gene duplication of the glucagon receptor lineage followed, giving rise to GLP-1R forms. As a result of further gene duplication of some of these early receptors, distinct glucagon, GIP, and GLP-1 receptors were thus created as indicated by our phylogenetic analysis. In particular, the GIP and glucagon receptors appear to be more closely related to each other than to GLP-1R. Based on our phylogeny, although GIP and the proglucagon peptides are from distinct genes and are structurally less related, the GIP-R is structurally more similar to the GCG-R while receptors for GLPs have branched out and evolved earlier. This is an example showing the independent evolvement of ligand-receptor pairs in evolution.

With the exception of fish GLP-1R-likes, all glucagon superfamily receptors can be grouped based on subtype. Our phylogenetic analyses indicate fish GLP-1R-likes to be closely related to fish GCG-Rs and are clustered together within the GCG-R group (Fig. 4). This intriguing observation was also taken note of in earlier studies.[5,56,58] It was previously hypothesized that as a result of 3R early in teleost evolution prior species diversification,[5] a duplicate GCG-R acquired GLP-1 binding ability while retaining glucagon binding ability, thus giving rise to receptors currently recognized as "fish GLP-1Rs" and are paralogous to GCG-Rs.[5] This can be supported by our chromosome synteny, which reveals the gene arrangement near the locus of "fish GLP-1Rs" sharing greater similarity to GCG-Rs than to GLP-1Rs. Furthermore, the original GLP-1R in fish was also thought to have been lost from the fish lineage.[5] However by extensive data mining of the fugu, pufferfish, and medaka genomes, we have been able to predict sequences resembling those of known

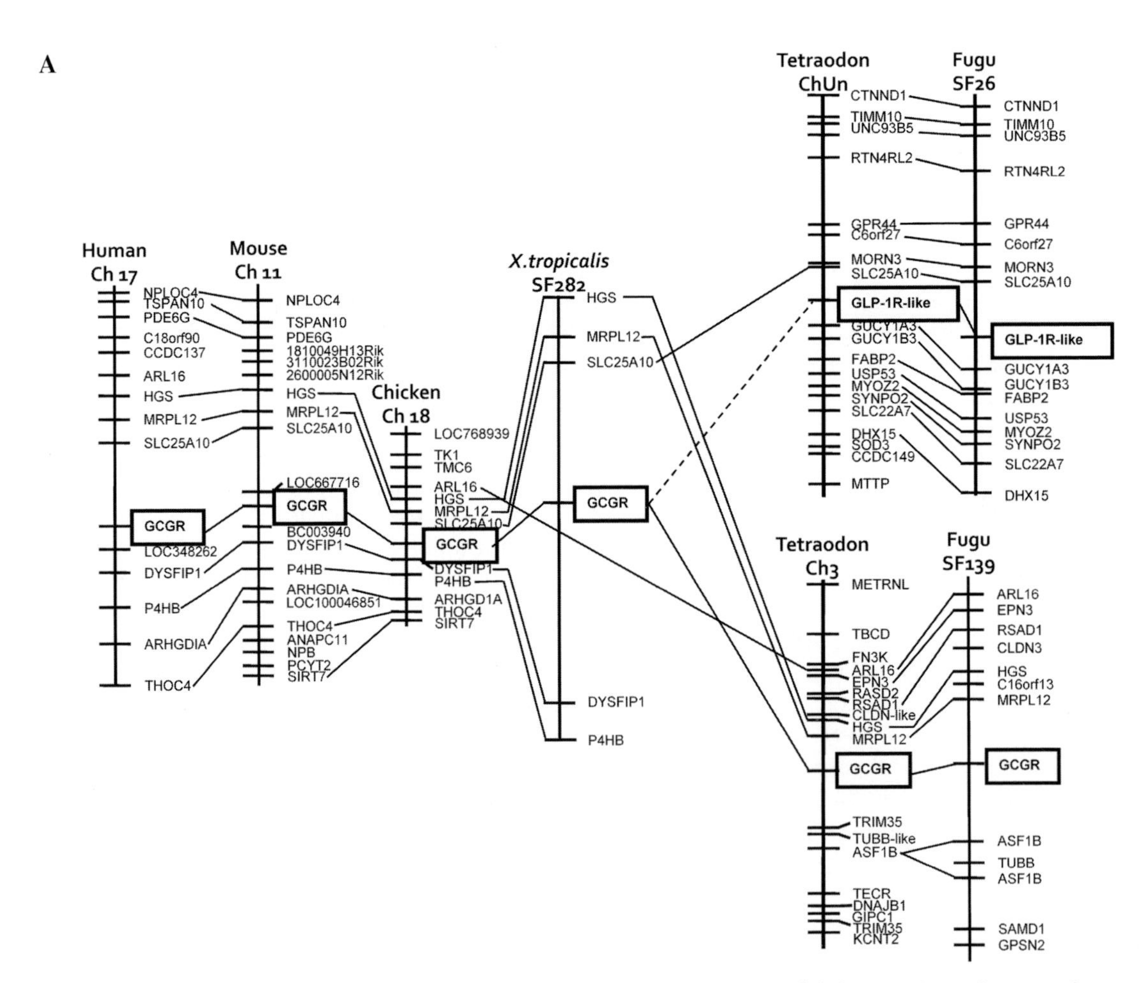

Figure 6. Chromosomal locations of (A) GCG-R, (B) GLP-1R, (C) GLP-2R, and (D) GIP-R in various vertebrate species. Genes adjacent to each of the receptors in different genomes are shown. The genes are named according to their annotation in the human genome. GCG-R, GLP-1R, GLP-2R, and GIP-R genes are boxed. *Continued.*

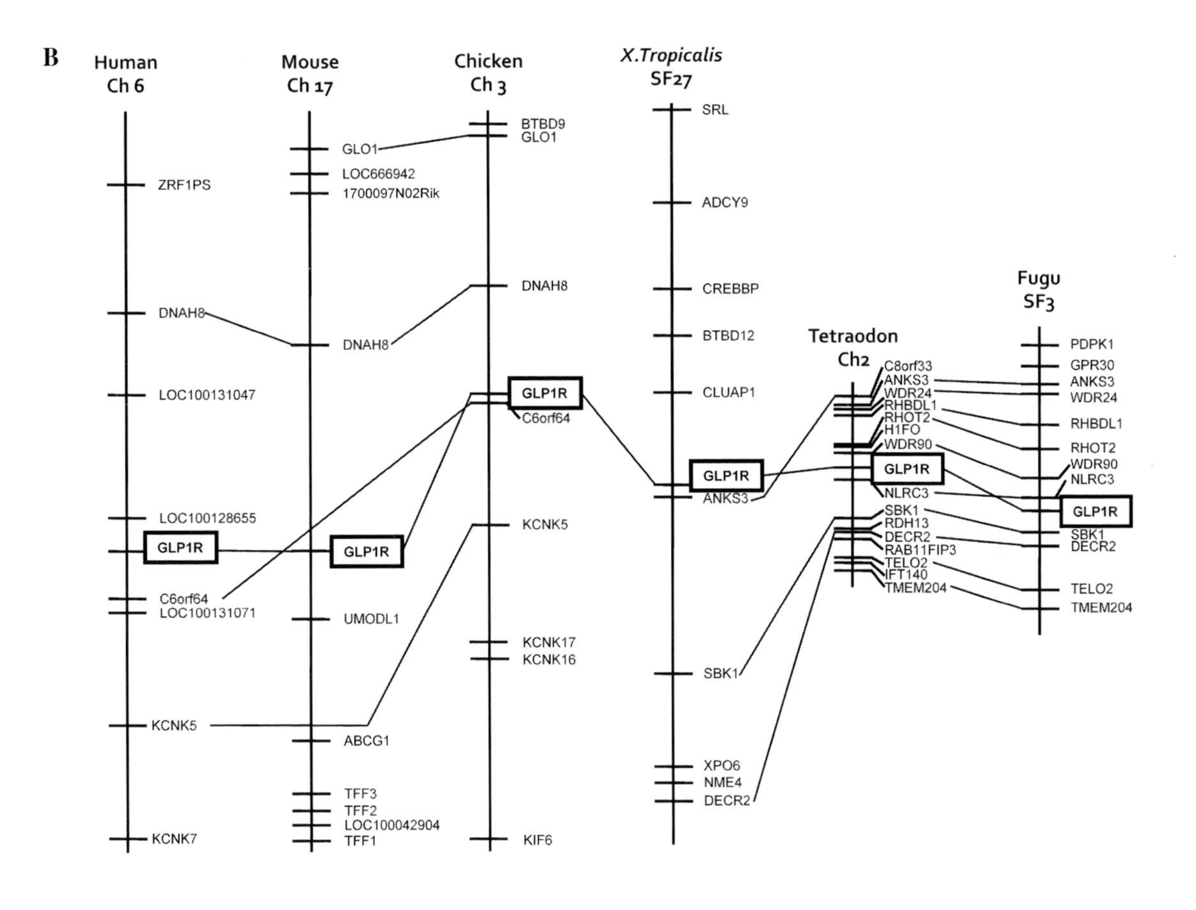

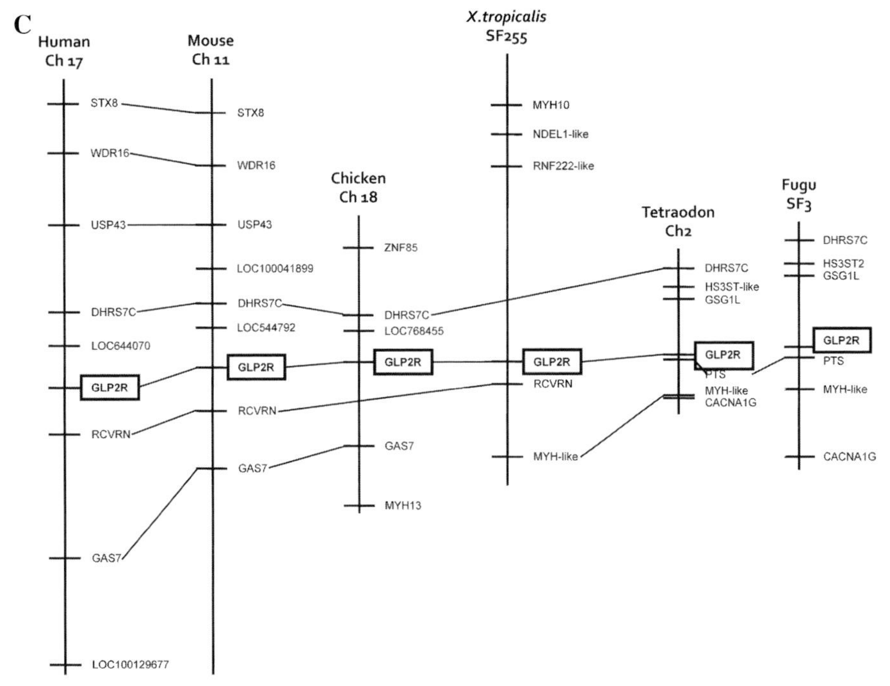

Figure 6. *Continued.*

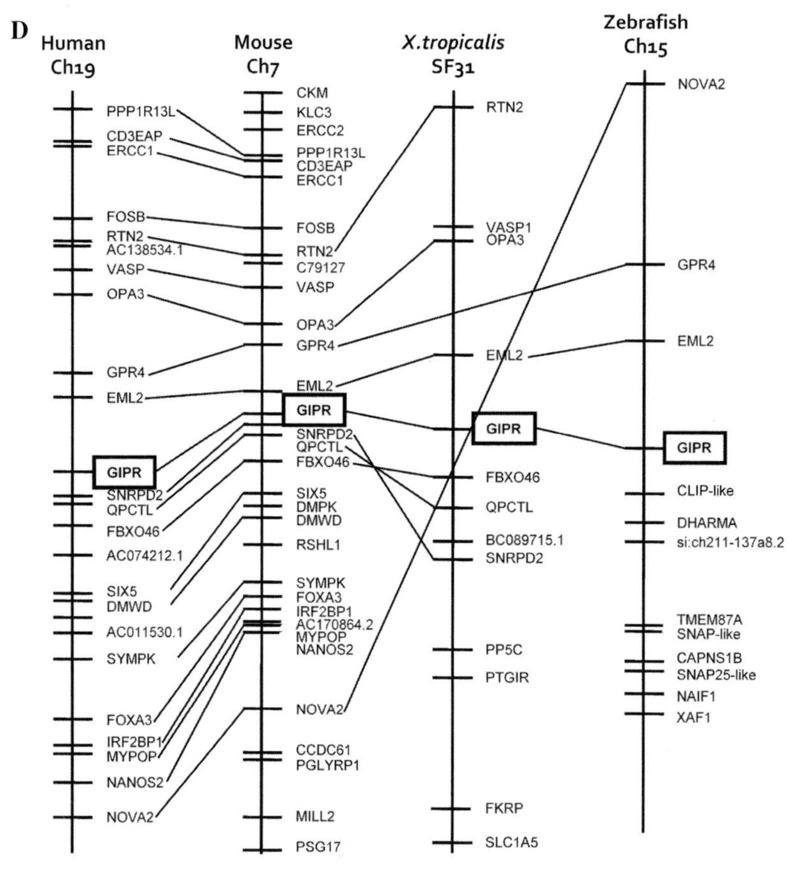

Figure 6. *Continued.*

GLP-1Rs, forming a separate GLP-1R group as supported by our phylogenetic analyses. Interestingly, the chromosomal locations of neighboring genes in these predicted GLP-1Rs do not correlate to those of mammalian nor avian orthologues (Fig. 6). Instead, these predicted GLP-1Rs are more similar to amphibian forms with neighboring genes such as ANKS3, SBK1, and DECR2 closely linked to the receptor's gene loci. Interestingly, several of the teleost neighboring genes such as WDR24, DECR2 could be found on the same chromosome as the mouse GLP-1R. These findings suggest that another GLP-1R in fish does indeed exist and were not lost. Therefore, receptors previously identified as "fish GLP-1Rs" should be referred to as "fish GLP-1R-likes."

Phylogeny by functional analysis

With respect to the bioactivity of glucagon and GLP-1, they are known to overlap, having gluconeogenic activity in fish. In knowing this, it would be speculated that glucagon and GLP-1 elicit their actions via the same receptor in fish. However evidence from hepatocyte binding experiments suggest otherwise, indicating separate receptors to be involved.[62] This is further confirmed by functionality tests based on goldfish receptors, where GCG-Rs bind exclusively with glucagon (IC_{50} = 0.6, 9 and 13 nmol/L for goldfish, zebrafish, and human, respectively) whereas GLP-1R-likes interact with both glucagon (EC_{50} = 0.53 and 1.2 nmol/L for goldfish and human, respectively) and GLP-1 (EC_{50} values = 0.18 and 0.9 nmol/L for goldfish and human, respectively)[58,59]. Fish GLP-1R-likes are able to accommodate a wider ligand spectrum and this includes extendin in addition to fish and mammalian glucagon and GLP-1s.[59] In contrast, mammalian GLP-1Rs preferably bind GLP-1s, whereas glucagon acts as a weak agonist.[63,64] Fish GLP-1R-likes are therefore structurally more flexible than mammalian GLP-1Rs, and this may be explained by its phylogenetic origin as demonstrated in our analysis. Fish GLP-1R-likes as shown are clustered with

GCG-Rs and are GCG-R duplicates that acquired GLP-1 binding ability. Therefore, a possible explanation to the gluconeogenic activity of fish GLP-1 is due to interaction with GLP-1R-likes that are paralogous to GCG-R as illustrated in our chromosome synteny. Based on this, we suspect the GLP-1R in fish may either be non-functional or function differently to known vertebrate GLP-1Rs having different expression patterns or signaling pathways, however this remains to be confirmed with cloning and functional tests.

Despite ligand binding differences, both GCG-Rs and GLP-1R-likes have maintained their locality in the liver and downstream effects in fish, further accounting for their homogeneous activity.[5,65] Based on glucagon and GLP-1's analogous roles, it is suggested that their peptide function did not diverge until after the divergence of fish and mammals.[42] To date, fish GLP-2's biological function remains unknown. However it is believed to behave differently to glucagon and GLP-1, suggesting divergence in function before divergence of fish and mammals.[42]

Unlike fish, frog GCG-R is unable to recognize a broad species range of glucagons. In particular, tiger frog GCG-R binds specifically with human glucagon but have no response to goldfish and zebrafish glucagons.[61] This binding specificity is attributed to by amino acids known to be catalytically important in binding of human glucagon to its receptor. Based on observation of several of these amino acids being substituted in fish glucagons, it is suggested that GCG-R's structure has become more stringently controlled with the emergence of frogs and subsequently maintained throughout vertebrate evolution.[61] As proposed, formation of a binding pocket in a receptor is a multistep process.[59] Therefore, sites in frog GCG-Rs, which were previously able to bind fish glucagons, may have been altered leading to recognition of non-fish forms only. To date, no amphibian GLP receptors have been isolated. However based on the duplicate glucagon receptor theory as previously proposed,[5] it can be speculated that amphibian GLP-1Rs bind specifically to GLP-1 only. This can be supported by GLP-1's insulinotropic and glycogenolytic activity in several frog species studied.[37,40,56]

With respect to the different binding abilities of fish and amphibian receptors as discussed, a similar approach can be used to predict the functional characteristics of GLP-2 and GIP receptors. Because fish glucagon and GLP-1 receptors are able to interact with both fish and nonfish peptide ligands, fish GLP-2 and GIP receptors may also likely have a binding pocket capable of recognizing cross-species' GLP-2 and GIP. Whereas amphibian receptors should have a narrower range, interacting with non-fish ligands as their binding pocket have likely been altered. However, these speculations remain to be confirmed, as the bioactivity of GLP-2 and GIP receptors is still understudied.

Independent evolution of proglucagon-derived peptides and receptors

When considering phylogenetic evolution, the proglucagon-derived peptides and receptors follow different evolutionary schemes with glucagon having most ancient origin for the peptides and GLP-2R for the receptors. This indicates that their pattern of evolution is independent of one another and ligand-receptor specificity likely evolved only much later after peptide and receptor subtype diversification. Although evolving differently, duplication either at the level of the gene or exon is demonstrated to be indispensible in providing the source of new materials from which ligand-receptor pair diversity has arisen. With regard to the proglucagon-derived peptides, all three hormones are triplication products of a common ancestral glucagon gene. In particular, fish and amphibians are interesting species to study, often possessing more than one copy of either the proglucagon gene or derived sequences. Examination of the various hormone sequences provides evidence for a series of gene and exon duplication events occurring throughout the fish and amphibian lineages, which is likely followed by species-specific loss of the duplicate copies. Within the fish lineage, a total of three rounds of gene duplication have likely occurred. These have resulted in duplicate proglucagon gene copies that are still evident in Agnathans and Osteichthyes, whereas loss or silencing of the duplicate gene may have occurred in the Chondrichthyes lineage. In amphibians, exon triplication specifically of the GLP-1 exon has resulted in multiple GLP-1 copies, which have been lost in some frog species by exon loss. Despite differences in gene or exon copies, the primary structure of the proglucagon-derived peptides has been well conserved in all vertebrate classes.

Throughout evolution, a peptide may undergo non-functionalization, sub-functionalization or neo-functionalization, and having a unique receptor is essential for eliciting of these functions. Indeed of the proglucagon peptide trio, GLP-1 is found to have acquired and subsequently lost gluconeogenic activity as illustrated by its contrasting functions in fish and mammals. As this is the case, this leads to possibility of either the peptide or receptor being responsible. In the case of the proglucagon-derived peptides, the high conservation of primary structure indicates strong evolutionary pressure to preserve function. Hence, the receptor may play a greater contributing factor as to why there was a change in biological activity. In fish, GLP-1R-likes bind both glucagon and GLP-1 whereas other vertebrate GLP-1Rs bind GLP-1 exclusively. GLP-1 in fish is able to have glucagon-like activity owing to interaction with fish GLP-1R-like that is speculated to be a GCG-R duplicate having acquired GLP-1 binding ability. In our study, we reveal that fish GLP-1Rs do exist, and this is supported by predicted sequences that cluster with known GLP-1Rs. Based on this finding, we believe previously identified "fish GLP-1Rs" should be termed as "fish GLP-1R-likes." Non-fish vertebrates have GLP-1Rs exhibiting high resemblance in both phylogenetic analysis and physiological function. Therefore, this may suggest that the fish GLP-1R does not function similarly to other vertebrate GLP-1Rs, thus resulting in GLP-1's unique insulinotropic activity in fish. If this holds true, then GLP-1R is likely to have acquired GLP-1 binding ability only after the fish lineage. We also believe that fish GLP-2 and GIP receptors have wide ligand binding range and amphibian counterparts at a lesser extent. Another noteworthy aspect is the existence of multiple frog GLP-1s, this raises the question as to whether the presence of these extra copies foreshadows an attempt to evolve new functions or if they have not yet been lost or silenced in the frog lineage following triplication. Lastly, identification of putative sequences from prochordates (such as the tunicate *Ciona intestinalis*[57,66]), which share some similarity with proglucagon receptors, further provides evidence for the presence of receptor genes even before vertebrate evolution and the importance of gene duplication events in driving the evolution of these receptors. Overall, the exact occurrence of the early evolutionary events still remains under debate and future molecular cloning of proglucagon-derived peptides and receptors for nonmammalian vertebrates is needed for clarification.

Conflicts of interest

The authors declare no conflicts of interest.

References

1. Irwin, D.M. 2001. Molecular evolution of proglucagon. *Regul. Pept.* **98:** 1–12.
2. De Duve, C. 1953. Glucagon; the hyperglycaemic glycogenolytic factor of the pancreas. *Lancet* **265:** 99–104.
3. Lund, P.K. *et al.* 1983. Anglerfish islet pre-proglucagon II. Nucleotide and corresponding amino acid sequence of the cDNA. *J. Biol. Chem.* **258:** 3280–3284.
4. Hasegawa, S. *et al.* 1990. Nucleotide sequence determination of chicken glucagon precursor cDNA. Chicken preproglucagon does not contain glucagon-like peptide II. *FEBS Lett.* **264:** 117–120.
5. Irwin, D.M. & K. Wong. 2005. Evolution of new hormone function: loss and gain of a receptor. *J. Hered.* **96:** 205–211.
6. Ohno, S. 1970. *Evolution by Gene Duplication.* Springer-Verlag. New York.
7. Venkatesh, B. 2003. Evolution and diversity of fish genomes. *Curr. Opin. Genet. Dev.* **13:** 588–592.
8. Kondrashov, F.A. & E.V. Koonin. 2001. Origin of alternative splicing by tandem exon duplication. *Hum. Mol. Genet.* **10:** 2661–2669.
9. Irwin, D.M. 2002. Ancient duplications of the human proglucagon gene. *Genomics* **79:** 741–746.
10. Hoyle, C.H. 1998. Neuropeptide families: evolutionary perspectives. *Regul. Pept.* **73:** 1–33.
11. Irwin, D.M., O. Huner & J.H. Youson. 1999. Lamprey proglucagon and the origin of glucagon-like peptides. *Mol. Biol. Evol.* **16:** 1548–1557.
12. Lopez, L.C. *et al.* 1984. Evolution of glucagon genes. *Mol. Biol. Evol.* **1:** 335–344.
13. Wriston, J.C., Jr. 1984. Comparative biochemistry of the guinea-pig: a partial checklist. *Comp. Biochem. Physiol. B.* **77:** 253–278.
14. Seino, S. *et al.* 1988. Appalachian spring: variations on ancient gastro-entero-pancreatic themes in New World mammals. *Horm. Metab. Res.* **20:** 430–435.
15. Irwin, D.M. 2009. Molecular evolution of mammalian incretin hormone genes. *Regul. Pept.* **155:** 121–130.
16. Irwin, D.M. & T. Zhang. 2006. Evolution of the vertebrate glucose-dependent insuliinotropic polypeptide (GIP) gene. *Comp. Biochem. Physiol.* 385–395.

17. Conlon, J.M. *et al.* 1995. Characterization of insulin, glucagon, and somatostatin from the river lamprey, Lampetra fluviatilis. *Gen. Comp. Endocrinol.* **100:** 96–105.
18. Kuraku, S. & S. Kuratani. 2006. Time scale for cyclostome evolution inferred with a phylogenetic diagnosis of hagfish and lamprey cDNA sequences. *Zoolog. Sci.* **23:** 1053–1064.
19. Conlon, J.M., P.F. Nielsen & J.H. Youson. 1993. Primary structures of glucagon and glucagon-like peptide isolated from the intestine of the parasitic phase lamprey Petromyzon marinus. *Gen. Comp. Endocrinol.* **91:** 96–104.
20. Wang, Y. *et al.* 1999. Multiple forms of glucagon and somatostatin isolated from the intestine of the southern-hemisphere lamprey *Geotria australis*. *Gen. Comp. Endocrinol.* **113:** 274–282.
21. Conlon, J.M. & L. Thim. 1985. Primary structure of glucagon from an elasmobranchian fish. *Torpedo marmorata*. *Gen. Comp. Endocrinol.* **60:** 398–405.
22. Conlon, J.M., L. O'Toole & L. Thim. 1987. Primary structure of glucagon from the gut of the common dogfish (*Scyliorhinus canicula*). *FEBS Lett.* **214:** 50–56.
23. Conlon, J.M., N. Hazon & L. Thim. 1994. Primary structures of peptides derived from proglucagon isolated from the pancreas of the elasmobranch fish, *Scyliorhinus canicula*. *Peptides* **15:** 163–167.
24. Conlon, J.M. *et al.* 1987. A glucagon-like peptide, structurally related to mammalian oxyntomodulin, from the pancreas of a holocephalan fish, *Hydrolagus colliei*. *Biochem. J.* **245:** 851–855.
25. Berks, B.C. *et al.* 1989. Isolation and structural characterization of insulin and glucagon from the holocephalan species *Callorhynchus milii* (elephantfish). *Biochem. J.* **263:** 261–266.
26. Hoegg, S. *et al.* 2004. Phylogenetic timing of the fish-specific genome duplication correlates with the diversification of teleost fish. *J. Mol. Evol.* **59:** 190–203.
27. Irwin, D.M. & J. Wong. 1995. Trout and chicken proglucagon: alternative splicing generates mRNA transcripts encoding glucagon-like peptide 2. *Mol. Endocrinol.* **9:** 267–277.
28. Conlon, J.M., S. Falkmer & L. Thim. 1987. Primary structures of three fragments of proglucagon from the pancreatic islets of the daddy Sculpin (Cottus scorpius). *Eur. J. Biochem.* **164:** 117–122.
29. Conlon, J.M. *et al.* 1988. Somatostatin-related and glucagon-related peptides with unusual structural features from the European eel (*Anguilla anguilla*). *Gen. Comp. Endocrinol.* **72:** 181–189.
30. Nguyen, T.M. *et al.* 1994. Characterization of insulins and proglucagon-derived peptides from a phylogenetically ancient fish, the paddlefish (*Polyodon spathula*). *Biochem. J.* **300**(Pt 2): 339–345.
31. Jaillon, O. *et al.* 2004. Genome duplication in the teleost fish *Tetraodon nigroviridis* reveals the early vertebrate proto-karyotype. *Nature* **431:** 946–957.
32. Kasahara, M. *et al.* 2007. The medaka draft genome and insights into vertebrate genome evolution. *Nature* **447:** 714–719.
33. Vandepoele, K. *et al.* 2004. Major events in the genome evolution of vertebrates: paranome age and size differ considerably between ray-finned fishes and land vertebrates. *Proc. Natl. Acad. Sci. USA* **101:** 1638–1643.
34. Amores, A. *et al.* 1998. Zebrafish hox clusters and vertebrate genome evolution. *Science* **282:** 1711–1714.
35. Aparicio, S. *et al.* 2002. Whole-genome shotgun assembly and analysis of the genome of Fugu rubripes. *Science* **297:** 1301–1310.
36. Robinson-Rechavi, M. *et al.* 2001. An ancestral whole-genome duplication may not have been responsible for the abundance of duplicated fish genes. *Curr. Biol.* **11:** 458–459.
37. Irwin, D.M. *et al.* 1997. The Xenopus proglucagon gene encodes novel GLP-1-like peptides with insulinotropic properties. *Proc. Natl. Acad. Sci. USA* **94:** 7915–7920.
38. Matutte, B. & J.M. Conlon. 2000. Characterization of insulin and atypically processed proglucagon-derived peptides from the surinam toad Pipa pipa (Anura:Pipidae). *Peptides* **21:** 1355–1360.
39. Yeung, C.M. & B.K. Chow. 2001. Identification of a proglucagon cDNA from Rana tigrina rugulosa that encodes two GLP-1s and that is alternatively spliced in a tissue-specific manner. *Gen. Comp. Endocrinol.* **124:** 144–151.
40. Conlon, J.M. *et al.* 1998. Purification and characterization of insulin, glucagon, and two glucagon-like peptides with insulin-releasing activity from the pancreas of the toad, Bufo marinus. *Endocrinology* **139:** 3442–3448.
41. Irwin, D.M. & P. Sivarajah. 2000. Proglucagon cDNAs from the leopard frog, Rana pipiens, encode two GLP-1-like peptides. *Mol. Cell Endocrinol.* **162:** 17–24.
42. Sivarajah, P., M.B. Wheeler & D.M. Irwin. 2001. Evolution of receptors for proglucagon-derived peptides: isolation of frog glucagon receptors. *Comp. Biochem. Physiol. B Biochem. Mol. Biol.* **128:** 517–527.
43. Cavanaugh, E.S., P.F. Nielsen & J.M. Conlon. 1996. Isolation and structural characterization of proglucagon-derived peptides, pancreatic polypeptide, and somatostatin from the urodele Amphiuma tridactylum. *Gen. Comp. Endocrinol.* **101:** 12–20.

44. Drucker, D.J. *et al.* 1996. Induction of intestinal epithelial proliferation by glucagon-like peptide 2. *Proc. Natl. Acad. Sci. USA.* **93:** 7911–7916.
45. Tang-Christensen, M. *et al.* 2000. The proglucagon-derived peptide, glucagon-like peptide-2, is a neurotransmitter involved in the regulation of food intake. *Nat. Med.* **6:** 802–807.
46. Tang-Christensen, M. *et al.* 2001. [Glucagon-like peptide 2, a neurotransmitter with a newly discovered role in the regulation of food ingestion]. *Ugeskr. Laeger.* **163:** 287–291.
47. Kieffer, T.J. & J.F. Habener. 1999. The glucagon-like peptides. *Endocr. Rev.* **20:** 876–913.
48. Rouille, Y., S. Martin & D.F. Steiner. 1995. Differential processing of proglucagon by the subtilisin-like prohormone convertases PC2 and PC3 to generate either glucagon or glucagon-like peptide. *J. Biol. Chem.* **270:** 26488–26496.
49. Dhanvantari, S., N.G. Seidah & P.L. Brubaker. 1996. Role of prohormone convertases in the tissue-specific processing of proglucagon. *Mol. Endocrinol.* **10:** 342–355.
50. Falkmer, S. & S. Van Noorden. 1983. *Ontogeny and Phylogeny of the Glucagon Cell.* Springer-Verlag. Heidelberg, Germany.
51. Plisetskaya, E.M. & T.P. Mommsen. 1996. Glucagon and glucagon-like peptides in fishes. *Int. Rev. Cytol.* **168:** 187–257.
52. Harmon, J.S. & M.A. Sheridan. 1992. Effects of nutritional state, insulin, and glucagon on lipid mobilization in rainbow trout, *Oncorhynchus mykiss. Gen. Comp. Endocrinol.* **87:** 214–221.
53. Albalat, A., J. Gutierrez & I. Navarro. 2005. Regulation of lipolysis in isolated adipocytes of rainbow trout (*Oncorhynchus mykiss*): the role of insulin and glucagon. *Comp. Biochem. Physiol. A Mol. Integr. Physiol.* **142:** 347–354.
54. Albalat, A. *et al.* 2005. Nutritional and hormonal control of lipolysis in isolated gilthead seabream (Sparus aurata) adipocytes. *Am. J. Physiol. Regul. Integr. Comp. Physiol.* **289:** R259–R265.
55. Silverstein, J.T. *et al.* 2001. Neuropeptide regulation of feeding in catfish, Ictalurus punctatus: a role for glucagon-like peptide-1 (GLP-1)? *Comp. Biochem. Physiol. B Biochem. Mol. Biol.* **129:** 623–631.
56. Mommsen, T.P., J.M. Conlon & D.M. Irwin. 2001. Amphibian glucagon family peptides: potent metabolic regulators in fish hepatocytes. *Regul. Pept.* **99:** 111–118.
57. Cardoso, J.C. *et al.* 2005. The secretin G-protein-coupled receptor family: teleost receptors. *J. Mol. Endocrinol.* **34:** 753–765.
58. Chow, B.K. *et al.* 2004. Identification and characterization of a glucagon receptor from the goldfish Carassius auratus: implications for the evolution of the ligand specificity of glucagon receptors in vertebrates. *Endocrinology* **145:** 3273–3288.
59. Yeung, C.M. *et al.* 2002. Isolation and structure-function studies of a glucagon-like peptide 1 receptor from goldfish Carassius auratus: identification of three charged residues in extracellular domains critical for receptor function. *Endocrinology* **143:** 4646–4654.
60. Mosjov, S. 2000. Glucagon-like peptide-1 (GLP-1) and the control of glucose metabolism in mammals and teleost fish. *Am. Zool.* **40:** 246–258.
61. Ngan, E.S. *et al.* 1999. Functional studies of a glucagon receptor isolated from frog Rana tigrina rugulosa: implications on the molecular evolution of glucagon receptors in vertebrates. *FEBS Lett.* **457:** 499–504.
62. Navarro, I. & T.W. Moon. 1994. Glucagon binding to hepatocytes isolated from two teleost fishes, the American eel and the brown bullhead. *J. Endocrinol.* **140:** 217–227.
63. Graziano, M.P. *et al.* 1993. Cloning and functional expression of a human glucagon-like peptide-1 receptor. *Biochem. Biophys. Res. Commun.* **196:** 141–146.
64. Fehmann, H.C. *et al.* 1994. Ligand-specificity of the rat GLP-I receptor recombinantly expressed in Chinese hamster ovary (CHO-) cells. *Z. Gastroenterol.* **32:** 203–207.
65. Roch, G.J., S. Wu & N.M. Sherwood. 2008. Hormones and receptors in fish: Do duplicates matter? *Gen. Comp. Endocrinol.*
66. Cardoso, J.C. *et al.* 2006. Evolution of secretin family GPCR members in the metazoa. *BMC Evol. Biol.* **6:** 108.

Ann. N.Y. Acad. Sci. ISSN 0077-8923

ANNALS OF THE NEW YORK ACADEMY OF SCIENCES
Issue: *Phylogenetic Aspects of Neuropeptides*

Oxytocin/vasopressin and gonadotropin-releasing hormone from cephalopods to vertebrates

Hiroyuki Minakata
Suntory Institute for Bioorganic Research, Osaka, Japan
Address for correspondence: Hiroyuki Minakata, Ph.D., 1-1-1, Wakayamadai, Shimamoto, Mishima, Osaka 618-8503, Japan. minakata@sunbor.or.jp

Recent advances in peptide search methods have revealed two peptide systems that have been conserved through metazoan evolution. Members of the oxytocin/vasopressin-superfamily have been identified from protostomian and deuterostomian animals, indicating that the oxytocin/vasopressin hormonal system represents one of the most ancient systems. In most protostomian animals, a single member of the superfamily shares oxytocin-like and vasopressin-like actions. Co-occurrence of two members has been discovered in modern cephalopods, octopus, and cuttlefish. We propose that cephalopods have developed two peptides in the molluscan evolutionary lineage like vertebrates have established two lineages in the oxytocin/vasopressin superfamily. The existence of gonadotropin-releasing hormone (GnRH) in protostomian animals was initially suggested by immunohistochemical analysis using chordate GnRH antibodies. A peptide with structural features similar to those of chordate GnRHs was originally isolated from octopus, and an identical peptide has been characterized from squid and cuttlefish. Novel forms of GnRH-like molecules from other molluscs, an annelid, arthropods, and nematodes demonstrate somewhat conserved structures at the N-terminal regions; but structures of the C-terminal regions critical to gonadotropin-releasing activity are diverse. These findings may be important for the study of the molecular evolution of GnRH in protostomian animals.

Keywords: oxytocin/vasopressin-superfamily peptides; gonadotropin-releasing hormone; peptide hormone; cephalopods

Introduction

Other molluscs are shell-protected, slowly moving scavengers that feed off plants, algae, or plankton. In contrast, cephalopods are agile and mobile predators that share their environment and way of life with fish or marine mammals and compete with them. Just as humans dominate the evolutionary lineage of vertebrates, cephalopods, "the primates in the sea," stand at the top of the evolutionary tree of marine invertebrates.

Recent advances in peptide hunting methods have revealed two peptide systems that are conserved in metazoan evolution. Peptide hormones such as oxytocin/vasopressin-superfamily peptides and gonadotropin-releasing hormone (GnRH) share structures and functions in vertebrates and in invertebrates.[1–7] As Gnathostomata and vertebrate groups above it (jawed vertebrates) possess at least two molecular forms of the oxytocin-family and vasopressin-family, two evolutionary lineages have been proposed.[1,2] In contrast, almost all invertebrates express only a single member of the oxytocin/vasopressin-superfamily of peptides.[6,8–11] Not surprisingly, a single member of the superfamily of proteins has both oxytocin-like and vasopressin-like actions.[12,13] In the evolution of molluscs, cephalopods have, like jawed vertebrates, secured two molecular forms with discriminative actions.[14–17]

The existence of GnRH in protostomian animals was initially suggested by immunohistochemical analysis using antibodies against mammalian and chicken GnRHs.[3–5] A peptide with structural features similar to those of chordate GnRHs was originally isolated from octopus,[18] and then an identical peptide was characterized from cuttlefish[19] and squid.[20] Novel forms of GnRH-like molecules from other molluscs have been characterized,[21,22] including annelids[22] and nematodes.[23] The

doi: 10.1111/j.1749-6632.2010.05569.x

structures of these GnRH-like peptides are somewhat conserved at the N-terminal regions, but those of the C-terminal regions, which are critical to their gonadotropin-releasing activity in vertebrates, are lacking or changed. Recently, genome sequencing was completed for an amphioxus, *Branchitostoma floridae*, the most basal chordate.[24] Characterization of amphioxus GnRH receptors led to the hypothesis that one receptor type leads to vertebrate GnRH receptors, whereas the other type, which related to the mollusc GnRH receptor, was lost in the vertebrate lineage.[25] The question was then raised as to how oxytocin/vasopressin and GnRH evolved in the molluscan lineage. Here, we review comparable structures and functions of the oxytocin/vasopressin superfamily of peptides and GnRH in modern cephalopods.

Results and discussion

Peptides of the molluscan oxytocin/vasopressin superfamily and their receptors

Members of the oxytocin/vasopressin superfamily of peptides are widely distributed among molluscs.[8,14–16,26–30] Lys-conopressin was isolated from a pond snail *Lymnaea stagnalis*[8] and also from a seaslug *Aplysia kurodai*.[26] Lys-conopressin possesses a positively charged amino acid, Lys, at position 8. Thus, based on the nomenclature in mammalian peptides it belongs to the vasopressin family. Lys-conopressin also potently induces contractions in the vas deferens, showing involvement in the control of ejaculation of semen during intromission in *Lymnaea*,[9] which corresponds to the role of oxytocin in mammals.[31]

The structural organization of the Lys-conopressin precursor is similar to that of the oxytocin precursor and the vasopressin precursor with a signal peptide, a nonapeptide, and a neurophysin domain.[1,2,8] The Lys-conopressin genome is composed of three exons and two introns at exactly the same locations as in the oxytocin genome and vasopressin genome of vertebrates.[8] The Lys-conopressin gene is a single copy gene in the *Lymnaea* genome, and related genes are absent.[9]

The venom of fish-hunting snails of the genus *Conus* is a rich source of neuroactive peptides (Table 1). Arg-conopressin S was isolated from the venom of *Conus striatus*,[27] and Lys-conopressin G was isolated from the venom of *Conus geographus* and *Conus imperialis*.[28] Recently, a novel bioactive peptide, conopressin-T, was isolated from *Conus tulipa* venom.[29] Conopressin-T, which possesses a Val instead of a conserved Gly at position 9, was found to act as a selective antagonist at the human V1a receptor and has partial agonist activity at the oxytocin receptor. [Val^9]oxytocin and [Val^9]vasopressin revealed that this position can function as an agonist/antagonist switch in the V1a receptor.[29] A novel peptide, designated γ-conopressin-vil, has been isolated from the venom of *Conus villepinii* (CLIQDCP-γ-G-NH_2, where γ-G is γ-carboxyglutamate).[30] γ-Conopressin-vil undergoes structural changes in the presence of Ca^{2+}. This suggests that the peptide binds Ca^{2+}, and that the Ca^{2+}-binding process is mediated by the γ-carboxyglutamate residue.[30]

Other members of the oxytocin/vasopressin superfamily of peptides in Protostomia have been found in arthropods[6,32] and annelids.[10,33–35] More thorough information on these peptides is available in a recent review by Salzet[36] and a report by Stafflinger *et al.*[6]

Cephalopods are the only invertebrates to differentiate their blood vessels into arteries and veins linked by a capillary network and can thus enjoy a high-pressure closed vessel system. A number of neurosecretory cells in the ventral median vasomotor lobe of the brain send a voluminous neuropil to the inner surface of the vena cava.[37] The nerves terminate in close contact with the blood forming the vena cava neurosecretory system. The terminals in the neuropil contain masses of dense granules. Vasopressin-like immunoreactivity occurred together with neurophysin-like immunostaining in the same terminals.[38] In 1992, Reich reported isolation of an oxytocin-like immunoreactive peptide from nerve terminals of the vena cava of *Octopus vulgaris*.[14] The peptide, named cephalotocin, exhibits 78% homology with mesotocin[14] (Table 1). However, Reich did not state its biological activities. Intriguingly, a novel peptide, octopressin, was isolated from octopus.[15] *Octopus* possesses two members of the oxytocin/vasopressin superfamily of peptides, as vertebrates do. This is an observation made for the first time for invertebrates.

Octopressin evoked contractions of smooth muscles such as rectum, oviduct, efferent branchial vessel, and anterior aorta.[15] In contrast, cephalotocin had no effect on these tissues.[15] The octopressin

Table 1. **Oxytocin/vasopressin-superfamily peptides, GnRHs, and GnRH-like peptides**

Name	Origin	Amino acid sequence
Oxytocin family peptides		
Oxytocin	Rat	Cys-Tyr-Ile-Gln-Asn-Cys-Pro-Leu-Gly-NH_2
Mesotocin	Chicken	Cys-Tyr-Ile-Gln-Asn-Cys-Pro-Ile-Gly-NH_2
Isotocin	Fish	Cys-Tyr-Ile-Ser-Asn-Cys-Pro-Ile-Gly-NH_2
Vasopressin family peptides		
Vasopressin	Rat	Cys-Tyr-Phe-Gln-Asn-Cys-Pro-Arg-Gly-NH_2
Vasotocin	Chicken	Cys-Tyr-Ile-Gln-Asn-Cys-Pro-Arg-Gly-NH_2
Oxytocin/Vasopressin superfamily peptides		
Lys-conopressin	Snail/sea slug	Cys-Phe-Ile-Arg-Asn-Cys-Pro-Lys-Gly-NH_2
Arg-conopressin S	Snail	Cys-Ile-Ile-Arg-Asn-Cys-Pro-Arg-Gly-NH_2
Conopressin-T	Snail	Cys-Tyr-Ile-Gln-Asn-Cys-Leu-Arg-Val-NH_2
γ-Conopressin-vil	Snail	Cys-Leu-Ile-Gln-Asp-Cys-Pro-g-carboxyGlu-NH_2
Cephalotocin	Octopus	Cys-Tyr-Phe-Arg-Asn-Cys-Pro-Ile-Gly-NH_2
Octopressin	Octopus	Cys-Phe-Trp-Thr-Ser-Cys-Pro-Ile-Gly-NH_2
GnRHs		
mGnRH	Rat	<Glu——His-Trp-Ser-Tyr-Gly-Leu-Arg-Pro-Gly-NH_2
cGnRH-I	Chicken	<Glu——His-Trp-Ser-Tyr-Gly-Leu-Gln-Pro-Gly-NH_2
cGnRH-II	Chicken	<Glu——His-Trp-Ser-His-Gly-Trp-Tyr-Pro-Gly-NH_2
sGnRH	Salmon	<Glu——His-Trp-Ser-Tyr-Gly-Trp-Leu-Pro-Gly-NH_2
Protostomian GnRHs and GnRH-like peptides		
oct-GnRH	Octopus	<Glu-Asn-Tyr-His-Phe-Ser-Asn-Gly-Trp-His-Pro-Gly-NH_2
oct-GnRH[a]	Squid/cuttlefish	-Gln-Asn-Tyr-His-Phe-Ser-Asn-Gly-Trp-His-Pro-Gly-Gly-Lys-Arg-
ap-GnRH[a]	Sea slug	-Gln-Asn-Tyr-His-Phe-Ser-Asn-Gly-Trp-Tyr-Ala-Gly-Lys-Arg-
ol-GnRH[a]	Owl limpet	-Gln-His-Tyr-His-Phe-Ser-Asn-Gly-Trp-Lys-Ser-Gly-Lys-Arg-
an-GnRH[a]	Annelid	-Gln-Ala-Tyr-His-Phe-Ser-His-Gly-Trp-Phe-Pro-Gly-Lys-Arg-
Ce-AKH-GnRH[a]	Nematode	-Gln-Met-Thr—Phe-Thr-Asp-Gln-Trp-Thr-Lys-Lys-Arg-
AKH	Locust	<Glu-Leu-Asn—Phe-Thr-Pro-Asn-Trp-Gly-Thr-NH_2
Dm-AKH	Fruitfly	<Glu-Leu-Thr—Phe-Ser-Pro-Asp-Trp-NH_2
Sp-GnRH[a,b]	Sea urchin	-Gln-Val-His-His-Arg-Phe-Ser-Gly-Trp-Arg-Pro-Gly-Gly-Lys-Lys-

Biochemically confirmed structures by Edman degradation, etc., and those predicted by in silico analysis are listed in this table. <Glu: pyroglutamic acid.

[a]Deduced structures from precursor cDNA were shown.

[b]Sea urchin (*Strongylocentrotus purpuraus*) is the Deuterostomia.

precursor and the cephalotocin precursor are composed of the typical structural units of the oxytocin/vasopressin superfamily precursors, a signal peptide, a nonapeptide, and a neurophysin domain.[1,2,15] Octopressin mRNA is expressed in the superior buccal lobe[15] which centrally controls feeding behavior, the inferior buccal ganglion, which does the same peripherally, and the gastric ganglion which regulates digestion.[37] This suggests that octopressin plays an important role in feeding behavior.[15] It is noteworthy that peripheral and central administration of oxytocin reduce feeding and act as a "satiety hormone" in rats.[2,39]

Specific expression of cephalotocin mRNA was observed in the ventral median vasomotor lobe, suggesting that cephalotocin synthesized in this lobe is transported to the vena cava and secreted into the blood.[15]

Oxytocin and vasopressin acting in the brain of mammals appear to be critical in neuronal processing for social recognition, maternal behaviors, and learning and memory.[2,40–43] Several million neuronal cells in the frontal and vertical lobes of the octopus brain create a series of matrices concerned with touch and visual learning.[44] These matrices are somewhat like the mammalian hippocampus.[44]

Octopressin mRNA is expressed in these lobes, suggesting its contribution to neuronal transmission and/or modulation in memory and learning systems.[15] Other lobes expressing octopressin mRNA are concerned with walking, swimming, changing color, respiration, etc.[37] Central administration of octopressin evoked hyperactivities of chromatophore cells, rapid respiration, and jetting water from the siphon to escape (unpublished results). Recently, Bardou et al. reported the cooccurrence of oxytocin-like and vasopressin-like immunoreactive signals in the central nervous system of the cuttlefish *Sepia officinalis.*[16] They used antibodies raised against mammalian oxytocin and vasopressin.[16] Several populations of oxytocin-like and vasopressin-like immunoreactive cell bodies and fibers were widely distributed in cerebral structures involved in learning processes, behavioral communication, feeding behavior, sexual activity, and metabolism, with no double immunostained cell bodies and fibres.[16] These results are somewhat similar to those of in situ hybridization experiments of mRNA expression of octopressin- and cephalotocin-precursors in the octopus.[15] Interestingly, intravenously injected of a low dose of octopressin enhanced long-term memory formation of a passive avoidance task, whereas a high dose attenuated it in *Sepia officinalis.*[17] Conversely, an enhancement of retention performance was observed at all doses of cephalotocin tested.[17]

As Gnathostomata and higher vertebrate (jawed vertebrates) possess at least two molecular forms of the oxytocin/vasopressin superfamily of peptides, two evolutionary lineages have been proposed: an isotocin-mesotocin-oxytocin line and a vasotocin-vasopressin line.[1,2] The jawless vertebrates (the Agnatha) have a single peptide belonging to the vasopressin family.[1,7] The oxytocin gene and the vasopressin gene may have arisen by duplication of a common ancestral gene after the radiation of the Agnatha, which occurred between 450 and 500 million years ago.[2] The presence of members of the oxytocin/vasopressin superfamily of peptides in invertebrates may go further back in time. The typical architecture of the precursors must have already been present in the Archaemetazoa, a stem group from which both vertebrates and invertebrates diverged about 640–760 million years ago.[2] It can be assumed that the octopus has developed two peptides in the molluscan evolutionary lineage as vertebrates have established two lineages of the oxytocin family and the vasopressin family.[15]

Genomes for cephalotocin and octopressin lack introns and consist of a single exon in their protein-coding regions,[45] which is unlike the three exon and two intron structures in *Lymnaea* Lys-conopressin.[8] The octopus may have lost introns during the evolutionary process in molluscs. Southern blot analysis revealed that a single copy of the octopressin gene and the cephalotocin gene was present in the genome.[45]

Three receptor genes cloned from the octopus brain show high homologies to those of the oxytocin/vasopressin superfamily of peptide receptors.[46,47] Two of them are subtypes of the cephalotocin receptor (CTR-1 and CTR-2) and one is the octopressin receptor (OPR). Expression of CTR-1 was strongly detected in the central and peripheral nervous systems, and slightly detected in peripheral tissues (Fig. 1). CTR-2 is mainly distributed in peripheral tissues, abundantly in the branchia and kidney (Fig. 1).[47] Significant synthesis of ammonia occurs within the kidney, and loss of ammonia takes place from the branchia.[48] Occurrence of a Na-dependent ATPase in the branchia of octopus was reported.[49] High expression of CTR-2 in the branchia and kidney may indicate that cephalotocin contributes ammonia excretion from the branchia like Arg-vasotocin does, which increases excretion of urea from the gill of toadfish,[50] and may control ionic balance coupled with Na/K-ATPase as it does in mammals.[51] Given that marine invertebrates including octopus are osmoconformers, water reabsorption activity, such as that induced by vasopressin may not be established. The OPR is widely expressed in both the central and peripheral nervous system and peripheral tissues, in which octopressin exhibits biological activities.[47] From the expressions (Fig. 1), a role for octopressin as a multifunctional neuropeptide has been suggested.[47]

The differences in the polarity of amino acid residues at position 8 are believed to confer on the oxytocin family and the vasopressin family the binding selectivity for their respective receptors.[52] Ile at position 8 is conserved between octopressin and cephalotocin, but amino acid residues at positions 2–5 are completely different. Cephalotocin has activity at CTR-1 and CTR-2, but not at the OPR. Conversely, octopressin activates only the OPR.[47] These results demonstrate that the amino acid residue at

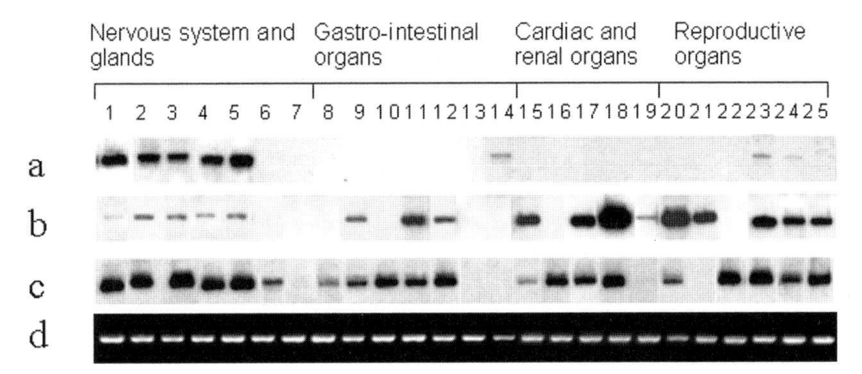

Figure 1. RT-PCR Southern-blot analysis of cephalotocin- (a: CTR1, b: CTR2), octopressin-receptors (c: OPR), and d: actin. 1: brain, 2: olfactory and peduncle lobes, 3: optic lobe, 4: buccal ganglion, 5: gastric ganglion, 6: optic gland, 7: white body, 8: esophagus, 9: crop, 10: stomach, 11: rectum, 12: salivary gland, 13: liver, 14: pancreas, 15: systemic heart, 16: aorta, 17: branchial vessel, 18: branchia, 19: kidney, 20: vas deferens, 21: testis, 22: egg, 23: ovary, 24: oviduct, and 25: oviducal gland.

position 8 is not responsible for the specific activity of octopressin and cephalotocin at the CTR and OPR, respectively, but the binding selectivity of octopressin and cephalotocin are dependent on amino acid residues at positions 2–5.[47] These specificities have been established in the evolutionary lineage of invertebrates, which is distinct from those of vertebrates.

Cephalopod GnRH and protostomian GnRH-like peptides

In vertebrates, a crucial role for regulating reproduction of gonadotropin-releasing hormone (GnRH) has been well defined.[3,53,54] Conversely, studies about molecular forms and roles of invertebrate GnRH-like peptides have only recently been launched.[18–22,55–62] The presence and distribution of GnRH-like immunoreactivity using multiple GnRH antibodies in neurons and reproductive tissues indicate that GnRH or GnRH variants exist throughout invertebrate phyla (Mollusca, Annelida, Arthropoda).[3–5,56–59] Neurons containing GnRH-like peptide stimulate egg-laying in the fresh water snails *Helisoma trivolvis* and *Lymnaea stagnalis*.[56] At least three isoforms of GnRH-like peptide are present in the central nervous system of the prawn *Macrobrachium rosenbergii*, and GnRH stimulates ovarian maturation in prawns.[58] A peptide with structural features similar to that of vertebrate GnRH was isolated from the brain of octopus (*Octopus vulgaris*).[18] The peptide showed luteinizing hormone-releasing activity on anterior pituitary cells of the Japanese quail *Coturnix coturnix*.[18] Thus, the peptide was named oct-GnRH. Oct-GnRH immunoreactivity was observed in the glandular cells of the mature optic gland, which is analogous to the anterior pituitary in the context of gonadal maturation.[60] Oct-GnRH stimulated the synthesis and release of sex steroids from the ovary and testis, and elicited contractions of the oviduct.[61] These results suggest that oct-GnRH in the octopus induces gonadal maturation and oviposition by regulating sex steroidogenesis and a series of egg-laying behaviors.[61] An immunohistochemical study using two different antibodies against vertebrate GnRHs suggested the presence of at least two forms of GnRH-containing neuronal projections in the spear-squid brain (*Loligo bleekeri*).[62] More comprehensive information on oct-GnRH, including the receptor involved in reproductive regulation of the octopus, is available in a recent review by Minakata *et al.*[63] Based on the precursor of octopus GnRH (oct-GnRH), the cDNA of a GnRH-like peptide from a sea slug, *Aplysia californica* (ap-GnRH), has been cloned.[21] Subsequently, searches within the expressed sequence tag (EST) database of Phyla Mollusca and Annelida have revealed two novel forms of GnRH-like molecules, one from an owl limpet *Lottia gigantea* (ol-GnRH), and one from a polychaete annelid *Capitella* sp. (an-GnRH) (Table 1).[22] The C-terminal-Pro-Gly-NH_2 structure critical to gonadotropin-releasing activity in vertebrate GnRHs is not conserved in these peptides.[21,22] Amino acid residues at the Pro position are diverse and the Gly residue necessary to convert it to a C-terminal -Gly-NH_2 is mutated to Lys or Arg.[21,22] Although

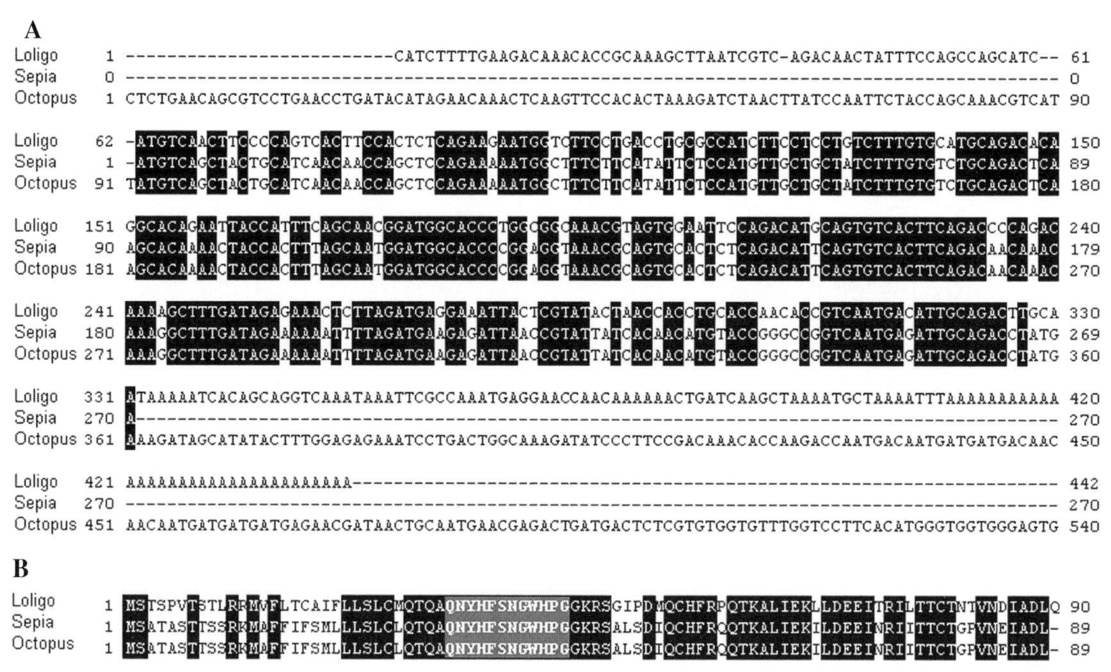

Figure 2. Nucleotide (A) and amino acid sequence (B) alignments of the GnRH precursor that originated from *Loligo edulis*, *Sepia officinalis*, and *Octopus vulgaris*. Oct-GnRH sequences are shown in bold face in a gray box in (B).

cloning has not been completed, a cDNA with exactly the same sequence as an oct-GnRH precursor from a European cuttlefish *Sepia officinalis* was reported and its expression was detected in brain and ovary.[19] The open reading frame of the cDNA cloned from a swordtip squid, *Loligo edulis*, is 80.5% similar to that of oct-GnRH[20] (Fig. 2). The amino acid structures of *Sepia* and *Loligo* GnRH are identical to those of oct-GnRH. Thus, the matured molecular form of cephalopod GnRH (oct-GnRH) is a dodecapeptide,[18–20] and those of other molluscan and annelidan GnRH-like peptides are undecapeptides.[21,22] Because the structures of the C-terminal regions of these peptides are different from that of vertebrate GnRH, it is questionable whether GnRH antibodies recognize these peptides. These data provide valuable support for the molecular evolution of protostomian GnRH, and also raise the important question as to why cephalopods have conserved a single GnRH molecule and homologous precursor genes. It is noteworthy that the deduced molecular form of the GnRH (sp-GnRH) of the deuterostomian Echinoderm *Strongylocentrotus purpuraus*[23] is a dodecapeptide that possesses a C-terminal-Pro-Gly-NH_2, like oct-GnRH, but Arg is inserted between His and Phe, and it lacks a residue between Ser and Gly (Table 1).

Tsai and Zhang recently proposed that chordate GnRH and protostomian GnRH-like peptides likely share a common ancestor, but that GnRH-like molecule has been lost in the ecdysozoan lineage (nematodes and arthropods) whereas it has been preserved in lophotrochozoans (annelids and molluscs).[22] GnRH receptor-like receptors have been retained in Ecdysozoa, but they now bind other ligands. The ligand of the GnRH-receptor orthologues of *Drosophila melanogaster* and *Bombyx mori* has been identified as adipokinetic hormone (AKH).[64] In the nematode *Caenorhabditis elegans* a receptor homologous to the AKH receptor of insects and the GnRH receptor of vertebrates (Ce-GnRHR) has been identified.[65] *Drosophila* AKH can activate Ce-GnRHR, whereas human GnRH and chicken GnRH-II cannot.[23] Lindemans et al. proposed that the ligand is analogous to AKH, but an AKH-GnRH signaling system is present in the nematode as indicated by its effects on reproductive functions.[23] Gene silencing experiments of Ce-GnRHR or its ligand or both result in a delay in egg laying.[23] However, the structure of the nematode peptide (Ce-AKH-GnRH) is far different from those of GnRHs (Table 1). It is debatable whether the nomenclature, AKH-GnRH, is appropriate.

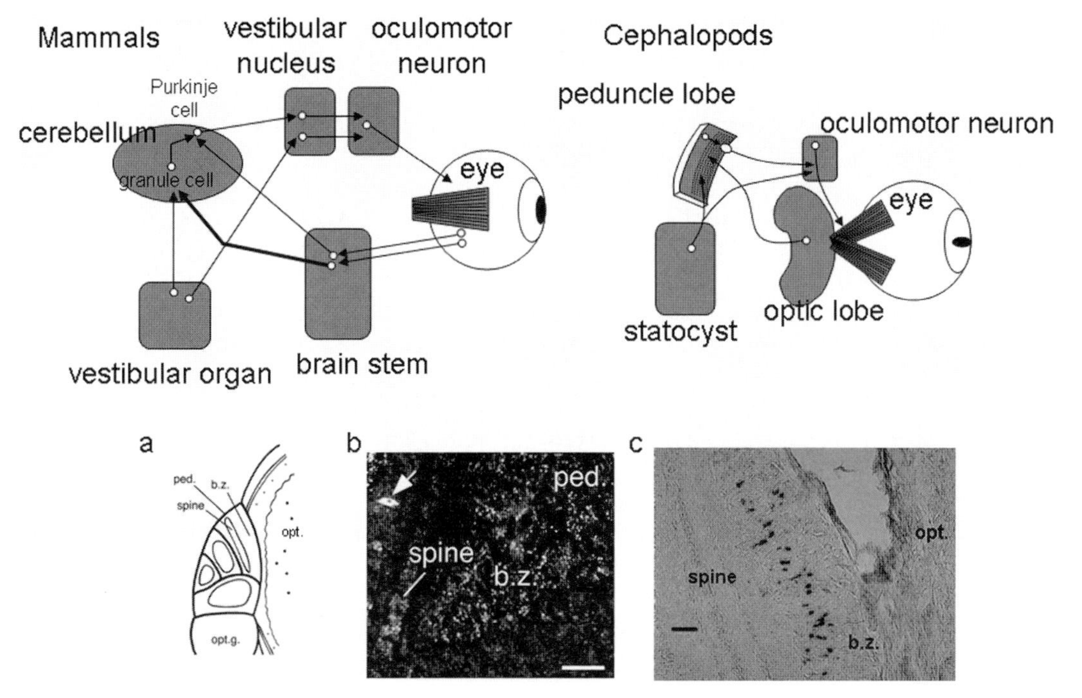

Figure 3. A cephalopod system similar to the cerebellum. The cephalopod system controls head and eye positions and movements. Modified from the illustration of Young's report (see Ref. 68). (A) Illustration of a horizontal section of the peduncle lobe. ped: peduncle lobe, b.z.: basal zone, opt.: optic lobe. (B) oct-GnRH-immunopositive fibers in the spine and the basal zone of the peduncle lobe. Scale bar = 50 μm. (C) oct-GnRH mRNA expression in the basal zone of the peduncle lobe. Scale bar = 100 μm.

Cephalopods possess well-developed eyes with large optic lobes, which conduct complex analysis of visual information. There are neurons that send commands to motor centers in the central brain and hence to the muscles of the mantle, funnel, fins, and arms.[37] Many of the neurons indirectly influence the motor system by pathways that involve the peduncle lobe of the optic tract[66] (Fig. 3A). Electrical stimulation in the peduncle lobe exhibits coordinated motor patterns and surgical lesions of the peduncle lobe result in various motor defects, resulting in jerky and less precise movements.[67] The diagram of the octopus system, including the peduncle lobe, the oculomotor neuron, the eye, the optic lobe, and the statocyst, corresponds to that of mammals[68] (Fig. 3). Oct-GnRH was injected into the blood sinus near the peduncle lobe of hatchlings of the octopus *Octopus bimaculoides* and caused several motor system abnormalities such as unstable crawling, up-side down postures, and abnormal movements of the arms.[63] Oct-GnRH-immunoreactive fibers in the spine of the peduncle lobe were detected[60] (Fig. 3B) as was expression of oct-GnRH mRNA in neuronal cell bodies in the basal zone (Fig. 3C). There are two types of cells in the basal zone cell layer: large motor cells and small cells.[66] Further experiments are necessary to confirm which type of cells express oct-GnRH. Transmitters involved in this system have not been clarified yet, but the presence of 5-HT[69] and GABA[70] has been reported. Administration of oct-GnRH into the blood sinus reaches the peduncle lobe and may cause disruption of signal transmission. Thus, oct-GnRH may be involved in signal transmission within the octopus "cerebellar" system. More information on the multifunctional oct-GnRH system is discussed in a previous review.[63]

Conclusion

The oxytocin/vasopressin superfamily of peptides and receptors has been well conserved in invertebrates. Moreover, the molecular structure, precursor organization, and actions mediated by these peptide receptors are comparable to those of vertebrates. Almost all invertebrates except for cephalopods possess only a single member of the superfamily, which has both oxytocin-like and vasopressin-like actions. In the evolution of molluscs, the most evolved cephalopods have secured two molecular forms with discriminative actions, as have all jawed vertebrate groups. The superfamily peptides of cephalopods may conserve mammalian oxytocin/vasopressin-like roles in the brain for learning and memory, an observation made for the first time in invertebrate species.

On the other hand, molecular forms of GnRH are not likely conserved in vertebrates and invertebrates. As proposed by Tsai *et al.* and Lindeman *et al.*, GnRH-receptor like receptors have been retained in metazoan animals, but they have recruited diverse ligands. What does the similarity between oct-GnRH and vertebrate GnRH mean? In addition to their endocrine activities, extrahypothalamic GnRHs in vertebrates are modulators of many physiological functions such as behavior, metabolism, immunity, etc.[71] The octopus with its advanced brain may enjoy a chordate-like multifunctional GnRH system. Taken together, the above data suggest that there are some parallels between the ligand-receptor coevolution of the cephalopod and the chordate systems as a result of adaptive evolution.

Conflict of interest

The author declares no conflict of interest.

References

1. Hoyle, C.H. 1998. Neuropeptide families: evolutionary perspectives. *Regul. Pept.* **73:** 1–33.
2. Gimpl, G. & F. Fahrenholz. 2001. The oxytocin receptor system: structure, function, and regulation. *Physiol. Rev.* **81:** 629–683.
3. Gorbman, A. & S.A. Sower. 2003. Evolution of the role of GnRH in animal (Metazoan) biology. *Gen. Comp. Endocrinol.* **134:** 207–213.
4. Tsai, P.S. 2006. Gonadotropin-releasing hormone in invertebrates: structure, function, and evolution. *Gen. Comp. Endocrinol.* **148:** 48–53.
5. Kah, O., C. Lethimonier, G. Somoza, *et al.* 2007. GnRH and GnRH receptors in metazoan: a historical, comparative, and evolutive perspective. *Gen. Comp. Endocrinol.* **153:** 346–364.
6. Stafflinger, E., K.K. Hansen, F. Hauser, *et al.* 2008. Cloning and identification of an oxytocin/vasopressin-like receptor and its ligand from insects. *Proc. Natl. Acad. Sci. USA* **105:** 3262–3267.
7. Gwee, P.C., B.H. Tay, S. Brenner & B. Venkatesh. 2009. Characterization of the neurohypophysial hormone gene loci in elephant shark and the Japanese lamprey: origin of the vertebrate neurohypophysial hormone genes. *BMC Evol. Biol.* **9:** 47.
8. van Kesteren, R.E., A.B. Smit, R.W. Dirks, *et al.* 1992. Evolution of the vasopressin/oxytocin superfamily: characterization of a cDNA encoding a vasopressin-related precursor, preproconopressin, from the mollusc *Lymnaea stagnalis*. *Proc. Natl. Acad. Sci. USA* **89:** 4593–4597.
9. van Kesteren, R.E., A.B. Smit, R.P. Lange de, *et al.* 1995. Structural and functional evolution of the vasopressin/oxytocin superfamily: vasopressin-related conopressin is the only member present in *Lymnaea*, and is involved in the control of sexual behavior. *J. Neurosci.* **15:** 5989–5998.
10. Oumi, T., K. Ukena, O. Matsushima, *et al.* 1994. Annetocin: an oxytocin-related peptide isolated from the earthworm, *Eisenia foetida*. *Biochem. Biophys. Res. Commun.* **198:** 393–399.
11. Kawada, T., T. Sekiguchi, Y. Itoh, *et al.* 2008. Characterization of a novel vasopressin/oxytocin superfamily peptide and its receptor from an ascidian, *Ciona intestinalis*. *Peptides.* **29:** 1672–1678.
12. van Kesteren, R.E., C.P. Tensen, A.B. Smit, *et al.* 1996. Co-evolution of ligand–receptor pairs in the vasopressin/oxytocin superfamily of bioactive peptides. *J. Biol. Chem.* **271:** 3619–3626.
13. Kawada, T., A. Kanda, H. Minakata, *et al.* 2004. Identification of a novel receptor for an invertebrate oxytocin/vasopressin superfamily peptide: molecular and functional evolution of the oxytocin/vasopressin superfamily. *Biochem. J.* **382:** 231–237.
14. Reich, G. 1992. A new peptide of the oxytocin/vasopressin family isolated from nerves of the cephalopod *Octopus vulgaris*. *Neurosci. Lett.* **134:** 191–194.
15. Takuwa-Kuroda, K., E. Iwakoshi-Ukena, A. Kanda & H. Minakata. 2003. Octopus, which owns the most advanced brain in invertebrates, has two members of vasopressin/oxytocin superfamily as in vertebrates. *Regul. Pept.* **115:** 139–419.
16. Bardou, I., E. Maubert, J. Leprince, *et al.* 2009. Distribution of oxytocin-like and vasopressin-like immunoreactivities within the central nervous system of the cuttlefish, *Sepia officinalis*. *Cell Tissue Res.* **336:** 249–266.
17. Bardou, I., J. Leprince, R. Chichery, *et al.* 2010. Vasopressin/oxytocin-related peptides influence long-term memory of a passive avoidance task in the cuttlefish, *Sepia officinalis*. *Neurobiol. Learn Mem.* **93:** 240–247.
18. Iwakoshi, E., K. Takuwa-Kuroda, Y. Fujisawa, *et al.* 2002. Isolation and characterization of a GnRH-like peptide

from *Octopus vulgaris. Biochem. Biophys. Res. Commun.* **291:** 1187–1193.

19. Di Cristo, C., E. De Lisa & A. Di Cosmo. 2009. GnRH in the brain and ovary of *Sepia officinalis. Peptides* **30:** 531–537.
20. Onitsuka, C., A. Yamaguchi, H. Kanamaru, *et al.* 2009. Molecular cloning and expression analysis of a GnRH-like dodecapeptide in the swordtip squid, *Loligo edulis. Zoolog. Sci.* **26:** 203–208.
21. Zhang, L., J.A. Tello, W. Zhang & P.S. Tsai. 2008. Molecular cloning, expression pattern, and immunocytochemical localization of a gonadotropin-releasing hormone-like molecule in the gastropod mollusc, *Aplysia californica. Gen. Comp. Endocrinol.* **156:** 201–209.
22. Tsai, P.S. & L. Zhang. 2008. The emergence and loss of gonadotropin-releasing hormone in protostomes: orthology, phylogeny, structure, and function. *Biol. Reproduc.* **9:** 798–805.
23. Lindemans, M., F. Liu, T. Janssen, *et al.* 2009. Adipokinetic hormone signaling through the gonadotropin-releasing hormone receptor modulates egg-laying in *Caenorhabditis elegans. Proc. Natl. Acad. Sci. USA* **106:** 1642–1647.
24. Holland, L.Z., R. Albalat, K. Azumi, *et al.* 2008. The amphioxus genome illuminates vertebrate origins and cephalochordate biology. *Genome Res.* **18:** 1100-1111.
25. Tello, J.A. & M.M. Sherwood. 2009. Amphioxus: beginning of vertebrate and end of invertebrate type GnRH receptor lineage. *Endocrinol.* **150:** 2847–2856.
26. McMaster, D., Y. Kobayashi & K. Lederis. 1992. A vasotocin-like peptide in *Aplysia kurodai* ganglia: HPLC and RIA evidence for its identity with Lys-conopressin G. *Peptides* **13:** 413–421.
27. Cruz, L.J., V. de Santos, G.C. Zafaralla, *et al.* 1987. Invertebrate vasopressin/oxytocin homologs. Characterization of peptides from *Conus geographus* and *Conus striatus* venoms. *J. Biol. Chem.* **262:** 15821–15824.
28. Nielsen, D.B., J. Dykert, J.E. Rivier & J.M. McIntosh. 1994. Isolation of Lys-conopressin-G from the venom of the worm-hunting snail, *Conus imperialis. Toxicon* **32:** 845–848.
29. Dutertre, S., D. Croker, N.L. Daly, *et al.* 2008. Conopressin-T from *Conus tulipa* reveals an antagonist switch in vasopressin-like peptides. *J. Biol. Chem.* **283:** 7100–7108.
30. Möller, C. & F. Marí. 2007. A vasopressin/oxytocin-related conopeptide with gamma-carboxyglutamate at position 8. *Biochem. J.* **404:** 413–419.
31. Todd, K. & S.L. Lightman. 1986. Oxytocin release during coitus in male and female rabbits: effect of opiate receptor blockade with nalaxone. *Psychoneuroendocrinology* **11:** 367–371.
32. Proux, J.P., C.A. Miller, J.P. Li, *et al.* 1987. Identification of an arginine vasopressin-like diuretic hormone from *Locusta migratoria. Biochem. Biophys. Res. Commun.* **149:** 180–186.
33. Oumi, T., K. Ukena, O. Matsushima, *et al.* 1996. Annetocin, an annelid oxytocin-related peptide, induces egg-laying behavior in the earthworm, *Eisenia foetida. J. Exp. Zool.* **276:** 151–156.
34. Fujino, Y., T. Nagahama, T. Oumi, *et al.* 1999. Possible functions of oxytocin/vasopressin-superfamily peptides in annelids with special reference to reproduction and osmoregulation. *J. Exp. Zool.* **284:** 401–406.
35. Satake, H., K. Takuwa, H. Minakata & O. Matsushima. 1999. Evidence for conservation of the vasopressin/oxytocin superfamily in Annelida. *J. Biol. Chem.* **274:** 5605–5611.
36. Salzet, M. 2006. Molecular aspect of annelid neuroendocrine system. In *Invertebrate Neuropeptides and Hormones: Basic Knowledge and Recent Advances.* H. Satake, Ed.: 17–35, Transworld Research Network. Kerala, India.
37. Young, J.Z. 1971. *The Anatomy of the Nervous System of Octopus vulgaris.* Oxford University Press. London.
38. Martin, R., D. Frosch & K.H. Voigt. 1980. Immunocytochemical evidence for melanotropin-and vasopressin-like material in a cephalopod neurohemal organ. *Gen. Comp. Endocrinol.* **42:** 235–243.
39. Arletti, R., A. Benelli & A. Bertolini. 1990. Oxytocin inhibits food and fluid intake in rats. *Physiol. Behav.* **48:** 825–830.
40. Rose, J.D. & F.L. Moore. 2002. Behavioral neuroendocrinology of vasotocin and vasopressin and the sensorimotor processing hypothesis. *Front. Neuroendocrinol.* **81:** 317–341.
41. Bielsky, I.F. & L.J. Young. 2004. Oxytocin, vasopressin, and social recognition in mammals. *Peptides* **25:** 1565–1574.
42. McEwen, B.B. 2004. General introduction to vasopressin and oxytocin: structure/metabolism, evolutionary aspects, neural pathway/receptor distribution, and functional aspects relevant to memory processing. *Adv. Pharmacol.* **50:** 1–50.
43. Caldwell, H.K. & W.S. Young. 2006. Oxytocin and vasopressin: Genetics and behavioral implications. In *Handbook of Neurochemistry and Molecular Neurobiology.* R. Lim, 3rd, Ed.: 573–607. Springer. New York.
44. Young, J.Z. 1995. Multiple matrices in the memory system of *Octopus.* In *Cephalopod Neurobiology.* N.J. Abbott, R. Williamson & L. Maddock, Eds.: 431–443. Oxford University Press. London.

45. Kanda, A., K. Takuwa-Kuroda, E. Iwakoshi-Ukena & H. Minakata. 2003. Single exon structures of the oxytocin/vasopressin superfamily peptides of octopus. *Biochem. Biophys. Res. Commun.* **309:** 743–748.
46. Kanda, A., K. Takuwa-Kuroda, E. Iwakoshi-Ukena, *et al.* 2003. Cloning of *Octopus* cephalotocin receptor, a member of the oxytocin/vasopressin superfamily. *J. Endocrinol.* **179:** 281–291.
47. Kanda, A., H. Satake, T. Kawada & H. Minakata. 2005. Novel evolutionary lineages of the invertebrate oxytocin/vasopressin superfamily peptides and their receptors in the common octopus (*Octopus vulgaris*). *Biochem. J.* **387:** 85–91.
48. Potts, W.T.W. 1965. Ammonia excretion in *Octopus dofleini*. *Comp. Biochem. Physiol.* **14:** 339–355.
49. Schoffeniels, E. 1962. Isolation of a sodium-dependent ATPase from the gills of *Octopus vulgaris* L. *Life Sci.* **9:** 437–440.
50. Mahlmann, S., W. Meyerhof, H. Hausmann, *et al.* 1994. Structure, function, and phylogeny of [Arg^8]vasotocin receptors from teleost fish and toad. *Proc. Natl. Acad. Sci. USA* **91:** 1342–1345.
51. Bertuccio, C.A., F.R. Ibarra, J.E. Toledo, *et al.* 2002. Endogenous vasopressin regulates Na-K-ATPase and Na^+-K^+-Cl^- cotransporter *rbsc-1* in rat outer medulla. *Am. J. Physiol. Renal Physiol.* **282:** F265–F270.
52. Barberis, C., B. Mouillac & T. Durroux. 1998. Structural bases of vasopressin/oxytocin receptor function. *J. Endocrinol.* **156:** 223–229.
53. Debeljuk, L., A. Arimura & A.V. Schally. 1973. Stimulation of release of FSH and LH by infusion of LH-RH and some of its analogues. *Neuroendocrinology*. **11:** 130–136.
54. Conn, P.M. & W.F. Crowley Jr. 1994. Gonadotropin-releasing hormone and its analogs. *Annu. Rev. Med.* **45:** 391–405.
55. Adams, B.A., J.A. Tello, J. Erchegyi, *et al.* 2003. Six novel gonadotropin-releasing hormones are encoded as triplets on each of two genes in the protochordate, *Ciona intestinalis*. *Endocrinology* **144:** 1907–1919.
56. Young, K.G., J.P. Chang & J.I. Goldberg. 1999. Gonadotropin-releasing hormone neuronal system of the freshwater snails *Helisoma trivolvis* and *Lymnaea stagnalis*: possible involvement in reproduction. *J. Comp. Neurol.* **404:** 427–437.
57. Nuurai, P., J. Poljaroen, Y. Tinikul, *et al.* 2009. The existence of gonadotropin-releasing hormone-like peptides in the neural ganglia and ovary of the abalone, *Haliotis asinina* L. *Acta Histochem.* In press.
58. Ngernsoungnern, A., P. Ngernsoungnern, S. Kavanaugh, *et al.* 2008. The identification and distribution of gonadotropin-releasing hormone-like peptides in the central nervous system and ovary of the giant freshwater prawn, *Macrobrachium rosenbergii*. *Invert. Neurosci.* **8:** 49–57.
59. Ngernsoungnern, P., A. Ngernsoungnern, S. Kavanaugh, *et al.* 2008. The presence and distribution of gonadotropin-releasing hormone-liked factor in the central nervous system of the black tiger shrimp, *Penaeus monodon*. *Gen. Comp. Endocrinol.* **155:** 613–622.
60. Iwakoshi-Ukena, E., K. Ukena, K. Takuwa-Kuroda, *et al.* 2004. Expression and distribution of octopus gonadotropin-releasing hormone in the central nervous system and peripheral organs of the octopus (*Octopus vulgaris*) by in situ hybridization and immunohistochemistry. *J. Comp. Neurol.* **477:** 310–323.
61. Kanda, A., T. Takahashi, H. Satake & H. Minakata. 2006. Molecular and functional characterization of a novel gonadotropin-releasing-hormone receptor isolated from the common octopus (*Octopus vulgaris*). *Biochem. J.* **395:** 125–135.
62. Amano, M., Y. Oka, Y. Nagai, *et al.* 2008. Immunohistochemical localization of a GnRH-like peptide in the brain of the cephalopod spear-squid, *Loligo bleekeri*. *Gen. Comp. Endocrinol.* **156:** 227–284.
63. Minakata, H., S. Shigeno, N. Kano, *et al.* 2009. Octopus gonadotrophin-releasing hormone: a multifunctional peptide in the endocrine and nervous systems of the cephalopod. *J. Neuroendocrinol.* **21:** 322–326.
64. Staubli, F., T.J. Jorgensen, G. Cazzamali, *et al.* 2002. Molecular identification of the insect adipokinetic hormone receptors. *Proc. Natl. Acad. Sci. USA* **99:** 3446–3451.
65. Vadakkadath, M.S., M.J. Gallego, R.J. Haasl, *et al.* 2006. Identification of a gonadotropin-releasing hormone receptor orthologue in *Caenorhabditis elegans*. *BMC Evol. Biol.* **6:** 103.
66. Woodhams, P.L. 1977. The ultrastructure of a cerebellar analogue in Octopus. *J. Comp. Neurol.* **174:** 329–346.
67. Messenger, J.B. 1967. The peduncle lobe: a visuo-motor centre in Octopus. *Proc. R. Soc. B.*, **167:** 225–251.
68. Young J.Z. 1976. The 'cerebellum' and the control of eye movements in the cephalopods. *Nature* **264:** 572–574.
69. Kito-Yamashita, T., C. Haga, K. Hirai, *et al.* 1990. Localization of serotonin immunoreactivity in cephalopod visual system. *Brain Res.* **521:** 81–88.
70. Cornwell, C.J., J.B. Messenger & R. Williamson. 1993. Distribution of GABA-like immunoreactivity in the octopus brain. *Brain Res.* **621:** 353–357.
71. Fernald, R.D. & R.B. White. 1999. Gonadotropin-releasing hormone genes: phylogeny, structure, and functions. *Front. Neuroendocrinol.* **20:** 224–240.

Ann. N.Y. Acad. Sci. ISSN 0077-8923

ANNALS OF THE NEW YORK ACADEMY OF SCIENCES
Issue: *Phylogenetic Aspects of Neuropeptides*

Somatostatin and its receptors from fish to mammals

Manuel D. Gahete, Jose Cordoba-Chacón, Mario Duran-Prado, María M. Malagón, Antonio J. Martinez-Fuentes, Francisco Gracia-Navarro, Raul M. Luque, and Justo P. Castaño

Department of Cell Biology, Physiology and Immunology, University of Córdoba, Instituto Maimónides de Investigación Biomédica de Córdoba (IMIBIC), and CIBER Fisiopatología de la Obesidad y la Nutrición (CIBERobn 06/03), Córdoba, Spain

Address for correspondence: Dr. Justo P. Castaño, Dpt. de Biologia Celular, Fisiologia e Inmunologia, Edif. Severo Ochoa. Planta 3, Campus Univ. de Rabanales, Universidad de Cordoba, 14014-Cordoba, Spain. justo@uco.es.

Somatostatin (SST) and its receptors (sst) make up a molecular family with unique functional complexity and versatility. Widespread distribution and frequent coexpression of sst subtypes underlies the multiplicity of (patho)physiological processes controlled by SST (central nervous system functions, endocrine and exocrine secretion, cell proliferation). This complexity is clearly reflected in the intricate evolutionary development of this molecular family. Recent studies postulate the existence of an ancestral somatostatin/urotensin II (SST/UII) gene, which originated two ancestral, SST and UII, genes by local duplication. Subsequently, segment duplication would have originated two diverging SST genes in both fish (SS1/SS2) and tetrapods [(SST/cortistatin(CST))]. SST/CST actions are mediated by a family of GPCRs (sst1–5) encoded by five different genes. sst1–4 sequences are highly conserved compared with sst5, suggesting unique evolutionary and functional relevance for the latter. Indeed, we recently identified novel truncated but functional sst5 variants in several species, which may help to explain part of the complexity of the SST/CST/sst family. Comparative and phylogenetic analysis of this molecular family would enhance our understanding of its paradigmatic evolutionary complexity and functional versatility.

Keywords: somatostatin; cortistatin; somatostatin receptors; evolution; vertebrates

Introduction

Somatostatin-14 (SST14) is the best-known member of a family of related peptides, which were likely originated during the process of tetraploidization that took place during vertebrate evolution. Indeed, genomes sequencing analyses suggest that two rounds of extensive gene duplications occurred in early vertebrate evolution, which resulted from two more or less complete genome doublings (2R hypothesis). In addition, all euteleosts share a common ancestor that underwent a third tetraploidization (3R). Nevertheless, this hypothesis does not explain all vertebrate gene duplications, suggesting that processes of local duplications have also occurred in vertebrate lineage.[1] Many families of neuroendocrine peptides, including that of the SST genes, have been expanded during this evolutive process,[1] and as result, the SST family is composed of several related peptides highly conserved during evolution of vertebrate lineage.[2]

Peptides that constitute this family exert their functions through binding and activation of five distinct seven trans-membrane domain (TMD) G-protein coupled receptors (GPCRs), the so-called somatostatin receptors (sst1–5). Components of the sst family are encoded by five different genes that also underwent the process of local/segmented duplication occurred in early vertebrates,[3] however, the sequence of sst subtypes has been highly conserved during vertebrate evolution. Although each sst subtype seems to preferentially mediate one or more specific SST actions, most tissues co-express several sst subtypes, suggesting that the precise actions of SST depend on the interaction of the ssts expressed in each cell.[4,5]

Since the discovery of SST in 1973 as an inhibitor factor of GH-release,[6] a vast amount of

doi: 10.1111/j.1749-6632.2010.05511.x

patho-physiological functions have been associated to this peptides family.[4] By and large, SST acts as a universal inhibitor of endocrine and exocrine secretions. Thus, SST inhibits GH secretion in all vertebrate species studied,[7–18] while its actions in other pituitary secretions vary in a species-dependent manner.[19–30] In addition, SST14 inhibits exocrine secretions[4,5,31] and modulates nervous system functions and digestive and immune system actions.[32,33] In contrast, other non-SST14 SST family peptides, such as CST, have been reported to exert different, and even opposite actions in a tissue- or species-dependent manner, probably by binding to sst subtypes with different affinities than SST-14 or by activating disparate intracellular pathways.[34–36]

In sum, the SST/ssts system represents a paradigm of a very complex, versatile, and functionally relevant neuroendocrine system, which has been highly conserved since its appearance, probably, in early vertebrate evolution.

Evolution of the SST gene

SST was originally discovered as a 14-residue peptide (SST14), with a disulfide bridge that allows its cyclic structure.[6] However, transcription of SST gene leads in fact to the synthesis of a pre-pro-hormone (PPSS1), which can codify various SST-related peptides. The most important one besides SST14 is an amino-terminally extended form of 28 residues, called SST28, which was isolated from the porcine gut and exhibits a distinct tissular distribution. Indeed, whereas SST14 is predominantly produced in the central nervous system (CNS) but also in many peripheral organs, SST28 is mainly synthesized by mucosal epithelial cells along the gastrointestinal tract (GIT).

As mentioned earlier, the processes of local and segmented duplication that occurred in the first stages of vertebrate evolution originated a large number of SST-related peptides encoded by up to four different genes.[1] Because SST14 is present in all vertebrate phyla,[37–40] except for holocephalan fish,[41] and its nucleotide-amino acidic sequence has been highly conserved across species, its coding gene, SS1, is considered the ancestral SST gene.[2] However, comparative genome organization studies suggest that SS1 gene shares with urotensin (UII) gene a common ancestor present in early vertebrates. Although both peptides show limited structural similarities and the nucleotide sequences of their precursors are not highly conserved, both SST and UII peptides exhibit a disulfide bridge, and a share common FWK motif, which is essential for their biological activity. Based on these evidences, Tostivint *et al.* have postulated the existence of an ancestral SST/UII gene, which underwent two rounds of gene duplication (2R), and thereby produced two SST- and two UII-related genes along the evolutionary process.[2]

In accordance with the 2R hypothesis, the tetraploidization that occurred in early vertebrates originated the mentioned SS1 gene and a new SST-related gene, termed somatostatin 2 (SS2) in lower vertebrates and cortistatin (CST) in mammals. In contrast with the high conservation of SST14 sequence along vertebrate lineage, SS2 and CST peptides show variable structure (despite sharing a unique proline residue at position 2), and this may explain that the relation between fish SS2 and human CST has remained unclear during years. Recently, genomic organization analyses have clarified it revealing that SS2 and CST derive from orthologous genes. SS2/CST peptidic precursor has lower similarities with that of SS1, but the C-terminal regions of both are highly conserved and, similar to that occurred in SS1, various peptides of diverse length are generated from the SS2 propeptide (CST14, CST17, CST28, CST29, SS2–22, SS2–25). Indeed, in human, CST17 and SST14 share 11 amino acids, including the FWK motif, which makes up the receptor binding core, as well as the two cysteine residues responsible for their conserved cyclic structure.[20] This sequence identity and structural homology between SST and CST could explain their close pharmacology.[42] Indeed, both peptides exhibit a comparable subnanomolar binding affinity to the sst subtypes. On the other hand, CST, unlike SST, has been reported to bind to the receptor for ghrelin (GHS-R)[43] and also to the Mrgx2 receptor,[44] although this latter receptor seems to be somewhat promiscuous and more specific for proadrenomedullin and its related peptides.

In fish lineage, two more SST-related genes have been described in literature. SS3, previously known as SSII, was initially found in anglerfish, trout, and goldfish. Most peptides derived from PSS3 have 25–28-aminoacids in length, with amino acids changes at [Tyr^7, Gly^{10}]SST14. The vicinity between SS3 and SS1 loci suggest that the two genes arose by a local duplication, and thus they are considered orthologous genes. In addition, in catfish and zebrafish, a SS4 gene, also named SSIII or atypical SSII, has been reported that encoded for SST variants containing Tyr, Ser, Arg, and Ala residues at positions 6, 10, 11, and 13. Although its origin is still unknown, it seems to be a paralogous gene that could have resulted from the 3R event that occurred in the early ray-finned fish lineage.[2]

Evolution of sst genes

As mentioned earlier, diversity of ssts, together with intrincated evolution of SST-related peptides, seems to be responsible for the multiplicity and complexity of SST actions. Nowadays, ssts are classified as Class A GPCRs and are encoded by five intronless genes, sst1–5, that give rise to six different receptor isoforms, including a carboxyterminal spliced variant of the sst2.[4,31] Although ssts have been described in all vertebrate classes studied to date, no data have been reported in invertebrates metazoans. According to their sequence identity and pharmacological properties, ssts can be subdivided into two groups, namely SRIF1, which includes sst2, sst3, and sst5, and SRIF2, which includes sst1 and sst4. Phylogenetic analysis of ssts suggests that, as other GPCRs, a putative sst precursor gene was duplicated and originated SRIF1 and SRIF2 receptors genes early in the vertebrate evolution, before tetrapods and teleosts split up. According with the 2R hypothesis, a second step consisted in an additional duplication event that derived into sst1, sst4, sst3, and an ancestral sst2/sst5. At last, the ancestral sst2/sst5 divided into the actual sst2 and sst5.[3] In fish, the scenario of sst subtypes is more complex due to polyploidization occurred during fish lineage evolution, which resulted in several and different copies of the same gene. Thus, some teleosts have two sst1 (A and B), two sst3 (A and B), and three sst5 (A, B, and C). In spite of this branched evolutive process, sst sequences are highly conserved between species, and within the sst subtype family, being mostly divergent at both, N- and C-terminal domains. Interestingly, the most conserved subtype between species is the sst1, whereas the most divergent is the sst5. For example, sst1A and B from *Carassius auratus* shows 76% similarity with their human counterpart. Moreover, within each species, sst5 seems to be also the most divergent subtype. Thus, human sst1 shows 64%, 62%, and 58% similarity with sst2, sst3, and sst4, respectively, whereas human sst5 shows 42%, 48%, 47%, and 46% with sst1, sst2, sst3, and sst4, respectively.[31] This lower similarity for the sst5 subtype suggest a unique evolutionary relevance for this receptor, which is supported by its involvement in certain atypical processes.[45] In line with this, and despite the difficulty to identify additional ssts in mammals, we have discovered several truncated variants of the human,[46] rat, mouse,[47] and pig sst5 subtype, which increase the complexity of the SST/ssts system.

SST functions from fish to mammals

Control of endocrine secretions

SST family peptides are one of the major negative regulatory systems of endocrine secretions in vertebrates, both by acting directly on endocrine tissues (pituitary, pancreas, liver, adrenal gland) or indirectly through regulatory centers (mainly hypothalamus).[4]

Pituitary

Actions on somatotropes. SST is the foremost negative regulator of GH secretion and its inhibitory function is conserved during vertebrate evolution. In teleost, although SST14 is a potent inhibitor of basal and stimulated GH secretion in several species,[20] the function of other SST peptides is less clear. In goldfish, mammalian SST14 and SST28 were equipotent in inhibiting GH release from the pituitary, whereas salmonid SS3–25 and catfish SS2–22 had no effect on GH release.[32] In addition, it has been reported that goldfish SS2–28 was a more potent inhibitor of GH than SST14.[32] In other nonmammalian vertebrate phyla, SST action on GH release is slightly different to that exerted in teleost. In amphibians and reptilians, SS1 by itself has no apparent effect on GH release, but inhibits the *in vitro* release of GH stimulated by TRH (the main GH-release stimulatory factor in amphibian and reptilian pituitary glands),[9,10] whereas both SS1 and SS2 are able to inhibit GHRH-induced rise of GH in frog.[10] In chicken, SST inhibits basal and GHRH-stimulated GH secretion.[14,16] In mammals, SST is essential to establish and maintain pulsatility of GH secretion; however, several studies have revealed that the precise role of SST, and also GHRH, in the control of GH secretion is species- and, even, gender-dependent, suggesting a complex relationship between these two peptides and additional factors.[48] Specifically, in human, SST and CST had an inhibitory effect on GH release both *in vivo* and *in vitro*, probably through activation of sst2 and sst5.[5,28] Taken together, these data indicate that, although the role of SST axis in regulation of GH secretion is well conserved in the vertebrate lineage, the precise peptides and receptors involved in this control are phyla- and species-dependent.

Actions on other pituitary cell types. SST has been also shown to inhibit most of pituitary secretions. Specifically, it inhibits prolactin (PRL) release *in vivo* and *in vitro* from fish pituitary cells[29] and, in earlier studies, from rat.[19] In human, *in vitro* experiments showed that sst2 is the main receptor involved in the inhibition of PRL secretion from fetal human pituitary cells, although, in most prolactinomas and somato-prolactinomas, SST and CST have been found to exert a suppressive effect on PRL secretion, mainly acting through sst5 subtype.[28] Surprisingly, SST seems to exert a stimulatory role in ACTH secretion in teleost,[49] whereas it has long been known to negatively regulate ACTH secretion *in vitro* in mammals.[25] Specifically, several studies indicate that sst2 and sst5 mediate the inhibitory response of corticotropes to SST.[23] In addition, it has been reported that SST and CST display a similar, strong inhibitory effect on ACTH hypersecretion in patients with Cushing's disease.[21] On the other hand, SST has rarely been found to modulate normal gonadotrope function. Thus, an isolated study on male rat anterior pituitary cells in culture showed that SST did not affect basal release of either LH or FSH, whereas it suppressed LHRH-induced release of LH, but not that of FSH.[30] Conversely, several studies have demonstrated that SST inhibits gonadotropin release from human pituitary adenomas (both clinically nonfunctioning adenomas and somatotrope adenomas cosecreting the alpha-subunit).[24] More recently, use of receptor-specific somatostatin analogs on clinically nonfunctioning pituitary adenoma cells cultured *in vitro* indicated that the regulation of alpha-subunit secretion by SST is mediated through the subtypes sst2A, sst3, and sst5.[50] SST has also been reported to inhibit stimulated TSH release *in vitro* from fetal human pituitary cells, as well as in cultures of pituitary cells from rats, birds, and amphibians.[27,29,51,52] In fact, in this latter species, it has been demonstrated that both SS1 and SS2 inhibited TSH release induced by PACAP-38 but did not alter spontaneous (basal) TSH secretion, suggesting a similar role for both neurohormones in the regulation of thyrotrope function *in vitro*.[27] When viewed together, the data obtained in normal and tumoral pituitary cells in culture from mammalian and nonmammalian vertebrates suggest that the pleiotropic regulatory/modulatory capability of SST axis on pituitary cell function is clearly dependent on the specific species, peptide, and receptor subtype considered and could have evolutionary implications.

Extrapituitary actions

Vertebrate pancreatic islets express all five sst subtypes, being sst1, 2, and 5 those predominantly expressed in human islets. Immunohistochemical colocalization of sst1–5 with insulin, glucagon, and SST has revealed that, in humans, β-cells are rich in sst1 and sst5, α-cells in sst2, and δ-cells in sst5, suggesting a receptor-dependent regulation of pancreatic secretions. In teleost, injection of SS2–25, but not SST14, reduced plasma insulin levels; whereas, injections of both equally depressed plasma glucagon levels,[53] demonstrating that SST isoforms may possess disparate physiological effects. In rats, pioneer studies showed selective suppression of glucagon by SST14 and of insulin by SST28.[4] In human, SST and CST displays the same *in vivo* inhibitory effects on insulin secretion in either physiological and pathological conditions (e.g., acromegaly or prolactinoma).[54] Current knowledge suggest that, multiple ssts would act in concert to convey the inhibitory signal of SST axis on insulin and glucagon release, through an action that is species-dependent, and where it is critical the specific combination and the levels of expression of the ssts involved.[55–61] Specifically, sst2 seems to be the main receptor mediating the inhibitory actions of SST in insulin and glucagon in humans, with sst1 and sst5 playing also a role in regulating insulin. In contrast, in rodents, sst5 appear as the predominant receptor mediating SST-induced inhibition of insulin release and sst2 seems to mediate glucagon inhibition.[61]

SST has also been shown to be able to regulate endocrine secretions in other peripheral organs such as ovary, thyroid, and adrenal gland. Specifically, in teleosts, SST plasma level are inversely correlated with thyroid hormone levels and SST inhibits TSH-stimulated release of T4, and T3 as well as that of calcitonin from murine parafollicular cells.[4] Furthermore, recent studies using human medullary thyroid carcinoma TT-cell line have shown that activation of specific ssts by its selective ligands regulates diverse cellular processes, including proliferation, calcitonin secretion, and RNA synthesis, in a mechanism involving sst2, sst5, and sst1.[62] In the ovary, SST acts as regulator of progesterone release, specifically by inhibiting basal, but not LH-stimulated, progesterone release from cultured human granulosa-luteal cells.[63] Finally, adrenal gland is a known target of SST action, which displays a rich concentration of sst2 and modest levels of sst1 and 3. Recently, SST has been shown to regulate also secretion of catecholamines from chromaffin cells. In this case, SST does not inhibit but enhances the release of catecholamines induced by a high concentration of acetylcholine in bovine adrenal medullary cells.[64]

Digestive actions

In gastrointestinal tract (GIT), SST is mainly secreted from stomach D cells into the extracellular space to act as a paracrine factor on nearby endocrine cells and as an autocrine factor to inhibit its own secretion; however, it is also present in the myenteric and submucosal nerve plexus of the gut, where it functions as a neurotransmitter and neuromodulator. These and other widespread sources of SST production in GIT make it to contain approximately 65% of the entire SST content in the body. In addition, although the results are quite variable depending on the technique used, variable expression of all ssts has been described in vertebrate stomachs and GIT.[5]

In mammals, SST is known to inhibit release of GIT peptides, including secretin, ghrelin, CCK, vasoactive intestinal polypeptide (VIP), gastric inhibitory peptide (GIP), motilin, enteroglucagon, gastric acid, peptide intrinsic factor, bile, and colonic fluids.[4] Not only GIT secretions are affected by SST, but other GIT functions such as bowel motility, gastric emptying, GIT transit time, small bowel segmentation, gallbladder contractility, bowel flow, splanchnic and liver blood flow, and intestinal absorption of carbohydrates, amino acids, calcium, and triacylglycerols are inhibited as well.[4] Although the above actions are mediated by the SST-14/sst2 tandem in mammals, other actions such as inhibition of peptide YY and glucagon-like peptide secretion from intestinal endocrine cells seem to be mediated by sst5 and SST-28.[4]

Although SST actions are well characterized in mammalian GIT, studies in nonmammalian vertebrates are scarce. Specifically, in amphibians, SST seems to inhibit pepsinogen secretion and, in teleost, the only digestive action reported for SS1–14 is reduction of basal and/or stimulated gastric acid secretion.[32]

Taken together, these results indicate that although GIT functions of SST are mainly mediated by SST-14 through sst2, more studies in nonmammalian vertebrates are needed to establish the precise role of SST axis in GIT functions of vertebrate lineage.

Neuronal actions

Comparative analyses developed in several vertebrate species indicate that SST and sst2A are widely distributed in the brain of vertebrates, although the other sst subtypes are also expressed in certain CNS regions, suggesting that the general organization of the somatostatinergic system has been well conserved during evolution.[33,65] Indeed, SST is able to act as a neurotransmitter, a neuropeptide, or a neuromodulatory agent in the CNS, mediating motor, cognitive, and sensory effects, acting mainly through sst2A and other subtypes coexpressed in different brain regions.[33] Acting as a neurotransmitter, SST inhibits presynaptically glutamate release in various mammalian, bird, and amphibian models by acting through sst1, sst2, or sst5, depending on the anatomical structure,[33] and also inhibits GABA release in cat and rat.[33] In mammals, intracerebroventricular administration of SST increases locomotor activity through activation of sst2 and sst4, whereas its depletion has been reported to induce learning and memory deficits. These actions in central and peripheral nervous system seems to be mediated by sst2.[33] Interestingly, SST and CST, which is highly expressed in CNS, share several functional properties, such as the depression of neuronal activity.[66] However, in nervous system, both peptides exert clearly distinct actions. Indeed, it has been found that CST at low doses decreased locomotor activity, an effect opposite to that observed for SST at the same doses. However, at a high doses SST and CST induce seizures and barrel rotation, suggesting that CST may act through ssts under saturating conditions. In addition, SST increases REM without affecting slow-wave sleep, while CST induces slow-wave sleep,[66] again suggesting a different role of SST family components in the control of relevant physiological processes.

Immune system actions

Accumulated evidence suggests that the immune and neuroendocrine systems cross-talk by sharing ligands and receptors. Certainly, hormones and neuropeptides produced by the neuroendocrine system often modulate the function of lymphoid organs and immune cells. Immunohistochemical analysis has revealed that SST is expressed in the thymus and other lymphoid organs of amphibians, birds, and mammals (mouse, rat, pig, and humans).[67] In addition, presence of CST, sst1, sst2, and sst3, as well as, MrgX2 has been reported in normal human thymus,[67] suggesting an autocrine or paracrine regulation in this organ. Interestingly, immune cells (T-cell, B-cell, monocytes, macrophages, dentritic cells, etc.) express CST and GHS-R but not SST, and the presence of different ssts seems to be species-dependent in these cell types; for example, sst2 and sst4 are predominantly expressed in rodents, whereas sst2 and sst3 are mainly expressed in human immune cell subsets.[67] Moreover, immune cells can alter their sst subtype expression panel during maturation/differentiation. Taken together, these data suggest that, at least in mammals, CST could mediate the autocrine or paracrine effect of SST axis in immune cells through activation of ssts, MrgX2, and/or GHS-R.

In mammals, SST has been reported to act in an inhibitory fashion on T-cell, B-cell, and monocyte–macrophage lineages. However, a biphasic response has been often observed, that is, inhibition of a certain response at low SST concentrations and stimulation of that same response at higher concentrations. Specifically, SST inhibits the proliferation of mouse and human T-lymphocyte through binding sst2 or possibly sst3.[67] In addition, SST acting through sst2 and/or sst5 inhibit the production of immunoglobulins by murine B cells, diminishes the number of rat antigen-specific plasmatic cells formed during a primary immune response, and reduces the *in vitro* differentiation of human peripheral blood B-lymphocytes into plasmatic cells.[67] However, the data concerning SST action on monocyte–macrophage lineage are contradictory. Indeed, SST has been reported to exert inhibitory and stimulatory effects on the chemotaxis of macrophages, and on the cytokine release and the effector functions of macrophages and monocytes.[67] However, because CST expression levels correlate with immune cell differentiation and activation,[68,69] and MrgX2 and GHS-R are widely expressed in the immune system, CST, rather than SST, has been proposed to be a major endogenous regulatory factor in the immune system.[36] Indeed, CST has been claimed as a therapeutic antiinflammatory peptide on arthritis and lethal endotoxemia treatment.[35,36]

Hence, similar to that occurred in SST digestive actions, more studies are needed in nonmammalian vertebrates to understand the immune role of SST axis in this lineage.

New ssts receptors

In addition to the remarkable versatility of the SST/sst family, new players have emerged in the SST physiology during recent years. Although sst are typically known as 7TMD receptors encoded by intronless genes, and only one splice variant of sst2 has been described in mammals,[4,31] our group has recently identified various truncated sst5 variants that possess less than 7TMD and are generated by cryptic introns in 3′-region of the gene.[46] Generation of splice variants with less than the canonical 7TMDs is a relatively common process in GPCR families, which seems to provide an additional step in the regulation of pathways controlled by these receptors. These truncated sst5 receptors have been identified in human[46] as well as in pig, rat, and mouse.[47] The fact that sst5 sequence is the less intraspecifically conserved among sst subtypes and that it is involved in certain atypical processes[45] together with the notion that it possess a recent common antecessor with sst2, led us to think as plausible to find new variants of sst5 gene.

Specifically, the complete CDS of human truncated sst5 variants comprise 822 and 651 nucleotides, which encode proteins of 273 and 216 amino acids, respectively. Analysis of their corresponding secondary structures (hydrophobicity profiles) indicated that these sequences would encode truncated sst5 variants with five and four putative TMDs, and thus the receptor variants were termed sst5TMD5 and sst5TMD4. Both receptors are functional in terms of mobilization of intracellular calcium, being these actions selective for SST14 and CST17, respectively. Truncated sst5 isoforms are expressed differentially in normal tissues and are also present in diverse types of pituitary tumors, where they are mostly coexpressed with native ssts, mainly sst2 and sst5. Studies are underway aimed at elucidating the precise pathophysiological relevance of the novel sst5 variants.

Concluding remarks

Available evidence suggests that the intrinsic evolutive process that took place during vertebrate lineage diversification underlies the wide structural and functional variety and complexity observed in the SST/ssts family. Indeed, SST-related peptides comprise a family of two genes that coded for, at least, four peptides in tetrapods and up to four genes in some teleost groups. In addition, these peptides exert their biological actions through binding to, at least, eight identified receptors, including the carboxyterminal spliced variant sst2B and the two truncated isoforms, sst5TMD5 and sst5TMD4, encoded by five independent genes. Consequently, the evolutive expansion process that experienced the ancestral SST and sst genes during the early history of vertebrate lineage imparted the SST/sst family a high complexity and versatility in controlling a wide range of relevant biological processes.

Acknowledgments

This work was supported by Research Grants BFU2007-60180/BFI and BFU2008-01136/BFI (Ministerio de Ciencia e Innovación), BIO-139 and CTS-1705 (Junta de Andalucía), Ayudas predoctorales de formación en investigación en salud del Fondo de Investigación Sanitaria (FIS, ISCIII: FI06/00804; to JCC), Programa Nacional de becas de FPU (FPU-AP20052473, to MDG) and Programa Ramón y Cajal del Ministerio de Educación y Ciencia (RYC-2007-00186, to R.M.L.), Spain, and IPSEN Pharmaceuticals (to J.P.C). CIBER is an initiative of Instituto de Salud Carlos III, Ministerio de Ciencia e Innovación, Spain.

Conflicts of interest

The authors declare no conflict of interest.

References

1. Larhammar, D., G. Sundstrom, S. Dreborg, *et al.* 2009. Major genomic events and their consequences for vertebrate evolution and endocrinology. *Ann. N.Y. Acad. Sci.* **1163:** 201–208.
2. Tostivint, H., L. Joly, I. Lihrmann, *et al.* 2006. Comparative genomics provides evidence for close evolutionary relationships between the urotensin II and somatostatin gene families. *Proc. Natl. Acad. Sci. USA* **103:** 2237–2242.
3. Moaeen-ud-Din, M. & L.G. Yang. 2009. Evolutionary history of the somatostatin and somatostatin receptors. *J. Genet*. **88:** 41–53.
4. Patel, Y.C. 1999. Somatostatin and its receptor family. *Front. Neuroendocrinol.* **20:** 157–198.
5. Moller, L.N., C.E. Stidsen, B. Hartmann & J.J. Holst. 2003. Somatostatin receptors. *Biochim. Biophys. Acta.* **1616:** 1–84.
6. Brazeau, P., W. Vale, R. Burgus, *et al.* 1973. Hypothalamic polypeptide that inhibits the secretion of immunoreactive pituitary growth hormone. *Science* **179:** 77–79.
7. Gahete, M.D., M. Duran-Prado, R.M. Luque, *et al.* 2009. Understanding the multifactorial control of growth hormone release by somatotropes: lessons from comparative endocrinology. *Ann. N.Y. Acad. Sci.* **1163:** 137–153.
8. Cook, A.F. & R.E. Peter. 1984. The effects of somatostatin on serum growth hormone levels in the goldfish, Carassius auratus. *Gen. Comp. Endocrinol.* **54:** 109–113.
9. Hall, T.R. & A. Chadwick. 1984. Effects of synthetic mammalian thyrotrophin releasing hormone, somatostatin and dopamine on the secretion of prolactin and growth hormone from amphibian and reptilian pituitary glands incubated in vitro. *J. Endocrinol.* **102:** 175–180.
10. Jeandel, L., A. Okuno, T. Kobayashi, *et al.* 1998. Effects of the two somatostatin variants somatostatin-14 and [Pro2, Met13]somatostatin-14 on receptor binding, adenylyl cyclase activity and growth hormone release from the frog pituitary. *J. Neuroendocrinol.* **10:** 187–192.
11. Le Bail, P.Y., J.P. Sumpter, J.F. Carragher, *et al.* 1991. Development and validation of a highly sensitive radioimmunoassay for chinook salmon (Oncorhynchus tshawytscha) growth hormone. *Gen. Comp. Endocrinol.* **83:** 75–85.
12. Melamed, P., N. Eliahu, B. Levavi-Sivan, *et al.* 1995. Hypothalamic and thyroidal regulation of growth hormone in tilapia. *Gen. Comp. Endocrinol.* **97:** 13–30.
13. Peng, C. & R.E. Peter. 1997. Neuroendocrine regulation of growth hormone secretion and growth in fish. *Zool. Stud.* **36:** 79–89.

14. Piper, M.M. & T.E. Porter. 1997. Responsiveness of chicken embryonic somatotropes to somatostatin (SRIF) and IGF-I. *J. Endocrinol.* **154:** 303–310.
15. Sheridan, M.A. & J.D. Kittilson. 2004. The role of somatostatins in the regulation of metabolism in fish. *Comp. Biochem. Physiol. B Biochem. Mol. Biol.* **138:** 323–330.
16. Spencer, G.S., S. Harvey, A.R. Audsley, *et al.* 1986. The effect of immunization against somatostatin on growth rates and growth hormone secretion in the chicken. *Comp. Biochem. Physiol. A.* **85:** 553–556.
17. Tannenbaum, G.S., J.C. Painson, A.M. Lengyel & P. Brazeau. 1989. Paradoxical enhancement of pituitary growth hormone (GH) responsiveness to GH-releasing factor in the face of high somatostatin tone. *Endocrinology* **124:** 1380–1388.
18. Very, N.M., D. Knutson, J.D. Kittilson & M.A. Sheridan. 2001. Somatostatin inhibits growth of rainbow trout. *J. Fish Biol.* **59:** 157–165.
19. Enjalbert, A., J. Epelbaum, S. Arancibia, *et al.* 1982. Reciprocal interactions of somatostatin with thyrotropin-releasing hormone and vasoactive intestinal peptide on prolactin and growth hormone secretion in vitro. *Endocrinology* **111:** 42–47.
20. Gahete, M.D., M. Duran-Prado, R.M. Luque, *et al.* 2008. Are somatostatin and cortistatin two siblings in regulating endocrine secretions? in vitro work ahead. *Mol. Cell Endocrinol.* **286:** 128–134.
21. Giordano, R., A. Picu, L. Bonelli, *et al.* 2007. The activation of somatostatinergic receptors by either somatostatin-14 or cortistatin-17 often inhibits ACTH hypersecretion in patients with Cushing's disease. *Eur. J. Endocrinol.* **157:** 393–398.
22. Grau, E.G., C.A. Ford, L.M. Helms, *et al.* 1987. Somatostatin and altered medium osmotic pressure elicit rapid changes in prolactin release from the rostral pars distalis of the tilapia, Oreochromis mossambicus, in vitro. *Gen. Comp. Endocrinol.* **65:** 12–18.
23. Hofland, L.J., J. Van Der Hoek, R. Feelders, *et al.* 2005. The multi-ligand somatostatin analogue SOM230 inhibits ACTH secretion by cultured human corticotroph adenomas via somatostatin receptor type 5. *Eur. J. Endocrinol.* **152:** 645–654.
24. Klibanski, A., J.M. Alexander, H.A. Bikkal, *et al.* 1991. Somatostatin regulation of glycoprotein hormone and free subunit secretion in clinically nonfunctioning and somatotroph adenomas in vitro. *J. Clin. Endocrinol. Metab.* **73:** 1248–1255.
25. Luque, R.M., M.D. Gahete, U. Hochgeschwender & R.D. Kineman. 2006. Evidence that endogenous SST inhibits ACTH and ghrelin expression by independent pathways. *Am. J. Physiol. Endocrinol. Metab.* **291:** 395–403.
26. Luque, R.M., M.D. Gahete, R.J. Valentine & R.D. Kineman. 2006. Examination of the direct effects of metabolic factors on somatotrope function in a non-human primate model, Papio anubis. *J. Mol. Endocrinol.* **37:** 25–38.
27. Okada, R., K. Yamamoto, Y. Ito, *et al.* 2006. Effects of pituitary adenylate cyclase-activating polypeptide, vasoactive intestinal polypeptide, and somatostatin on the release of thyrotropin from the bullfrog pituitary. *Ann. N.Y. Acad. Sci.* **1070:** 474–480.
28. Rubinfeld, H., M. Hadani, G. Barkai, *et al.* 2006. Cortistatin inhibits growth hormone release from human fetal and adenoma pituitary cells and prolactin secretion from cultured prolactinomas. *J. Clin. Endocrinol. Metab.* **91:** 2257–2263.
29. Shimon, I., J.E. Taylor, J.Z. Dong, *et al.* 1997. Somatostatin receptor subtype specificity in human fetal pituitary cultures. Differential role of SSTR2 and SSTR5 for growth hormone, thyroid-stimulating hormone, and prolactin regulation. *J. Clin. Invest.* **99:** 789–798.
30. Yu, W.H., M. Kimura & S.M. McCann. 1997. Effect of somatostatin on the release of gonadotropins in male rats. *Proc. Soc. Exp. Biol. Med.* **214:** 83–86.
31. Olias, G., C. Viollet, H. Kusserow, *et al.* 2004. Regulation and function of somatostatin receptors. *J. Neurochem.* **89:** 1057–1091.
32. Klein, S.E. & M.A. Sheridan. 2008. Somatostatin signaling and the regulation of growth and metabolism in fish. *Mol. Cell Endocrinol.* **286:** 148–154.
33. Viollet, C., G. Lepousez, C. Loudes, *et al.* 2008. Somatostatinergic systems in brain: networks and functions. *Mol. Cell Endocrinol.* **286:** 75–87.
34. de Lecea, L., J.R. Criado, O. Prospero-Garcia, *et al.* 1996. A cortical neuropeptide with neuronal depressant and sleep-modulating properties. *Nature* **381:** 242–245.
35. Gonzalez-Rey, E., A. Chorny, R.G. Del Moral, *et al.* 2007. Therapeutic effect of cortistatin on experimental arthritis by downregulating inflammatory and Th1 responses. *Ann. Rheum. Dis.* **66:** 582–588.
36. Gonzalez-Rey, E., A. Chorny, G. Robledo & M. Delgado. 2006. Cortistatin, a new antiinflammatory peptide with therapeutic effect on lethal endotoxemia. *J. Exp. Med.* **203:** 563–571.
37. Tostivint, H., I. Lihrmann, C. Bucharles, *et al.* 1996. Occurrence of two somatostatin variants in the frog brain: characterization of the cDNAs, distribution of the mRNAs, and receptor-binding affinities of the peptides. *Proc. Natl. Acad. Sci. USA* **93:** 12605–12610.

38. Tostivint, H., D. Vieau, N. Chartrel, *et al.* 2002. Expression and processing of the [Pro(2),Met(13)]somatostatin-14 precursor in the intermediate lobe of the frog pituitary. *Endocrinology* **143:** 3472–3481.
39. Trabucchi, M., H. Tostivint, I. Lihrmann, *et al.* 2002. Polygenic expression of somatostatin in the sturgeon Acipenser transmontanus: molecular cloning and distribution of the mRNAs encoding two somatostatin precursors. *J. Comp. Neurol.* **443:** 332–345.
40. Vaudry, H., N. Chartrel & J.M. Conlon. 1992. Isolation of [Pro2,Met13]somatostatin-14 and somatostatin-14 from the frog brain reveals the existence of a somatostatin gene family in a tetrapod. *Biochem. Biophys. Res. Commun.* **188:** 477–482.
41. Conlon, J.M. 1990. [Ser5]-somatostatin-14: isolation from the pancreas of a holocephalan fish, the Pacific ratfish (Hydrolagus colliei). *Gen. Comp. Endocrinol.* **80:** 314–320.
42. de Lecea, L. & J.P. Castano. 2006. Cortistatin: not just another somatostatin analog. *Nat. Clin. Pract. Endocrinol. Metab.* **2:** 356–357.
43. Deghenghi, R., M. Papotti, E. Ghigo & G. Muccioli. 2001. Cortistatin, but not somatostatin, binds to growth hormone secretagogue (GHS) receptors of human pituitary gland. *J. Endocrinol. Invest.* **24:** 1–3.
44. Robas, N., E. Mead & M. Fidock. 2003. MrgX2 is a high potency cortistatin receptor expressed in dorsal root ganglion. *J. Biol. Chem.* **278:** 44400–44404.
45. Luque, R.M., M. Duran-Prado, S. Garcia-Navarro, *et al.* 2006. Identification of the somatostatin receptor subtypes (sst) mediating the divergent, stimulatory/inhibitory actions of somatostatin on growth hormone secretion. *Endocrinology* **147:** 2902–2908.
46. Duran-Prado, M., M.D. Gahete, A.J. Martinez-Fuentes, *et al.* 2009. Identification and characterization of two novel truncated but functional isoforms of the somatostatin receptor subtype 5 differentially present in pituitary tumors. *J. Clin. Endocrinol. Metab.* **94:** 2634–2643.
47. Córdoba-Chacón, J., M.D. Gahete, M. Duran-Prado, *et al.* 2009. Identification and characterization of new functional truncated variants of somatostatin receptor subtype 5 in rodents. *Cell Mol. Life Sci.* **67:** 1147–1163.
48. Goldenberg, N. & A. Barkan. 2007. Factors regulating growth hormone secretion in humans. *Endocrinol. Metab. Clin. North Am.* **36:** 37–55.
49. Langhorne, P. 1986. Somatostatin stimulates ACTH release in brown trout (Salmo trutta L.). *Gen. Comp. Endocrinol.* **61:** 71–75.
50. Pawlikowski, M., H. Lawnicka, H. Pisarek, *et al.* 2007. Effects of somatostatin-14 and the receptor-specific somatostatin analogs on chromogranin A and alpha-subunit (alpha-SU) release from "clinically nonfunctioning" pituitary adenoma cells incubated in vitro. *J. Physiol. Pharmacol.* **58:** 179–188.
51. Geris, K.L., B. De Groef, E.R. Kuhn & V.M. Darras. 2003. In vitro study of corticotropin-releasing hormone-induced thyrotropin release: ontogeny and inhibition by somatostatin. *Gen. Comp. Endocrinol.* **132:** 272–277.
52. Dieguez, C., S.M. Foord, J.R. Peters, *et al.* 1984. Interactions among epinephrine, thyrotropin (TSH)-releasing hormone, dopamine, and somatostatin in the control of TSH secretion in vitro. *Endocrinology* **114:** 957–961.
53. Eilertson, C.D. & M.A. Sheridan. 1993. Differential effects of somatostatin-14 and somatostatin-25 on carbohydrate and lipid metabolism in rainbow trout Oncorhynchus mykiss. *Gen. Comp. Endocrinol.* **92:** 62–70.
54. Grottoli, S., V. Gasco, F. Broglio, *et al.* 2006. Cortistatin-17 and somatostatin-14 display the same effects on growth hormone, prolactin, and insulin secretion in patients with acromegaly or prolactinoma. *J. Clin. Endocrinol. Metab.* **91:** 1595–1599.
55. Singh, V., M.D. Brendel, S. Zacharias, *et al.* 2007. Characterization of somatostatin receptor subtype-specific regulation of insulin and glucagon secretion: an in vitro study on isolated human pancreatic islets. *J. Clin. Endocrinol. Metab.* **92:** 673–680.
56. Efendic, S., R. Luft & V. Grill. 1974. Effect of somatostatin on glucose induced insulin release in isolated perfused rat pancreas and isolated rat pancreatic islets. *FEBS Lett.* **42:** 169–172.
57. Gerich, J.E., R. Lovinger & G.M. Grodsky. 1975. Inhibition by somatostatin of glucagon and insulin release from the perfused rat pancreas in response to arginine, isoproterenol and theophylline: evidence for a preferential effect on glucagon secretion. *Endocrinology* **96:** 749–754.
58. Brown, M., J. Rivier & W. Vale. 1976. Biological activity of somatostatin and somatostatin analogs on inhibition of arginine-induced insulin and glucagon release in the rat. *Endocrinology* **98:** 336–343.
59. Strowski, M.Z., M.P. Dashkevicz, R.M. Parmar, *et al.* 2002. Somatostatin receptor subtypes 2 and 5 inhibit corticotropin-releasing hormone-stimulated adrenocorticotropin secretion from AtT-20 cells. *Neuroendocrinology* **75:** 339–346.
60. Strowski, M.Z., M. Kohler, H.Y. Chen, *et al.* 2003. Somatostatin receptor subtype 5 regulates insulin secretion and glucose homeostasis. *Mol. Endocrinol.* **17:** 93–106.

61. Strowski, M.Z., R.M. Parmar, A.D. Blake & J.M. Schaeffer. 2000. Somatostatin inhibits insulin and glucagon secretion via two receptors subtypes: an in vitro study of pancreatic islets from somatostatin receptor 2 knockout mice. *Endocrinology* **141:** 111–117.
62. Zatelli, M.C., F. Tagliati, D. Piccin, *et al.* 2002. Somatostatin receptor subtype 1-selective activation reduces cell growth and calcitonin secretion in a human medullary thyroid carcinoma cell line. *Biochem. Biophys. Res. Commun.* **297:** 828–834.
63. Holst, N., M.B. Jacobsen, E. Haug, *et al.* 1995. Somatostatin in physiological concentrations inhibits basal and enhances luteinizing hormone-stimulated progesterone release from human granulosa-luteal cells. *Hum. Reprod.* **10:** 1363–1366.
64. Ribeiro, L., F. Martel & I. Azevedo. 2006. The release of 3H-1-methyl-4-phenylpyridinium from bovine adrenal chromaffin cells is modulated by somatostatin. *Regul. Pept.* **137:** 107–113.
65. Laquerriere, A., P. Leroux, B.J. Gonzalez, *et al.* 1989. Distribution of somatostatin receptors in the brain of the frog Rana ridibunda: correlation with the localization of somatostatin-containing neurons. *J. Comp. Neurol.* **280:** 451–467.
66. de Lecea, L. 2008. Cortistatin—functions in the central nervous system. *Mol. Cell Endocrinol.* **286:** 88–95.
67. van Hagen, P.M., V.A. Dalm, F. Staal & L.J. Hofland. 2008. The role of cortistatin in the human immune system. *Mol. Cell Endocrinol.* **286:** 141–147.
68. Dalm, V.A., P.M. van Hagen, P.M. van Koetsveld, *et al.* 2003. Expression of somatostatin, cortistatin, and somatostatin receptors in human monocytes, macrophages, and dendritic cells. *Am. J. Physiol. Endocrinol. Metab.* **285:** E344–E353.
69. Dalm, V.A., P.M. van Hagen, P.M. van Koetsveld, *et al.* 2003. Cortistatin rather than somatostatin as a potential endogenous ligand for somatostatin receptors in the human immune system. *J. Clin. Endocrinol. Metab.* **88:** 270–276.

Ann. N.Y. Acad. Sci. ISSN 0077-8923

ANNALS OF THE NEW YORK ACADEMY OF SCIENCES
Issue: *Phylogenetic Aspects of Neuropeptides*

Urotensin II, from fish to human

Hubert Vaudry,[1,2,3] Jean-Claude Do Rego,[4] Jean-Claude Le Mevel,[5] David Chatenet,[1,3,6] Hervé Tostivint,[1,7] Alain Fournier,[3,6] Marie-Christine Tonon,[1,2,3] Georges Pelletier,[8] J. Michael Conlon,[9] and Jérôme Leprince[1,2,3]

[1]Laboratory of Cellular Neuroendocrinology, INSERM U413, European Institute for Peptide Research (IFRMP 23), University of Rouen, Mont-Saint-Aignan, France. [2]Regional Platform for Cell Imaging (PRIMACEN), Mont-Saint-Aignan, France. [3]International Associated Laboratory Samuel de Champlain, Mont-Saint-Aignan, France. [4]Laboratory of Experimental Neuropsychopharmacology, EA 4359, IFRMP 23, University of Rouen, Rouen, France. [5]Laboratory of Neurophysiology, INSERM U650, University of Bretagne Occidentale, Brest, France. [6]INRS, Institut Armand Frappier, Laval, Quebec, Canada. [7]Laboratory Evolution of Endocrine Regulation, CNRS UMR 7221, Muséum d'Histoire Naturelle, Paris, France. [8]Laboratory of Molecular Endocrinology, Oncology and Genetics, Laval University Medical Center, Quebec, Canada. [9]Faculty of Medicine and Health Science, UAE University, Al Ain, United Arab Emirates

Address for correspondence: Hubert Vaudry, Laboratory of Cellular and Molecular Neuroendocrinology, INSERM U413, IFRMP 23, University of Rouen, 76821 Mont-Saint-Aignan, France. hubert.vaudry@univ-rouen.fr

The cyclic peptide urotensin II (UII) was originally isolated from the urophysis of teleost fish on the basis of its ability to contract intestinal smooth muscle. The UII peptide has subsequently been isolated from frog brain and, later on, the pre-proUII cDNA has been characterized in mammals, including humans. A UII paralog called urotensin II-related peptide (URP) has been identified in the rat brain. The UII and URP genes originate from the same ancestral gene as the somatostatin and cortistatin genes. In the central nervous system (CNS) of tetrapods, UII is expressed primarily in motoneurons of the brainstem and spinal cord. The biological actions of UII and URP are mediated through a G protein–coupled receptor, termed UT, that exhibits high sequence similarity with the somatostatin receptors. The UT gene is widely expressed in the CNS and in peripheral organs. Consistent with the broad distribution of UT, UII and URP exert a large array of behavioral effects and regulate endocrine, cardiovascular, renal, and immune functions.

Keywords: urotensin II; URP; UT; evolution; behavioral effects; cardiovascular system

Introduction

The caudal neurosecretory system of teleost fish consists of a population of magnocellular neurons (Dahlgren cells), linearly arranged in the ventral region of the spinal cord, that project their axons into a lobular organ called the urophysis (Fig. 1). The observation that urophysial extracts exert prominent myotropic activities not only in teleosts[1] but also in various vertebrate groups[1,2] has prompted several teams to identify the regulatory substances that were responsible for these pharmacological effects.

Among other factors, the urophysis contains large amounts of acetylcholine,[3] identifying Dahlgren cells as a subset of cholinergic neurons that share many similarities (size, localization, and neurochemical phenotype) with motoneurons. In addition, the urophysis harbors several biologically active peptides that have been termed urotensins. The distinguished groups of Karl Lederis and Howard Bern have concurrently purified and characterized two of these regulatory neuropeptides. Urotensin I (UI), initially isolated from the urophysis of the white sucker *Catostomus commersoni* as a 41-amino acid peptide,[4] turned out to be a paralog of corticotropin-releasing hormone, whereas urotensin II (UII), isolated from the urophysis of the goby *Gillichthys mirabilis* is a cyclic, 12-amino acid peptide that exhibits some structural similarities with somatostatin.[5] Another biologically active peptide isolated from the urophysis of the milkfish *Chanos chanos* and the rainbow trout *Salmo gairdneri* was indistinguishable from arginine vasotocin.[6]

doi: 10.1111/j.1749-6632.2010.05514.x

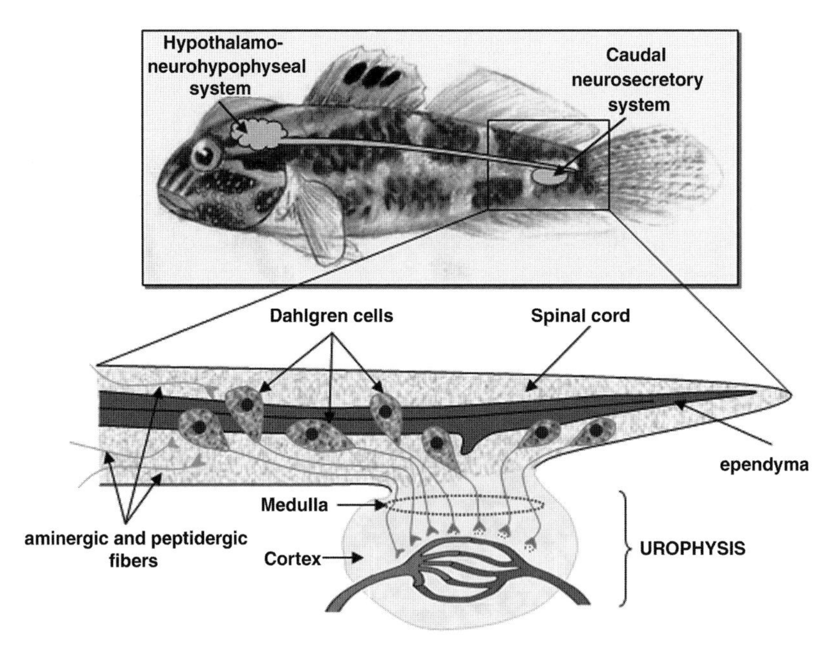

Figure 1. Schematic representation of the organization of the caudal neurosecretory system and its vascular connections in the urophysis of teleost fish.

Because the urophysis is a peculiarity of teleosts, it had long been thought that urotensins were oddities produced exclusively by the fish caudal neurosecretory system. Several observations, however, suggested that this view might not be correct. Firstly, UII-immunoreactive neurons had been localized in the anterior (extraurophyseal) region of the spinal cord and in the brain of fish[7,8] and UII-like immunoreactivity had even been described in the cerebral ganglia of gastropods.[9] Second, UII had been characterized in brain extracts from the long-nose skate and rainbow trout.[10] Third, binding studies using radioiodinated goby UII had revealed the existence of functional UII receptors in rat vascular tissue.[11] Fourth, pharmacological studies had shown that fish UII provoked relaxation of the mouse anococcygeus muscle,[12] caused endothelium-dependent relaxation of rat aorta strips[13] and lowered blood pressure in the anesthetized rat.[14] These observations provided many clues for the existence of a UII-like peptide in extraurophysial organs, including mammalian tissues.[15]

Discovery of UII in tetrapods

The concentration of neuropeptides in the brain of amphibians is usually much higher than in the brain of mammals. For instance, the concentrations of neuropeptide Y and pituitary adenylate cyclase-activating polypeptide are two orders of magnitude higher in the brain of frogs than in the brain of mammals.[16,17] Taking advantage of this situation, we have been able to isolate in pure form UII from an extract of the whole brain of the European green frog *Rana ridibunda* (now renamed *Pelophylax ridibundus*). Unlike previously characterized fish UII peptides that all contain 12-amino acids, frog UII encompasses 13-amino acid residues.[18]

Degenerate oligonucleotides were then designed from the C-terminal amino acid sequence of frog UII and used to screen a frog brain cDNA library. The full-length frog UII cDNA encodes a 127-amino acid precursor protein encompassing a 16-amino acid signal peptide, a 98-amino acid N-terminal flanking peptide, and the 13-amino acid UII peptide.[19] Subsequently, prepro UII cDNAs or genomic sequences were characterized in humans,[19,20] chimpanzee (XP_001157678), cynomolgus monkey,[21] pig,[22] mouse,[23,21] rat,[23] and chicken,[24] and more recently in zebra finch (XP_002195510.1) (Fig. 2). The organization of the cDNAs of mammalian UII is identical to that of frog UII, but the overall sequence identity is very low except for the C-terminal octapeptide that has been strongly preserved across vertebrate species.[15] The length of the predicted

Species	Origin	Sequence
U-II		
Lamprey	brain	H-Asn-Asn-Phe-Ser-Asp-**Cys-Phe-Trp-Lys-Tyr-Cys**-Val-OH
Fugu	spinal cord cDNA	H-Thr-Gly-Asn-Asn-Glu-**Cys-Phe-Trp-Lys-Tyr-Cys**-Val-OH
Skate	brain	H-Asn-Asn-Phe-Ser-Asp-**Cys-Phe-Trp-Lys-Tyr-Cys**-Val-OH
Dogfish	spinal cord	H-Asn-Asn-Phe-Ser-Asp-**Cys-Phe-Trp-Lys-Tyr-Cys**-Val-OH
Sturgeon	spinal cord	H-Gly-Ser-Thr-Ser-Glu-**Cys-Phe-Trp-Lys-Tyr-Cys**-Val-OH
Paddlefish	spinal cord	H-Gly-Ser-Thr-Ser-Glu-**Cys-Phe-Trp-Lys-Tyr-Cys**-Val-OH
Goby	urophysis	H-Ala-Gly-Thr-Ala-Asp-**Cys-Phe-Trp-Lys-Tyr-Cys**-Val-OH
Zebrafish α	urophysis cDNA, spinal cord cDNA	H-Gly-Gly-Gly-Ala-Asp-**Cys-Phe-Trp-Lys-Tyr-Cys**-Val-OH
Zebrafish β	urophysis cDNA, spinal cord cDNA	H-Gly-Ser-Asn-Thr-Glu-**Cys-Phe-Trp-Lys-Tyr-Cys**-Val-OH
Sucker A	urophysis	H-Gly-Ser-Gly-Ala-Asp-**Cys-Phe-Trp-Lys-Tyr-Cys**-Val-OH
Sucker B	urophysis	H-Gly-Ser-Asn-Thr-Glu-**Cys-Phe-Trp-Lys-Tyr-Cys**-Val-OH
Carp α	urophysis, spinal cord cDNA	H-Gly-Gly-Gly-Ala-Asp-**Cys-Phe-Trp-Lys-Tyr-Cys**-Val-OH
Carp β1	urophysis	H-Gly-Gly-Asn-Thr-Glu-**Cys-Phe-Trp-Lys-Tyr-Cys**-Val-OH
Carp β2	urophysis	H-Gly-Ser-Asn-Thr-Glu-**Cys-Phe-Trp-Lys-Tyr-Cys**-Val-OH
Carp γ	urophysis, spinal cord cDNA	H-Gly-Gly-Gly-Ala-Asp-**Cys-Phe-Trp-Lys-Tyr-Cys**-Ile-OH
Flounder	urophysis, urophysis cDNA	H-Ala-Gly-Thr-Thr-Glu-**Cys-Phe-Trp-Lys-Tyr-Cys**-Val-OH
Trout	brain	H-Gly-Gly-Asn-Ser-Glu-**Cys-Phe-Trp-Lys-Tyr-Cys**-Val-OH
Grouper	cDNA	H-Ala-Gly-Asn-Ser-Glu-**Cys-Phe-Trp-Lys-Tyr-Cys**-Val-OH
Frog	brain, spinal cord cDNA	H-Ala-Gly-Asn-Leu-Ser-Glu-**Cys-Phe-Trp-Lys-Tyr-Cys**-Val-OH
Chicken	cDNA	H-Gly-Asn-Leu-Ser-Glu-**Cys-Phe-Trp-Lys-Tyr-Cys**-Val-OH
Zebra finch	predicted	H-Gly-Asn-Leu-Ser-Glu-**Cys-Phe-Trp-Lys-Tyr-Cys**-Val-OH
Mouse	spinal cord cDNA	<Gln-His-Lys-Gln-His-Gly-Ala-Ala-Pro-Glu-**Cys-Phe-Trp-Lys-Tyr-Cys**-Ile-OH
Rat	spinal cord cDNA	<Gln-His-Gly-Thr-Ala-Pro-Glu-**Cys-Phe-Trp-Lys-Tyr-Cys**-Ile-OH
Porcine A	spinal cord	H-Gly-Pro-Thr-Ser-Glu-**Cys-Phe-Trp-Lys-Tyr-Cys**-Val-OH
Porcine B	spinal cord	H-Gly-Pro-Pro-Ser-Glu-**Cys-Phe-Trp-Lys-Tyr-Cys**-Val-OH
Cattle	predicted	H-Gly-Pro-Ser-Ser-Glu-**Cys-Phe-Trp-Lys-Tyr-Cys**-Val-OH
Monkey	spinal cord cDNA	H-Glu-Thr-Pro-Asp-**Cys-Phe-Trp-Lys-Tyr-Cys**-Val-OH
Chimpanzee	cDNA	H-Glu-Thr-Pro-Asp-**Cys-Phe-Trp-Lys-Tyr-Cys**-Val-OH
Human	spinal cord cDNA	H-Glu-Thr-Pro-Asp-**Cys-Phe-Trp-Lys-Tyr-Cys**-Val-OH
URP		
Zebrafish	cDNA	H-Val-**Cys-Phe-Trp-Lys-Tyr-Cys**-Ser-Gln-Asn-OH
Chicken	spinal cord cDNA	H-Ala-**Cys-Phe-Trp-Lys-Tyr-Cys**-Ile-OH
Mouse	brain cDNA	H-Ala-**Cys-Phe-Trp-Lys-Tyr-Cys**-Val-OH
Rat	brain cDNA	H-Ala-**Cys-Phe-Trp-Lys-Tyr-Cys**-Val-OH
Horse	predicted	H-Ala-**Cys-Phe-Trp-Lys-Tyr-Cys**-Val-OH
Chimpanzee	predicted	H-Ala-**Cys-Phe-Trp-Lys-Tyr-Cys**-Val-OH
Human	brain cDNA	H-Ala-**Cys-Phe-Trp-Lys-Tyr-Cys**-Val-OH

Figure 2. Comparison of the primary structures of UII and URP.

mature peptides is also variable, ranging from 11 amino acids in humans to 17 amino acids in the mouse (Fig. 2). It should be noted however that, in most species of tetrapods, the sequence of UII has only been inferred from the potential cleavage sites and that the sequence of native UII has only been determined in frog[18] and pig.[21] In both cases, processing of the precursor occurs at a Lys–Lys–Arg cleavage site.

Immunohistochemical and *in situ* hybridization studies indicate that, in tetrapods, the UII gene is expressed primarily in motoneurons of the brainstem and spinal cord.[19,22,25–28] Specifically, in the mouse and rat central nervous systems (CNS), the UII gene is expressed in cranial nerve nuclei including the motor trigeminal nucleus, facial nucleus, ambiguous nucleus,[22,29] and in the ventral horn of the spinal cord (Fig. 3).[27,29,30]

Identification of the UII receptor

Soon after the identification of UII in humans,[19] four teams independently and almost simultaneously identified, through a reverse pharmacology approach GPR14 (an orphan receptor previously characterized in bovine[31] and rat[32]) as being the UII receptor.[20,21,33,34] The UII receptor, now renamed UT, shares high sequence identity with somatostatin receptors, and the UT gene is physically linked with the gene encoding one of the somatostatin receptors, sst3, at 17q23,[24] suggesting that these two receptors arose by local duplication. UT is expressed in various human tissues, notably in the cardiovascular system including the left atrium, ventricle, coronary artery, and aorta,[20,35,36] and in the kidney, skeletal muscle, and bladder.[20,33,36] UT is also widely expressed in the CNS, notably

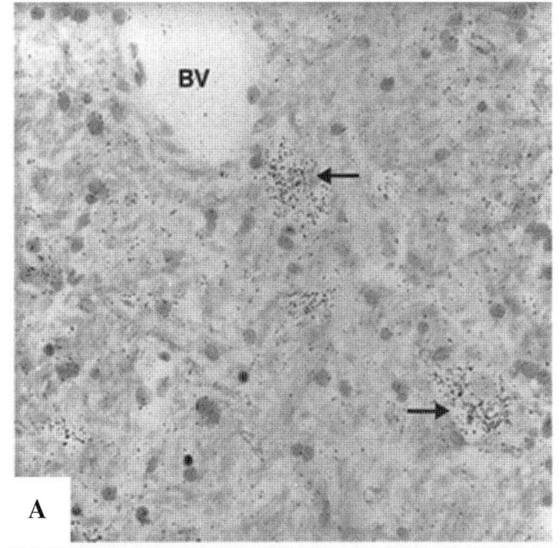

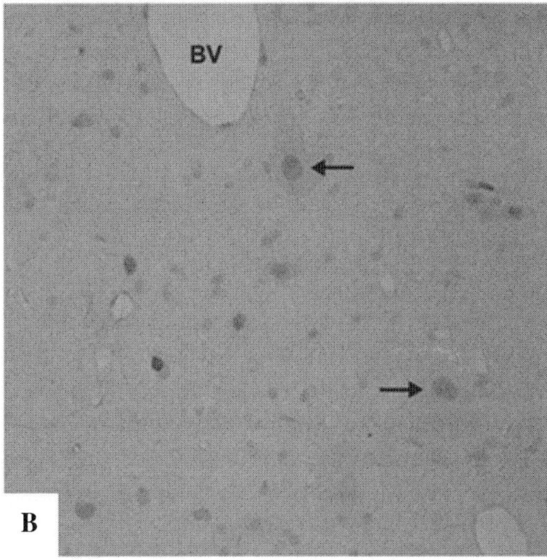

Figure 3. Bright-field micrographs showing the colocalization of prepro UII mRNA and immunoreactive androgen receptors in consecutive sections through the ventral horn of the spinal cord. (A) Accumulation of silver grains indicating the presence of prepro UII mRNA is observed over a few cells. (B) Large nuclei of motoneurons as well as smaller nuclei are immunostained for androgen receptors. The *arrows* point out two identified motoneurons expressing both prepro UII mRNA and androgen receptors. BV, blood vessel; magnification 420×. (Adapted from Ref. 30.)

in the telencephalon (entorhinal and piriform cortex, hippocampal formation, amygdala), diencephalon (arcuate, supraoptic, and suprachiasmatic nuclei, anterior hypothalamic area, circumventricular organs), mesencephalon (interpeduncular, red and pontine nuclei, substantia nigra), rhombencephalon (tegmental nuclei, locus coeruleus), cerebellum (Purkinje and granule cell layers), medulla oblongata (motor nuclei, raphe complex, nucleus of the solitary tract, dorsal motor nucleus of the vagus, nucleus ambigus), and spinal cord.[37,38] In the rat brain, UT is expressed both in neurons and glial cells.[39,40]

Studies conducted in GPR14-transfected cells,[23,33,34,41] isolated rabbit thoracic aorta,[42] and cultured rat glial cells[43] have shown that UT is primarily coupled to the phospholipase C/protein kinase C pathway: UII stimulates polyphosphoinositide turnover and provokes mobilization of Ca^{2+} from intracellular stores. However, UT can activate several other signaling pathways, including the phospholipase A2,[13,44,45] the MAP kinases ERK1/2,[46] the NO synthase,[47] and the Rho A/Rho kinase pathways.[48]

Urotensin II-related peptide

More than a decade after the initial characterization of UII in the frog brain,[18] a paralogous peptide was isolated from a rat brain extract and designated UII-related peptide (URP).[49] URP is a cyclic octapeptide whose sequence differs by only one amino acid from that of the C-terminal region of human and rodent UII (Fig. 2). The primary structure of URP has been fully conserved from rodents to humans, while chicken URP exhibits a substitution (Val → Ile) at its C-terminal extremity.[24] The UII and URP peptides both activate UT with similar potency.[50,51] The URP cDNA has been cloned and the expression patterns of the URP gene have been investigated in human, rat, and mouse.[29,49,52] The differential distribution of UII and URP mRNAs suggests that the peptides may exert complementary activities.

In humans, the UII gene and the cortistatin gene are located in close proximity on 1p36, and the URP gene and somatostatin gene are located in close proximity on 3q28,[24] suggesting that each pair of genes arose from tandem duplication. Chromosome synteny analysis indicates that the UII and URP genes actually originate from the same ancestral gene as the somatostatin and cortistatin genes.[24] The fact that UT possesses a high degree of sequence identity with somatostatin/cortistatin receptors provides additional evidence that UII/URP and somatostatin/cortistatin belong to the same superfamily. As a matter of fact, UII and URP can activate the somatostatin receptor subtypes 2 and 5,[53] and *vice versa* somatostatin and cortistatin stimulate to a certain extent UT.[33]

Conformation analysis and structure-activity relationships

The three-dimensional structures of UII and URP have been investigated by molecular modeling under NMR constraints. In micelles, human UII adopts two distinct conformations in equilibrium: the predominant one exhibits a well-defined II' β-turn whereas the minor one appears to be more flexible.[54] The structure of URP consists of an inverse γ-turn centered on the Trp^4–Lys^5–Tyr^6 sequence.[51] The interactions of UII with extracellular and transmembrane domains of UT have been investigated by surface plasmon resonance. Only extracellular loops 2 and 3 of UT adopt an α-helical conformation that is favorable to interaction with UII and URP.[55] Specifically, it has been proposed that the Asp^4 residue of UII comes close to the Leu^{288} and Ala^{289} residues of UT and that the cyclic portion of the peptide establishes hydrophobic bonds with the aliphatic side-chains of UT-TMVII residues.[56]

Sequential deletion of human UII residues has revealed that the minimal sequence required for retaining full biological activity corresponds to $UII_{(4-11)}$, and that $UII_{(5-10)}$ is about 200- to 1000-fold less potent than the parent peptide.[50,57–59] Any change in the cyclic core substantially impairs the recognition of the peptide by UT (Fig. 4). In particular, Ala- or D-substitution of the Phe^6, Trp^7, Lys^8, or Tyr^9 residues in UII, and the Phe^3, Trp^4, Lys^5, or Tyr^6 residues in URP almost totally suppresses the ability of the analogs to constrict rat aortic rings.[50,51,58,60] Substitution of the two cysteines by isosteric Ser or Ala residues abolishes the contractile activity of human UII and UII-NH_2, respectively, indicating that the integrity of the disulfide bridge is essential for the biological activity of the peptides.[50] However, the design of cystine-free cyclic hexapeptide analogs that behave as agonists of human UT has been recently reported.[61] Reduction of the size of the cyclic region of UII also impairs the binding of the analogs on UT.[59] The monoiodinated analog [3-iodo-Tyr^9]$UII_{(4-11)}$ is five times more potent

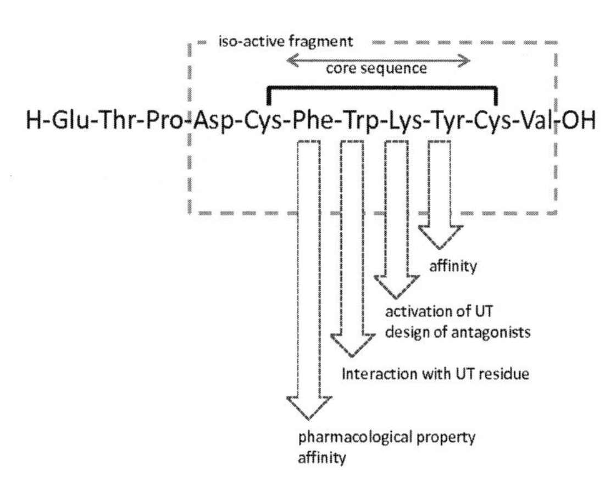

Figure 4. Schematic representation of the structure-activity relationships of human UII.

than the noniodinated peptide in the rat aortic ring assay,[50] suggesting that a modification of either the steric hindrance of the aromatic side-chain or the electronic density enhances the affinity of the UII analog for its receptor. Interestingly, it has been recently found that di-iodination of this Tyr moiety generates a potent partial agonist.[62] Substitution of the Lys^8 and Lys^5 residues of UII and URP, respectively, by an ornithine moiety generates antagonists of UT.[51,63–67] These observations have led to the design of a novel peptidic UII analog called urantide ([Pen^5,D-Trp^7,Orn^8]$UII_{(4-11)}$), which behaves as a pure antagonist in the rat aorta strip assay.[66,67] It should be noticed, however, that urantide acts as an agonist in human UT-transfected cells,[67] and is able, like UII, to induce a significant increase in systolic and mean arterial blood pressure after infusion in healthy human volunteers,[68] indicating that rat and human UT exhibit distinct structural features. In fact, a palmitoylation site, that is present in the cytoplasmic tail of human UT, but not in rodent UT, might serve as an anchor for the creation of an eighth transmembrane domain that probably changes the topology of the binding and/or activation site, and thus its signaling.[69,70]

Biological activities of UII in fish

UII has been initially partially purified on the basis of its vasoconstrictor effect in teleost fish.[71,72] This vasoactivity is one of the consequences of the general spasmogenic properties of UII, which can induce smooth muscle contraction in various tissues including urinary bladder, intestine, oviduct, and sperm duct.[73]

There is strong evidence that UII exerts osmoregulatory functions in fish. In particular, UII directly affects epithelial ion transport across the isolated skin, intestine, urinary bladder, and gill.[74–79] UII also stimulates cAMP formation in the isolated kidney from seawater-adapted flounders,[80] suggesting that the peptide may regulate water reabsorption in the fish nephron. In addition, UII may exert indirect actions on hydromineral homeostasis via stimulation of the secretion of prolactin,[81,82] a major osmoregulatory hormone in fish, and cortisol.[83,84] Supporting a physiologically important osmoregulatory role of UII in fish, *in vivo* studies have shown that prepro-UII mRNA levels in the caudal spinal cord and plasma UII concentrations fluctuate in accordance with external salinity in euryhaline fish.[85,86]

Intraarterial administration of UII in trout provokes a dose-dependent increase in arterial blood pressure associated with a decrease in heart rate[87] (Fig. 5) whereas intracerebroventricular (ICV) injection of UII produces a long-lasting hypertensive response without substantial modification of heart rate (Fig. 6).[88] Consistent with these observations, UII causes *in vitro* concentration-dependent contraction of dogfish vascular rings.[89] ICV injection of UII induces a gradual augmentation of the amplitude and frequency of ventilatory movements as well as a long-lasting increase in locomotor activity.[90,91]

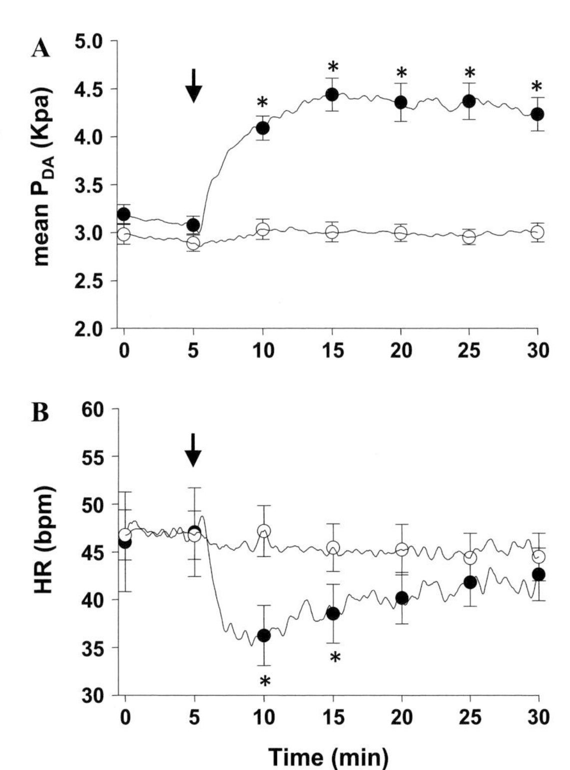

Figure 5. Time-course of the effect of intra-arterial injection of 50 pmol trout UII (filled circles) and vehicle (open circles) on mean dorsal aortic blood pressure (A, mean P_{DA}) and heart rate (B, HR) in 14 unanesthetized trout. The *arrows* indicate the time of injection. * $P < 0.05$ versus preinjection value (ANOVA followed by Dunnett's test).

Biological activities of UII in amphibians

The effects of UII on frog smooth muscle contraction have been studied *in vivo* and *in vitro*, using the homologous peptide. Bolus intraarterial injection of UII in the bullfrog *Rana catesbeiana* (now renamed *Lithobates catesbeianus*) provokes a rapid increase in blood pressure and a profound decrease in blood flow without any modification of heart rate.[45] Perifusion of rings from the left and right systemic arches with graded concentrations of UII induces a dose-dependent contraction.[45] In frog, UII also provokes an increase of the frequency of the contractions of the urinary bladder and an increase of the amplitude of the contractions of the ileum.[92] The contractile response is mimicked by arachidonic acid and blocked by indomethacin, indicating that the effect of UII on frog smooth muscle preparations is mediated through the phospholipase A2/prostaglandin-signaling pathway.[45,92]

In frog, the UII gene is primarily expressed in motoneurons of the brain stem and spinal cord, and these neurons appear to be innervated by recurrent fibers,[26] suggesting that UII may participate to the synchronization of motoneuron activity. However, a possible role of UII in locomotor activity in amphibians has not yet been investigated.

Biological activities of UII in mammals

Consistent with the widespread distribution of UT in the CNS of mammals, central administration of UII and/or URP provokes multiple behavioral effects (Fig. 6).[93] ICV injection of UII in the lateral

Effects	Species	Peptide	Reference
Hyperlocomotion	Mouse	mUII	Do Rego *et al.* 2005 *Psychopharmacology* **183:** 103-117
	Rat	hUII	Gartlon *et al.* 2001 *Psychopharmacology* **155:** 426-433
	Trout	tUII	Lancien *et al.* 2004 *Brain Res.* **1023:** 167-174
Anxiogenic-like effects	Mouse	bUII	Matsumoto *et al.* 2004 *Neurosci. Lett.* **358:** 99-102
		mUII	Do Rego *et al.* 2005 *Psychopharmacology* **183:** 103-117
Depressant-like effects	Mouse	mUII	Do Rego *et al.* 2005 *Psychopharmacology* **183:** 103-117
Orexigenic and dipsigenic effects	Mouse	mUII	Do Rego *et al.* 2005 *Psychopharmacology* **183:** 103-117
Increased REM sleep duration	Rat	rUII	Huitron-Resendiz *et al.* 2005 *J. Neurosci.* **25:** 5465-5474
Increased cerebral blood flow Exacerbation of post-infarct brain damage	Rat	rUII	Chuquet *et al.* 2008 *Exp. Neurol.* **210:** 577-784
Chronotropic and inotropic effects	Sheep	hUII	Watson *et al.* 2003 *Hypertension* **42:** 373-379
Hypertensive effect	Rat	rUII	Lin *et al.* 2003 *J. Hypertens.* **21:** 159-165
	Trout	tUII	Le Mevel *et al.* 2008 *Peptides* **29:** 830-837
Hyperventilatory effect	Trout	tUII	Lancien *et al.* 2004 *Brain Res.* **1023:** 167-174
Stimulation of prolactin and thyrotropin release	Rat	hUII	Gartlon *et al.* 2001 *Psychopharmacology* **155:** 426-433
Stimulation of adrenaline and ACTH release	Sheep	hUII	Watson *et al.* 2003 *Hypertension* **42:** 373-379

Figure 6. **Biological effects of intracerebroventricular injection of urotensin II. bUII, bovine urotensin II; hUII, human urotensin II; mUII, mouse urotensin II; rUII, rat urotensin II; tUII, trout urotensin II.**

ventricle induces a dose-dependent increase in locomotor activity in rat[94] and mouse.[95] ICV injection of UII also provokes orexigenic and dipsigenic responses in mouse.[95] Neuroanatomical studies reveal the occurrence of UT mRNA in the lateral septum and the bed nucleus of the lamina terminalis,[29,39] suggesting that UII and/or URP could exert anxiogenic effects. Consistent with this proposal, ICV administration of UII causes anxiety-like behavior in mouse and rat.[95,96] In the forced swimming test and the tail suspension tests, UII causes an increase in the immobility time of mice,[95] indicating that the peptide exerts a depressant-like effect. High concentrations of UT mRNA are found in the tegmental area[38,39] where cholinergic neurons controlling rapid eye movements (REM) are located.[97] Indeed, microinjection of UII into the pedunculopontine tegmental nucleus markedly increases REM sleep duration in rat (Fig. 6).[98] UT mRNA is also actively expressed in large cells of the locus coeruleus,[39] suggesting that UII may regulate sleep-wakefulness. In support of this hypothesis, *in vitro* studies indicate that UII stimulates the release of norepinephrine from rat cerebrocortical slices.[99] Interestingly, URP mRNA is abundant in catecholaminergic neurons of the locus coeruleus,[29] which suggests that the peptide may exert an autocrine effect on norepinephrine release. Collectively, these observations provide strong evidence for a role of UII and URP in the regulation of paradoxical sleep.

In the endocrine system, UII regulates the secretion of various hormones. In particular, central administration of UII in rat induces an increase in plasma prolactin and thyroid-stimulating hormone levels (Fig. 6).[94] ICV injection of UII in sheep elicits a dose-dependent elevation of plasma adrenocorticotropic hormone, adrenaline, and glucose (Fig. 6).[100]

UII inhibits glucose-induced insulin release from the perfused rat pancreas.[101,102] The UT antagonists palosuran and urantide potentiate glucose-evoked insulin output,[103] suggesting that endogenous UII exerts an inhibitory tone on β-cell secretory activity. The insulinostatic effect of UII can be ascribed to its inhibitory action on adenylyl cyclase through a pertusis toxin-sensitive G protein.[104]

Several studies suggest that UII may play a role in the pathogenesis of diabetes. Thus, diabetic mice exhibit higher concentrations of UII and UT mRNAs in skeletal muscle.[105] Plasma UII levels are also elevated in diabetic patients.[106,107] Finally, single nucleotide polyporphism in the UII gene is associated with higher risk to develop type 2 diabetes.[108] However, blockage of the urotensinergic system does not induce any variation in insulin secretion, sensitivity, or daily blood glucose levels in diet-treated patients with type 2 diabetes.[109]

Immunoblot analysis shows the occurrence of the UII precursor and UT in human adrenocortical tissue.[110] Infusion of UII in rat elevates plasma aldosterone levels[110] whereas, *in vitro*, UII causes a dose-dependent inhibition of corticosterone from cultured rat adrenocortical cells.[111]

The role of UII in the physiology of the cardiovascular system of mammals had long been suspected since early studies had revealed that fish UII exerts various effects on mammalian vascular smooth muscle. In particular, it had been reported that goby UII produces endothelium-dependent relaxation and endothelium-independent contraction of rat aorta rings[13] and increases arterial blood pressure in anesthetized rats.[14] There is now clear evidence that UII and UT are involved in the control of cardiovascular activity. *In vitro*, UII and URP exert a potent contractile effect on endothelium-denuded rat thoracic aorta, monkey coronary artery, and human pulmonary artery.[20,50,51,112] In contrast, in intact rat small mesenteric arteries, UII causes endothelium-dependent relaxation that is mediated through NO production.[113] It should be noted however that the vasoactive profile of UII is species-dependent and, within each single species, varies substantially according to the type of blood vessel considered.[114] *In vivo*, intravenous injection of UII[115,116] or URP[49] in the rat produces a long-lasting reduction of blood pressure whereas chronic infusion of UII has no substantial effect on blood pressure.[117] In humans, intraarterial injection of UII in the forearm induces local vasoconstriction[118] or has no apparent effect.[68,119]

Consistent with the expression of UT in the atrium and ventricle,[20,35,112,120] UII causes a concentration-dependent increase in the contractile force of human cardiac muscle *in vitro*.[121] The positive inotropic effect of UII does not depend on endothelin, serotonin (5-HT4), or β-adrenergic receptors.[121]

ICV administration of UII provokes an increase in heart rate and arterial blood pressure in both normotensive[122] and hypertensive rats (Fig. 6).[40] Microinjections of UII into the rat paraventricular nucleus or arcuate nucleus also increase heart rate and blood pressure.[123] In contrast, microinjection of UII into the A1 area (a region of the medulla oblongata that contains noradrenergic neurons) causes dose-dependent depressor and bradycardiac responses.[123] These data indicate that UII plays differential roles in brain regions involved in the control of cardiovascular activity.

There is now strong evidence that UII is implicated in cardiovascular diseases. In particular, UII, URP, and UT mRNA levels are increased in the atrium and/or ventricle of rats with congestive heart failure.[124] UII, URP, and UT mRNA expression is higher in cardiovascular organs[125] and kidney[126] of spontaneous hypertensive rats than in normotensive rats. Increased expression of UII mRNA is also observed in the myocardium of patients with congestive heart failure.[127] Circulating levels of UII are elevated in essential hypertension,[35] renal hypertension,[36] portal hypertension,[128] and heart failure.[129,130] These observations indicate that UII and URP may participate in the pathophysiology of cardiac, renal, and liver diseases. UII acts synergistically with mildly oxidized low-density lipoprotein to induce vascular smooth muscle cell proliferation,[131] suggesting a possible role of UII in the pathophysiology of atherosclerosis. Consistent with this notion, increased expression of UII and UT is observed in atherosclerotic lesions in the human aorta[132] and coronary arteries.[133,134] In addition, UT is expressed in rat and human peripheral blood mononuclear cells, and UII acts as a chemoattractant for human monocytes.[135] UII also activates migration and differentiation of adventitial fibroblasts from rat aorta.[136] These observations provide further evidence for the involvement of the UII system in the pathogenesis of atherosclerosis.[137]

However, recent data suggest that UII and URP may exert a cardioprotective effect. For instance, high concentrations of UII seem to have a beneficial effect in patients with end-stage renal disease.[138] In the ischemic heart, both rat UII and URP are able to reduce cardiac ischemia–reperfusion injury by increasing flow through the coronary circulation, by reducing contractility and therefore myocardial energy demand, and by inhibiting reperfusion myocardial damage.[139] Moreover, in a clinical study, using the Kaplan–Meier survival curve, it has been reported that patients with heart failure presenting high UII plasma concentrations have significantly better clinical prognostic.[140] Altogether these data highlight a potential cardioprotective role for both UII and URP and suggest that a degree of caution should be taken regarding the use of UT antagonists in different pathologies.

Acknowledgments

The preparation of this review and the experimental work described were supported by grants from the Institut National de la Santé et de la Recherche Médicale (U413), the European Institute for Peptide Research (IFRMP 23), Servier Laboratories, and the Région Haute-Normandie. H.V. is Affiliated Professor at the Institut National de la Recherche Scientifique – Institut Armand-Frappier, Montreal, and Associated Researcher at the Research Center in Molecular Endocrinology, Oncology and Genetics, Laval University, Quebec, Canada.

Conflicts of interest

The authors declare no conflict of interest.

References

1. Zelnik, P.R. & K. Lederis. 1973. Chromatographic separation of urotensins. *Gen. Comp. Endocrinol.* **20:** 392–400.
2. Chan, D.K.O. & H.A. Bern. 1976. The caudal neurosecretory system. *Cell Tissue Res.* **174:** 339–354.
3. Ichikawa, T. 1978. Acetylcholine in the urophysis of several species of teleosts. *Gen. Comp. Endocrinol.* **35:** 226–233.
4. Lederis, K., A. Letter, D. McMaster, *et al.* 1982. Complete amino acid sequence of urotensin I, a hypotensive and corticotrophin-releasing neuropeptide from catostomus. *Science* **218:** 162–164.
5. Pearson, D., J.E. Shively, B.R. Clark, *et al.* 1980. Urotensin II: a somatostatin-like peptide in the caudal neurosecretory system of fishes. *Proc. Natl. Acad. Sci. USA* **77:** 5021–5024.
6. Lacanilao, F. 1972. The urophysial hydroosmotic factor of fishes. II. Chromatographic and pharmacologic indication of similarity to arginine vasopressin. *Gen. Comp. Endocrinol.* **14:** 413–420.
7. Yulis, C.R. & K. Lederis. 1986. Extraurophyseal distribution of urotensin II immunoreactive neuronal perikarya and their process. *Proc. Natl. Acad. Sci. USA* **83:** 7079–7083.
8. Yulis, C.R. & K. Lederis. 1988. Occurrence of an anterior spinal, cerebrospinal fluid-contacting, urotensin II neuronal system in various fish species. *Gen. Comp. Endocrinol.* **70:** 301–311.
9. Gonzalez, G.C., M. Martinez-Padron, K. Lederis & K. Lukowiak. 1992. Distribution and coexistence of urotensin I and urotensin II peptides in the cerebral ganglia of *Aplysia californica*. *Peptides* **13:** 695–703.
10. Waugh, D. & J.M. Conlon. 1993. Purification and characterization of urotensin II from the brain of a teleost (trout, *Oncorhynchus mykiss*) and an elasmobranch (skate, *Raja rhina*). *Gen. Comp. Endocrinol.* **92:** 419–427.
11. Itoh, H., D. McMaster & K. Lederis. 1988. Functional receptors for fish neuropeptide urotensin II in major rat arteries. *Eur. J. Pharmacol.* **149:** 61–66.
12. Gibson, A., H.A. Bern, M. Ginsburg, *et al.* 1984. Neuropeptide-induced contraction and relaxation of the mouse anococcygeus muscle. *Proc. Natl. Acad. Sci. USA* **81:** 625–629.
13. Gibson, A. 1987. Complex effects of Gillichthys urotensin II on rat aortic strips. *Br. J. Pharmacol.* **91:** 205–212.
14. Gibson, A., P. Wallace & H.A. Bern. 1986. Cardiovascular effects of urotensin II in anesthetized and pithed rats. *Gen. Comp. Endocrinol.* **64:** 435–439.
15. Conlon, J.M. 2008. "Liberation" of urotensin II from the teleost urophysis: an historical overview. *Peptides* **29:** 651–657.
16. Chartrel, N., J.M. Conlon, J.M. Danger, *et al.* 1991. Characterization of melanotropin release-inhibiting factor (melanostatin) from frog brain: homology with human neuropeptide Y. *Proc. Natl. Acad. Sci. USA* **88:** 3862–3866.

17. Chartrel, N., M.C. Tonon, H. Vaudry, *et al.* 1991. Primary structure of frog pituitary adenylate cyclase-activating polypeptide (PACAP) and effects of ovine PACAP on frog pituitary. *Endocrinology* **129:** 3367–3371.
18. Conlon, J.M., F. O'Harte, D.D. Smith, *et al.* 1992. Isolation and primary structure of urotensin II from the brain of a tetrapod, the frog *Rana ridibunda. Biochem. Biophys. Res. Commun.* **188:** 578–583.
19. Coulouarn, Y., I. Lihrmann, S. Jégou, *et al.* 1998. Cloning of the cDNA encoding the urotensin II precursor in frog and human reveals intense expression of the urotensin II gene in motoneurons of the spinal cord. *Proc. Natl. Acad. Sci. USA* **95:** 15803–15808.
20. Ames, R.S., H.M. Sarau, J.K. Chambers, *et al.* 1999. Human urotensin-II is a potent vasoconstrictor and agonist for the orphan receptor GPR14. *Nature* **401:** 282–286.
21. Elshourbagy, N.A., S.A. Douglas, U. Shabon, *et al.* 2002. Molecular and pharmacological characterization of genes encoding urotensin II-related peptides and their cognate G-protein-coupled receptors from the mouse and monkey. *Br. J. Pharmacol.* **136:** 9–22.
22. Mori, M., T. Sugo, M. Abe, *et al.* 1999. Urotensin II is the endogenous ligand of a G-protein-coupled orphan receptor, SENR (GPR14). *Biochem. Biophys. Res. Commun.* **265:** 123–129.
23. Coulouarn, Y., S. Jégou, H. Tostivint, *et al.* 1999. Cloning, sequence analysis and tissue distribution of the mouse and the rat urotensin II precursors. *FEBS Lett.* **457:** 28–32.
24. Tostivint, H., L. Joly, I. Lihrmann, *et al.* 2006. Comparative genomics provides evidence for close evolutionary relationships between the urotensin II and somatostatin gene families. *Proc. Natl. Acad. Sci. USA* **103:** 2237–2242.
25. Chartrel, N., J.M. Conlon, F. Collin, *et al.* 1996. Urotensin II in the central nervous system of the frog *Rana ridibunda*: immunohistochemical localization and biochemical characterization. *J. Comp. Neurol.* **364:** 324–339.
26. Coulouarn, Y., C. Fernex, S. Jégou, *et al.* 2001. Specific expression of the urotensin II gene in sacral motoneurons of developing rat spinal cord. *Mech. Dev.* **101:** 187–190.
27. Dun, S.L., G.C. Brailoiu, J. Yang, *et al.* 2001. Urotensin II-immunoreactivity in the brainstem and spinal cord of the rat. *Neurosci. Lett.* **305:** 9–12.
28. Chartrel, N., J. Leprince, C. Dujardin, *et al.* 2004. Biochemical characterization and immunohistochemical localization of urotensin II in the human brainstem and spinal cord. *J. Neurochem.* **91:** 110–118.
29. Dubessy, C., D. Cartier, B. Lectez, *et al.* 2008. Characterization of urotensin II, distribution of urotensin II, urotensin II-related peptide and UT receptor mRNAs in mouse: evidence of urotensin II at the neuromuscular junction. *J. Neurochem.* **107:** 361–374.
30. Pelletier, G., I. Lihrmann & H. Vaudry. 2002. Role of androgens in the regulation of urotensin II precursor mRNA expression in the rat brainstem and spinal cord. *Neuroscience* **115:** 525–532.
31. Tal, M., D.A. Ammar, M. Karpuj, *et al.* 1995. A novel putative neuropeptide receptor expressed in neural tissue, including sensory epithelia. *Biochem. Biophys. Res. Commun.* **209:** 752–759.
32. Marchese, A., M. Heiber, T. Nguyen, *et al.* 1995. Cloning and chromosomal mapping of three novel genes, GPR9, GPR10, and GPR14, encoding receptors related to interleukin 8, neuropeptide Y, and somatostatin receptors. *Genomics* **29:** 335–344.
33. Liu, Q., S.S. Pong, Z. Zeng, *et al.* 1999. Identification of urotensin II as the endogenous ligand for the orphan G-protein-coupled receptor GPR14. *Biochem. Biophys. Res. Commun.* **266:** 174–178.
34. Nothacker, H.P., Z. Wang, A.M. McNeill, *et al.* 1999. Identification of the natural ligand of an orphan G-protein-coupled receptor involved in the regulation of vasoconstriction. *Nat. Cell Biol.* **1:** 383–385.
35. Matsushita, M., M. Shichiri, T. Imai, *et al.* 2001. Co-expression of urotensin II and its receptor (GPR14) in human cardiovascular and renal tissues. *J. Hypertens.* **19:** 2185–2190.
36. Totsune, K., K. Takahashi, Z. Arihara, *et al.* 2001. Role of urotensin II in patients on dialysis. *Lancet* **359:** 810–811.
37. Maguire, J.J., R.E. Kuc, M.J. Kleinz & A.P. Davenport. 2008. Immunocytochemical localization of the urotensin-II receptor, UT, in rat and human tissues: relevance to function. *Peptides* **29:** 735–742.
38. Clark, S.D., H.P. Nothacker, Z. Wang, *et al.* 2001. The urotensin II receptor is expressed in the cholinergic mesopontine tegmentum of the rat. *Brain Res.* **923:** 120–127.
39. Jégou, S., D. Cartier, C. Dubessy, *et al.* 2006. Localization of the urotensin-II receptor in the rat central nervous system. *J. Comp. Neurol.* **495:** 21–36.
40. Lin, Y., T. Tsuchihashi, K. Matsumura, *et al.* 2003. Central cardiovascular action of urotensin II in spontaneously hypertensive rats. *Hypertens. Res.* **26:** 839–845.

41. Rossowski, W.J., B.L. Cheng, J.E. Taylor, *et al.* 2002. Human urotensin II-induced aorta ring contractions are mediated by protein kinase C, tyrosine kinases, and Rho-kinase: inhibition by somatostatin receptor antagonists. *Eur. J. Pharmacol.* **438:** 159–170.
42. Saetrum Opgaard, O., H. Nothacker, F.J. Ehlert, *et al.* 2000. Human urotensin II mediates vasoconstriction via an increase in inositol phosphates. *Eur. J. Pharmacol.* **406:** 265–271.
43. Castel, H., M. Diallo, D. Chatenet, *et al.* 2006. Biochemical and functional characterization of high-affinity urotensin II receptors in rat cortical astrocytes. *J. Neurochem.* **99:** 582–595.
44. Gibson, A., S. Conyers, H.A. Bern, *et al.* 1988. The influence of urotensin II on calcium flux in rat aorta. *J. Pharm. Pharmacol.* **40:** 893–895.
45. Yano, K., J.W. Hicks, H. Vaudry, *et al.* 1995. Cardiovascular actions of frog urotensin II in the frog *Rana catesbeiana*. *Gen. Comp. Endocrinol.* **97:** 103–110.
46. Ziltener, P., C. Mueller, B. Haenig, *et al.* 2002. Urotensin II mediates ERK1/2 phosphorylation and proliferation in GPR14-transfected cell lines. *J. Recept. Signal Transduct. Res.* **22:** 155–168.
47. Katano, Y., A. Ishihata, T. Aita, *et al.* 2000. Vasodilatator effect of urotensin II, one of the most potent vasoconstricting factors, on rat coronary arteries. *Eur. J. Pharmacol.* **402:** 209–211.
48. Sauzeau, V., E. Le Mellionnec, J. Bertoglio, *et al.* 2001. Human urotensin II-induced contraction and arterial smooth muscle cell proliferation are mediated by RhoA and Rho-kinase. *Circ. Res.* **88:** 1102–1104.
49. Sugo, T., Y. Murakami, Y. Shimomura, *et al.* 2003. Identification of urotensin II-related peptide as the urotensin II-immunoreactive molecule in the rat brain. *Biochem. Biophys. Res. Commun.* **310:** 860–868.
50. Labarrère, P., D. Chatenet, J. Leprince, *et al.* 2003. Structure-activity relationships of human urotensin II and related analogues on rat aortic ring contraction. *J. Enz. Inh. Med. Chem.* **18:** 77–88.
51. Chatenet, D., C. Dubessy, J. Leprince, *et al.* 2004. Structure-activity relationships and structural conformation of a novel urotensin II-related peptide. *Peptides* **25:** 1819–1830.
52. Pelletier, G., I. Lihrmann, C. Dubessy, *et al.* 2005. Androgenic down-regulation of urotensin II precursor, urotensin II-related peptide precursor and androgen receptor mRNA in the mouse spinal cord. *Neuroscience* **132:** 689–696.
53. Malagon, M.M., M. Molina, M.D. Gahete, *et al.* 2008. Urotensin II and urotensin II-related peptide activate somatostatin receptor subtype 2 and 5. *Peptides* **29:** 711–720.
54. Carotenuto, A., P. Grieco, P. Campiglia, *et al.* 2004. Unraveling the active conformation of urotensin II. *J. Med. Chem.* **47:** 1652–1661.
55. Boivin, S., L. Guilhaudis, I. Milazzo, *et al.* 2006. Characterization of urotensin II receptor structural domains involved in the recognition of UII, URP, and urantide. *Biochemistry* **45:** 5993–6002.
56. Boivin, S., I. Ségalas-Milazzo, L. Guilhaudis, *et al.* 2008. Solution structure of urotensin-II receptor extracellular loop III and characterization of its interaction with urotensin-II. *Peptides* **29:** 700–710.
57. Perkins, T.D.J., S. Bansal & D.J. Barlow. 1990. Molecular modelling and design of analogues of the peptide hormone urotensin II. *Biochem. Soc. Trans.* **18:** 918–919.
58. Flohr, S., M. Kurz, E. Kostenis, *et al.* 2002. Identification of nonpeptidic urotensin II receptor antagonists by virtual screening based on a pharmacophore model derived from structure-activity relationships and nuclear magnetic resonance studies on urotensin II. *J. Med. Chem.* **45:** 1799–1805.
59. Kinney, W.A., H.R. Almond, J. Qi, *et al.* 2002. Structure-function analysis of urotensin II and its use in the construction of a ligand-receptor working model. *Angew. Chem. Int. Ed.* **41:** 2940–2944.
60. Brkovic, A., A. Hattenberger, E. Kostenis, *et al.* 2003. Functional and binding characterizations of urotensin II-related peptides in human and rat urotensin II-receptor assay. *J. Pharmacol. Exp. Ther.* **306:** 1200–1209.
61. Foister, S., L.L. Taylor, J.-J. Feng, *et al.* 2006. Design and synthesis of potent cystine-free cyclic hexapeptide agonists at the human urotensin receptor. *Org. Lett.* **8:** 1799–1802.
62. Batuwangala, M., V. Camarda, J. McDonald, *et al.* 2009. Structure-activity relationship study on Tyr^9 of urotensin-II(4–11): identification of a partial agonist of the UT receptor. *Peptides* **30:** 1130–1136.
63. Camarda, V., A. Rizzi, G. Calo, *et al.* 2002. Effects of human urotensin II in isolated vessels of various species; comparison with other vasoactive agents. *Naunyn. Schmiedebergs. Arch. Pharmacol.* **365:** 141–149.
64. Camarda, V., R. Guerrini, E. Kostenis, *et al.* 2002. A new ligand for the urotensin II receptor. *Br. J. Pharmacol.* **137:** 311–314.
65. Vergura, R., V. Camarda, A. Rizzi, *et al.* 2004. Urotensin II stimulates plasma extravasation in mice via UT receptor activation. *Naunyn. Schmiedebergs. Arch. Pharmacol.* **370:** 347–352.

66. Patacchini, R., P. Santicioli, S. Giuliani, *et al.* 2003. Urantide: an ultrapotent urotensin II antagonist peptide in the rat aorta. *Br. J. Pharmacol.* **140:** 1155–1158.
67. Camarda, V., W. Song, E. Marzola, *et al.* 2004. Urantide mimics urotensin-II induced calcium release in cells expressing recombinant UT receptors. *Eur. J. Pharmacol.* **498:** 83–86.
68. Cheriyan, J., T.J. Burton, T.J. Bradley, *et al.* 2009. The effects of urotensin II and urantide on forearm blood flow and systemic haemodynamics in humans. *Br. J. Clin. Pharmacol.* **68:** 518–523.
69. Weiss, J., S.-J. Han, S.-K. Kim, *et al.* 2008. Conformational changes involved in G-protein-coupled-receptor activation. *Trends Pharmacol. Sci.* **29:** 616–625.
70. Huynh, J., W.G. Thomas, M.-I. Aguilar, *et al.* 2009. Role of helix 8 in G protein-coupled receptors based on structure-function studies on the type 1 angiotensin receptor. *Mol. Cell. Endocrinol.* **302:** 118–127.
71. Bern, H.A. & K. Lederis. 1969. A reference preparation for the study of active substances in the caudal neurosecretory system of teleost. *J. Endocrinol.* **45:** xi–xii.
72. Chan, D.K.O., I. Chester Jones & S. Ponniah. 1969. Studies on the pressor substances in the caudal neurosecretory system of teleost fish: bioassay and fractionation. *J. Endocrinol.* **45:** 151–159.
73. Bern, H.A., D. Pearson, B.A. Larson, *et al.* 1985. Neurohormones from fish tails: the caudal neurosecretory system. Urophysiology and the caudal neurosecretory system of fishes. *Rec. Prog. Horm. Res.* **41:** 533–552.
74. Marshall, W.S. & H.A. Bern. 1979. Teleostean urophysis: urotensin II and ion transport across the isolated skin of marine teleost. *Science* **204:** 519–521.
75. Marshall, W.S. & H.A. Bern. 1981. Active chloride transport by the skin of marine teleost is stimulated by urotensin I and inhibited by urotensin II. *Gen. Comp. Endocrinol.* **43:** 484–491.
76. Mainoya, J.R. & H.A. Bern. 1982. Effects of teleost urotensins on intestinal absorption of water and NaCl in tilapia, *Sarotherodon mossambicus*, adapted to fresh water or seawater. *Gen. Comp. Endocrinol.* **47:** 54–58.
77. Mainoya, J.R. & H.A. Bern. 1984. Influence of vasoactive intestinal peptide and urotensin II on the absorption of water and NaCl by the anterior intestine of the Tilapia, *Sarotherodon mossambicus*. *Zool. Sci.* **1:** 100–105.
78. Loretz, C.A., R.W. Freel & H.A. Bern. 1983. Specificity of response of intestinal ion transport systems to a pair of natural peptide hormone analogs: somatostatin and urotensin II. *Gen. Comp. Endocrinol.* **52:** 198–206.
79. Loretz, C.A. & H.A. Bern. 1981. Stimulation of sodium transport across the teleost urinary bladder by urotensin II. *Gen. Comp. Endocrinol.* **43:** 325–330.
80. Stenhouse, A.A. & R.J. Balment. 1998. Urotensin II stimulates cAMP production in the kidney of seawater-adapted but not freshwater-adapted flounder, *Platichthys flesus*. *Ann. N.Y. Acad. Sci.* **839:** 389–391.
81. Grau, E.G., R.S. Nishioka & H.A. Bern. 1982. Effects of somatostatin and urotensin II on tilapia pituitary prolactin release and interactions between somatostatin, osmotic pressure Ca^{++}, and adenosine 3′,5′-monophosphate in prolactin release *in vitro*. *Endocrinology* **110:** 910–915.
82. Rivas, R.J., R.S. Nishioka & H.A. Bern. 1986. *in vitro* effects of somatostatin and urotensin II on prolactin and growth hormone secretion in tilapia, *Oreochromis mossambicus*. *Gen. Comp. Endocrinol.* **63:** 245–251.
83. Arnold Reed D.E. & R.J. Balment. 1994. Peptide hormones influence *in vitro* interrenal secretion of cortisol in the trout, *Oncorhynchus mykiss*. *Gen. Comp. Endocrinol.* **96:** 85–91.
84. Kelsall, C.J. & R.J. Balment. 1998. Native urotensins influence cortisol secretion and plasma cortisol concentration in the euryhaline flounder, *Platichthys flesus*. *Gen. Comp. Endocrinol.* **112:** 210–219.
85. Bond, H., M.J. Winter, J.M. Warne, *et al.* 2002. Plasma concentrations of arginine vasotocin and urotensin II are reduced following transfer of the euryhaline flounder (*Platichthys flesus*) from seawater to fresh water. *Gen. Comp. Endocrinol.* **125:** 113–120.
86. Lu, W., M. Greenwood, L. Dow, *et al.* 2006. Molecular characterization and expression of urotensin II and its receptor in the flounder (*Platichthys flesus*): a hormone system supporting body fluid homeostasis in euryhaline fish. *Endocrinology* **147:** 3692–3708.
87. Le Mevel, J.C., F. Lancien & N. Mimassi. 2008. Central and peripheral cardiovascular, ventilatory and motor effects of trout urotensin II in the trout. *Peptides* **29:** 830–837.
88. Le Mevel, J.C., K.R. Olson, D. Conklin, *et al.* 1996. Cardiovascular actions of trout urotensin II in the conscious trout, *Oncorhynchus mykiss*. *Am. J. Physiol.* **271:** R1335–R1343.
89. Hazon, N., C. Bjenning & J.M. Conlon. 1993. Cardiovascular actions of dogfish urotensin II in the dogfish *Scyliorhinus canicula*. *Am. J. Physiol.* **265:** R573–R576.
90. Lancien, F., J. Leprince, N. Mimassi, *et al.* 2004. Central effects of native urotensin II on motor activity, ventilatory movements, and heart rate in the trout *Oncorhynchus mykiss*. *Brain Res.* **1023:** 167–174.

91. Lancien, F., J. Leprince, N. Mimassi, *et al.* 2005. Time-course effects of centrally administered native urotensin-II on motor and cardioventillatory activity in trout. *Ann. N.Y. Acad. Sci.* **1040:** 371–374.
92. Yano, K., H. Vaudry & J.M. Conlon. 1994. Spasmogenic actions of frog urotensin II on the bladder and ileum of the frog, *Rana catesbeiana. Gen. Comp. Endocrinol.* **96:** 412–419.
93. Do Rego, J.C., J. Leprince, E. Scalbert, *et al.* 2008. Behavioral actions of urotensin-II. *Peptides* **29:** 838–844.
94. Gartlon, J., F. Parker, D.C. Harrison, *et al.* 2001. Central effects of urotensin-II following ICV administration in rats. *Psychopharmacology* **155:** 426–433.
95. Do Rego, J.C., D. Chatenet, M.H. Orta, *et al.* 2005. Behavioral effects of urotensin-II centrally administered in mice. *Psychopharmacology* **183:** 103–117.
96. Matsumoto, Y., M. Abe, T. Watanabe, *et al.* 2004. Intracerebroventricular administration of urotensin II promotes anxiogenic-like behaviors in rodents. *Neurosci. Lett.* **358:** 99–102.
97. Quattrochi, J.J., A.N. Mamelak, R.D. Madison, *et al.* 1989. Mapping neuronal inputs to REM sleep induction sites with carbachol-fluorescent microspheres. *Science* **245:** 984–986.
98. De Lecea, L. & P. Bourgin. 2008. Neuropeptide interactions and REM sleep: a role for urotensin II? *Peptides* **29:** 845–851.
99. Ono, T., Y. Kawaguchi, M. Kudo, *et al.* 2008. Urotensin II evokes neurotransmitter release from rat cerebrocortical slices. *Neurosci. Lett.* **440:** 275–279.
100. Watson, A.M., G.W. Lambert, K.J. Smith & C.N. May. 2003. Urotensin II acts centrally to increase epinephrine and ACTH release and cause potent inotropic and chronotropic actions. *Hypertension* **42:** 373–379.
101. Silvestre, R.A., J. Rodríguez-Gallardo, E.M. Egido & J. Marco. 2001. Inhibition of insulin release by urotensin II—a study on the perfused rat pancreas. *Horm. Metab. Res.* **33:** 379–381.
102. Silvestre, R.A., E.M. Egido, R. Hernández, *et al.* 2004. Urotensin-II is present in pancreatic extracts and inhibits insulin release in the perfused rat pancreas. *Eur. J. Endocrinol.* **151:** 803–809.
103. Marco, J., E.M. Egido, R. Hernández & R.A. Silvestre. 2007. Evidence for endogenous urotensin-II as an inhibitor of insulin secretion. Study in the perfused rat pancreas. *Peptides* **29:** 852–858.
104. Silvestre, R.A., E.M. Egido, R. Hernández & J. Marco. 2009. Characterization of the insulinostatic effect of urotensin II: a study in the perfused rat pancreas. *Regul. Pept.* **153:** 37–42.
105. Wang, H.X., X.J. Zeng, Y. Liu, *et al.* 2009. Elevated expression of urotensin II and its receptor in skeletal muscle of diabetic mouse. *Regul. Pept.* **154:** 85–90.
106. Totsune, K., K. Takahashi, Z. Arihara, *et al.* 2003. Increased plasma urotensin II levels in patients with diabetes mellitus. *Clin. Sci.* **104:** 1–5.
107. Suguro, T., T. Watanabe, S. Kodate, *et al.* 2008. Increased plasma urotensin-II levels are associated with diabetic retinopathy and carotid atherosclerosis in Type 2 diabetes. *Clin. Sci.* **115:** 327–334.
108. Wenyi, Z., S. Suzuki, M. Hirai, *et al.* 2003. Role of urotensin II gene in genetic susceptibility to Type 2 diabetes mellitus in Japanese subjects. *Diabetologia* **46:** 972–976.
109. Sidharta, P.N., K. Rave, L. Heinemann, *et al.* 2009. Effect of the urotensin-II receptor antagonist palosuran on secretion of and sensitivity to insulin in patients with type 2 diabetes mellitus. *Br. J. Clin. Pharmacol.* **68:** 502–510.
110. Giuliani, L., L. Lenzini, M. Antonello, *et al.* 2009. Expression and functional role of urotensin-II and its receptor in the adrenal cortex and medulla: novel insights for the pathophysiology of primary aldosteronism. *J. Clin. Endocrinol. Metab.* **94:** 684–690.
111. Albertin, G., V. Casale, A. Ziolkowska, *et al.* 2006. Urotensin-II and UII-receptor expression and function in the rat adrenal cortex. *Int. J. Mol. Med.* **17:** 1111–1115.
112. Maguire, J.J., R.E. Kuc & A.P. Davenport. 2000. Orphan-receptor ligand human urotensin II: receptor localization in human tissues and comparison of vasoconstrictor responses with endothelin-1. *Br. J. Pharmacol.* **131:** 441–446.
113. Bottrill, F.E., S.A. Douglas, C.R. Hiley & R. White. 2000. Human urotensin-II is an endothelium-dependent vasodilator in rat small arteries. *Br. J. Pharmacol.* **130:** 1865–1870.
114. Douglas, S.A., A.C. Sulpizio, V. Piercy, *et al.* 2000. Differential vasoconstrictor activity of human urotensin-II in vascular tissue isolated from the rat, mouse, dog, pig, marmoset and cynomolgus monkey. *Br. J. Pharmacol.* **131:** 1262–1274.
115. Gardiner, S.M., J.E. March, P.A. Kemp, *et al.* 2001. Depressor and regionally-selective vasodilator effects of human and rat urotensin II in conscious rats. *Br. J. Pharmacol.* **132:** 1625–1629.
116. Abdelrahman, A.M. & C.C Pang. 2002. Involvement of the nitric oxide/L-arginine and sympathetic nervous systems on the vasodepressor action of human urotensin II in anesthetized rats. *Life Sci.* **71:** 819–825.

117. Kompa, A.R., W.G. Thomas, F. See, *et al.* 2004. Cardiovascular role of urotensin II: effect of chronic infusion in the rat. *Peptides* **25:** 1783–1788.
118. Böhm, F. & J. Pernow. 2002. Urotensin II evokes potent vasoconstriction in humans in vivo. *Br. J. Pharmacol.* **135:** 25–27.
119. Wilkinson, I.B., J.T. Affolter, S.L. De Haas, *et al.* 2002. High plasma concentrations of human urotensin II do not alter local or systemic hemodynamics in man. *Cardiovasc. Res.* **53:** 341–347.
120. Leonard, A.D., J.P. Thompson, E.L. Hutchinson, *et al.* 2009. Urotensin II receptor expression in human right atrium and aorta: effects of ischaemic heart disease. *Br. J. Anaesth.* **102:** 477–484.
121. Russell, F.D., P. Molenaar & D.M. O'Brien. 2001. Cardiostimulant effects of urotensin-II in human heart in vitro. *Br. J. Pharmacol.* **132:** 5–9.
122. Lin, Y., T. Tsuchihashi, K. Matsumura, *et al.* 2003. Central cardiovascular action of urotensin II in conscious rats. *J. Hypertens.* **21:** 159–165.
123. Lu, Y., C.J. Zou, D.W. Huang & C.S. Tang. 2002. Cardiovascular effects of urotensin II in different brain areas. *Peptides* **23:** 1631–1635.
124. Nakayama, T., T. Hirose, K. Totsune, *et al.* 2008. Increased gene expression of urotensin II-related peptide in the heart of rats with congestive heart failure. *Peptides* **29:** 801–808.
125. Hirose, T., K. Takahashi, N. Mori, *et al.* 2009. Increased expression of urotensin II, urotensin II-related peptide and urotensin II receptor mRNAs in the cardiovascular organs of hypertensive rats: comparison with endothelin-1. *Peptides* **30:** 1124–1129.
126. Mori, N., T. Hirose, T. Nakayama, *et al.* 2009. Increased expression of urotensin II-related peptide and its receptor in kidney with hypertension or renal failure. *Peptides* **30:** 400–408.
127. Douglas, S.A., L. Tayara, E.H. Ohlstein, *et al.* 2002. Congestive heart failure and expression of myocardial urotensin II. *Lancet* **359:** 1990–1997.
128. Heller, J., M. Schepke, M. Neef, *et al.* 2002. Increased urotensin II plasma levels in patients with cirrhosis and portal hypertension. *J. Hepatol.* **37:** 767–772.
129. Richards, A.M., M.G. Nicholls, J.G. Lainchbury, *et al.* 2002. Plasma urotensin II in heart failure. *Lancet* **360:** 545–546.
130. Ng, L.L., I. Loke, R.J. O'Brien, *et al.* 2002. Plasma urotensin in human systolic heart failure. *Circulation* **106:** 2877–2880.
131. Watanabe, T., R. Pakala, T. Katagiri & C.R. Benedict. 2001. Synergistic effect of urotensin II with mildly oxidized LDL on DNA synthesis in vascular smooth muscle cells. *Circulation* **104:** 16–18.
132. Bousette, N., L. Patel, S.A. Douglas, *et al.* 2004. Increased expression of urotensin II and its cognate receptor GPR14 in atherosclerotic lesions of the human aorta. *Atherosclerosis* **176:** 117–123.
133. Maguire, J.J., R.E. Kuc, K.E. Wiley, *et al.* 2004. Cellular distribution of immunoreactive urotensin-II in human tissues with evidence of increased expression in atherosclerosis and a greater constrictor response of small compared to large coronary arteries. *Peptides* **25:** 1767–1774.
134. Hassan, G.S., S.A. Douglas, E.H. Ohlstein & A. Giaid. 2005. Expression of urotensin-II in human coronary atherosclerosis. *Peptides* **26:** 2464–2472.
135. Segain, J.P., M. Rolli-Derkinderen, N. Gervois, *et al.* 2007. Urotensin II is a new chemotactic factor for UT receptor-expressing monocytes. *J. Immunol.* **179:** 901–909.
136. Zhang, Y.G., J. Li, Y.G. Li & R.H. Wei. 2008. Urotensin II induces phenotypic differentiation, migration, and collagen synthesis of adventitial fibroblasts from rat aorta. *J. Hypertens.* **26:** 1119–1126.
137. Pakala R. 2008. Role of urotensin II in atherosclerotic cardiovascular diseases. *Cardiovasc. Revasc. Med.* **9:** 166–178.
138. Zoccali, C., F. Mallamaci, F.A. Benedetto, *et al.* 2008. Urotensin II and cardiomyopathy in end-stage renal disease. *Hypertension* **51:** 326–333.
139. Prosser, H.C.G., M.E. Forster, A.M. Richards & C.J. Pemberton. 2008. Urotensin II and urotensin II-related peptide (URP) in cardiac ischemia-reperfusion injury. *Peptides* **29:** 770–777.
140. Khan, S.Q., S.S. Bhandari, P. Quinn, *et al.* 2007. Urotensin II is raised in acute myocardial infarction and low levels predict risk of adverse clinical outcome in humans. *Int. J. Cardiol.* **117:** 323–328.

Ann. N.Y. Acad. Sci. ISSN 0077-8923

ANNALS OF THE NEW YORK ACADEMY OF SCIENCES
Issue: *Phylogenetic Aspects of Neuropeptides*

Molecular coevolution of kisspeptins and their receptors from fish to mammals

Haet Nim Um,[1] Ji Man Han,[1] Jong-Ik Hwang,[1] Sung In Hong,[2] Hubert Vaudry,[3] and Jae Young Seong[1]

[1]Laboratory of G Protein-Coupled Receptors, Graduate School of Medicine, Korea University, Seoul, Republic of Korea. [2]Department of East-West Medicine, Graduate School of East-West Medical Science, Kyung Hee University, Seoul, Republic of Korea. [3]INSERM U413, Laboratory of Cellular and Molecular Neuroendocrinology, European Institute for Peptide Research (IFRMP 23), University of Rouen, Mont-Saint-Aignan, France

Address for correspondence: Jae Young Seong, Laboratory of G Protein Coupled Receptors, Graduate School of Medicine, Korea University, Seoul 136-705, Republic of Korea. jyseong@korea.ac.kr

Kisspeptin and its receptor, GPR54, play a pivotal role in vertebrate reproduction. Recent advances in bioinformatic tools combined with comparative genomics have led to the identification of a large number of kisspeptin and GPR54 genes in a variety of vertebrate species. Genome duplications may have produced at least two isoforms of both ligand (KiSS1 and KiSS2) and receptor (GPR54-1 and GPR54-2). Additional isoforms of kisspeptin (KiSS1b) and GPR54 (GPR54-1b) have been found in an amphibian species, *Xenopus* (*Silurana*) *tropicalis*. Here, we describe the evolutionary lineages of these kisspeptin and GPR54 isoforms using genome synteny and phylogenetic analyses, and possible molecular interactions between kisspeptin and GPR54 subtypes based on ligand-receptor selectivity. Together, kisspeptin and GPR54 provide an excellent model for understanding molecular coevolution of the peptide ligand and GPCR pairs.

Keywords: kisspeptin; G protein–coupled receptor; GPR54, evolution; genome synteny; phylogentic analysis

Introduction

Neuropeptides and their G protein–coupled receptors (GPCRs) have become diversified through evolutionary mechanisms, such as gene/chromosome duplication and gene modification, resulting in the generation of families of related yet distinct peptides and receptors.[1,2] Two rounds of large-scale genome duplication are believed to have occurred in an early vertebrate ancestor, resulting in up to four copies of each gene in vertebrates. These events have produced paralogous chromosomal regions, or paralogons,[3] providing a basis for using comparative genomics to understand gene relationships and origins. Rapid and vast accumulation of genome sequence information for many vertebrate species, together with advances in bioinformatic tools, has allowed large-scale genome comparisons. Comparison of genome or gene structure makes it possible us to discover conserved regions (functionally important) within the genome/gene structure. Alternatively, tracing back the conserved regions provides crucial clues to discovering orthologous or paralogous genes from different species.[4]

The neuropeptide kisspeptin (KiSS1) and its receptor GPR54 play a key role in vertebrate reproduction. The *KiSS1* gene was originally isolated from human melanoma and breast cancer cells, and identified as a tumor suppressor gene.[5,6] In 2001, kisspeptin, the product of the *KiSS1* gene, was characterized as an endogenous peptide ligand for an orphan GPCR, GPR54.[7] Kisspeptin was initially called metastin in consideration of its suppressive effects on tumor growth and metastasis.[7] In 2003, two independent groups reported that loss-of-function mutations in *GPR54* is responsible for human idiopathic hypogonadotropic hypogonadism (IHH).[8,9] Subsequently, it was shown that knockout of either the *KiSS1* or *GPR54* genes leads to impairment of sexual development and reproductive

doi: 10.1111/j.1749-6632.2010.05508.x
 © 2010 New York Academy of Sciences.

function,[8,10] suggesting that this ligand receptor pair plays a pivotal role in mammalian reproduction. Recently, using bioinformatics combined with molecular cloning tools, *KiSS1* and its isoform, *KiSS2*, have been identified in many vertebrates.[4,11–16] Simultaneously, *GPR54* isoforms (*GPR54-2*) have been characterized.[4,11,13,15–20] Thus, kisspeptins and GPR54s may provide an excellent model to understand molecular coevolution of peptide and receptor genes. This article will discuss the possible evolutionary history of the kisspeptin and GPR54 pair, using genome synteny and phylogenetic analyses, and molecular interactions between kisspeptin mature forms and GPR54s.

Kisspeptin genes in vertebrates

The human *KiSS1* gene encodes a 145-amino acid precursor that is enzymatically cleaved into peptides of 54, 14, 13, or 10 amino acids.[7] The C-terminal decapeptide (kisspeptin-10, metastin 45-54) has been shown to be the minimal sequence required to retain potency as high as that of native kisspeptins.[7,21,22] Comparison of mammalian kisspeptin-10 reveals a high degree of conservation, except for the C-terminal amidated amino acids (Phe for human and Tyr for rodents). In 2008, a few *KiSS1* genes were isolated in fish species.[11,12] Interestingly, comparison of the C-terminal decapeptide of zebrafish and medaka KiSS1 with the rodent decapeptide revealed only one amino acid substitution at position 3, while comparison with the decapeptides of fugu and tetraodon revealed three amino acid substitutions at positions 1, 3, and 5.[11] This relatively high variation within fish species suggests that *KiSS1s* in fugu and tetraodon are isoforms distinct from the *KiSS1s* in zebrafish, medaka and mammals. In support of this hypothesis, we have recently found a cDNA encoding a fugu- and tetraodon-like KiSS1 in the African clawed frog (*Xenopus laevis*) and have shown that a synthetic *Xenopus* KiSS1 dodecapeptide with C-terminal Tyr-amidation is more potent than mammalian kisspeptin-10 in activating the bullfrog kisspeptin receptor, bfGPR54, expressed in CV-1 cells.[13]

Following this report, the presence of the *KiSS1* isoform, *KiSS2*, was demonstrated in many vertebrate species including lamprey, elephant shark, zebrafish, medaka, see bass, goldfish, African clawed frog, western clawed frog (*Xenopus (or Silurana)*

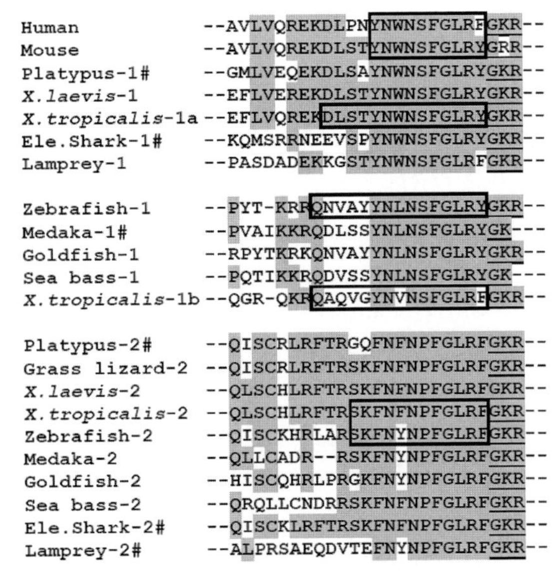

Figure 1. Alignment of amino acid sequences flanking kisspeptins in different species. The conserved amino acid residues are shaded. Amide donor glycine and dibasic cleavage sites are underlined. The potent kisspeptin forms that have been experimentally proven[4] are boxed.

tropicalis), anole lizard, and platypus, a mammalian monotreme species.[4,14–16] In addition, another isoform, *KiSS1b*, was isolated in *X. tropicalis*.[4] Amino acid sequence alignment of either *KiSS1* or *KiSS2* reveals considerable variation in the precursor portion but high conservation in the C-terminal decapeptide across vertebrate species. Thus, classification of *KiSS1* and *KiSS2* has been mainly based on the amino acid sequence of the C-terminal decapeptide.[4] The KiSS1 decapeptides possess an α-amidated Tyr or Phe residue at the C-terminal end and exhibit variation at position 3 (Leu for fish species, Val for *X. tropicalis KiSS1b*, and Trp for other species, including *X. tropicalis* KiSS1a).[4,14] In addition, sequences proximal to the decapeptide in fish KiSS1 and *X. tropicalis* KiSS1b are different from those of other species. Fish KiSS1 and *X. tropicalis* KiSS1b precursors have a conserved dibasic site five amino acids upstream of the decapeptide. These dibasic amino acids are followed by conserved Gln, indicating that fish *KiSS1* and *X. tropicalis KiSS1b* genes may produce a mature peptide of 15 amino acids, with N-terminal pyroglutamylation and C-terminal Tyr amidation (Fig. 1). Indeed, these pentadecapeptides are the most potent activators of fish and *X. tropicalis* GPR54-1.[4]

The amino acid sequences of the KiSS2 decapeptide differ from those of the KiSS1 decapeptide by three amino acids at positions 1, 3, and 5. In addition, a basic amino acid, Arg, occurs three amino acids upstream of the decapeptide, indicating that each KiSS2 cDNA encodes a novel putative peptide with 12 amino acids of which the C-terminal Phe can be amidated.[4] Indeed, the presence of the KiSS2 dodecapeptide has been demonstrated in the *X. laevis* brain using immunoaffinity purification.[4]

Accordingly, conserved synteny for both the *KiSS1*- or *KiSS2*-containing regions in vertebrate genomes has been observed. The *KiSS1* gene is usually positioned in genomic regions containing common loci, including PLEKHA6, GOLT1A, and REN.[4,14,16] These synteny analysis data, together with BLAST search results, suggest that some species, such as anole lizard (reptilian), chicken (avian), zebra finch (avian), and stickleback (fish) do not contain the *KiSS1* gene, while its neighboring genes are present (Fig. 2A). A BLAST search shows the presence of a *KiSS1*-like sequence in a region between *REN* and *GOLT1A* in the anole lizard; however, this gene may not produce a mature peptide form due to mutations in cleavage sites (unpublished data). Thus, a *KiSS1*-like sequence in the anole lizard is likely a pseudogene. Both *KiSS1* and *GOLT1A* are lost in the region between *REN* and *PLEKHA6* in the avian species chicken and zebra finch. The *KiSS1* gene is localized between *GOLT1A* and *CACNA1S* in some fish, but is absent from this region in stickleback. The *X. tropicalis KiSS1a* gene is positioned between the *REN* and *PLEKHA6*, suggesting that it is the ortholog of the fish and mammalian *KiSS1* genes. The *KiSS1b* gene is only present in *X. tropicalis* and is located in genome fragments containing *ALDH16A1*, *PIH1D1*, *PTH2*, *KRAS*, and *TEAD2*. While these neighboring genes are also found in human and anole lizard genomes, *KiSS1b* orthologs are absent in these species (Fig. 2B). These neighboring genes are separated on different chromosomes in fish, making synteny analysis difficult (Fig. 2B).

The *KiSS2* gene resides within a gene cluster including *GOLT1B*, *C12orf39*, *GYS2*, and *LDHB* in fish and anole lizards. This gene cluster is well conserved for mammalian and avian species; however, the *KiSS2* gene is absent in this genomic region of these species (Fig. 2C). Thus, neither the *KiSS1* nor the *KiSS2* gene is found in avian species. While the *KiSS2* genes for *X. tropicalis* and platypus have been identified,[4] genome synteny analysis cannot be performed due to insufficient information on the genome sequence of these species.

The structural organization of the *KiSS1* and *KiSS2* genes is very similar: they both contain two coding exons, one encoding both the signal peptide and part of the kisspeptin precursor, the other encoding the remainder of the precursor including the kisspeptin-10 sequence. This observation indicates that *KiSS1* and *KiSS2* originate from a common ancestral gene by genome duplication. In support of this hypothesis, large-scale synteny analysis reveals that some paralogous genes such as *GOLT1A/GOLT1B*, *PLEKHA5/PLEKHA6*, *PIK3C2B/PIK3CG*, and *ETNK1/ETNK2* are located in regions surrounding *KiSS1* and *KiSS2*, which may contribute to an ancestral paralogon.[16] The presence of an additional isoform for *KiSS1* in an amphibian species (*KiSS1b*) indicates that one more round of gene/chromosome duplication has occurred during evolution. The *KiSS1* gene is absent in some fish, avian and reptilian species, probably due to gene loss during evolution. The mammalian platypus contains two forms of the kisspeptin gene (*KiSS1* and *KiSS2*), but humans and rodents apparently possess only one kisspeptin form (*KiSS1*) (Fig. 2D).

GPR54 genes in vertebrates

The presence of two or three isoforms for the *kisspeptin* genes suggests the concurrent existence of multiple forms of cognate receptors. Multiple forms of *GPR54* have been recently identified from several vertebrate species, including zebrafish, medaka, goldfish, *X. tropicalis*, and platypus.[4,15,18] Fish and monotreme platypus genomes may possess genes for two receptors, GPR54-1 and GPR54-2, while *X. tropicalis* genomes may encode genes for three receptors, GPR54-1a, GPR54-1b, and GPR54-2. Genome and cDNA analyses reveal that all *GPR54* genes contain five coding exons with highly conserved splicing junctions,[4] indicating that these genes originate from a common ancestor. Inter-subclass and inter-species comparisons of the amino acid sequences reveal that the members of the GPR54-2 subfamily exhibit a high degree (74–81%) of sequence similarity with each other. In contrast, many GPR54-1 members showed a relatively low degree (45–55%) of amino acid sequence

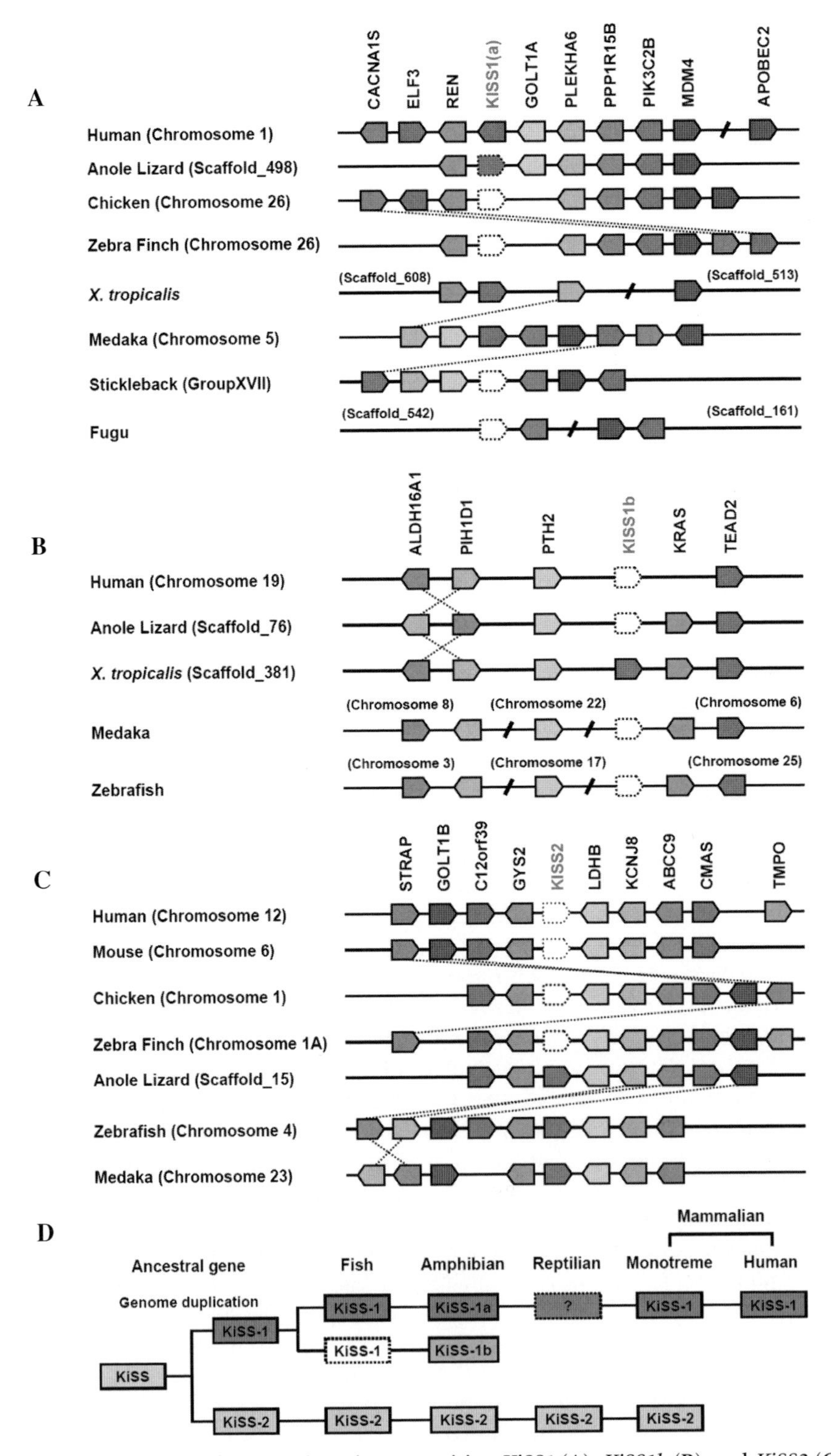

Figure 2. Conserved synteny for the genomic region comprising *KiSS1* (**A**), *KiSS1b* (**B**), and *KiSS2* (**C**) genes and proposed evolutionary lineage of kisspeptin genes (**D**). Gene loci organization in the genomic region containing the kisspeptin gene were obtained from the Ensembl Genome Browser (http://www.ensembl.org).

identity with each other, such that the classification of multiple GPR54 proteins by amino acid sequence identity does not fully explain the phylogenetic history of the GPR54 lineages.[4,15] Thus, additional data from genome synteny analysis are necessary to fully explore the evolutionary history of the GPR54 lineages.

The human *GPR54* gene is surrounded by the *PALM*, *C19orf21*, *PTBP1*, *PRTN3*, *MED16*, and *WDR18* genes on chromosome 19. Similar gene locations are observed for platypus, chicken, and *X. tropicalis*. However, this gene cluster is separated in fish species (Fig 3A). This synteny analysis shows that the ortholog for human *GPR54* is present in platypus and *X. tropicalis*, but not in chicken and fish (Fig. 3A). It is noteworthy that the fish *GPR54-1* is located in genomic regions containing common loci, including *PSAT1* and *ZCCHC6*. The ortholog for the fish *GPR54-1* gene is found in the *X. tropicalis* genome, but not in the human genome (Fig. 3B). Thus, it is likely that *X. tropicalis GPR54-1a* is a human *GPR54* ortholog, while *X. tropicalis GPR54-1b* is a fish *GPR54-1* ortholog.

The locus of fish *GPR54-2* is located between the *IPO13* and *ATP6V0B* loci. Likewise, the *X. tropicalis GPR54-2* gene is found in the genomic region containing the *ARTN*, *IPO13*, and *DPH2* loci. However, a fish and *X. tropicalis GPR54-2* ortholog in chicken and human has not been found in the corresponding genomic region containing *ARTN*, *IPO13*, *DPH2*, and *ATP6V0B* loci (Fig. 3C). Interestingly, the mammalian platypus and anole lizard *GPR54-2* genes reside in genomic regions containing the *RETN* locus and the *RETN*, *STXBP2*, *XAB2*, *KIAA1543*, *NOTCH3*, and *EPHX3* loci, respectively. The corresponding genomic region comprising the *STXBP2*, *XAB2*, *KIAA1543*, *NOTCH3* and *EPHX3* loci is found in human, *X. tropicalis*, and fish. The *GPR54-2* gene is absent in the genomic regions of these species (Fig. 3D). Thus, it is likely that *GPR54-2* genes in reptilian and mammalian monotreme originate from genomic regions distinct from those containing the *GPR54-2* of fish and amphibian.

Many vertebrate *GPR54* genes have been previously identified. However, the nomenclature for these genes has been confusing because their classifications are mainly dependent on phylogenetic analysis. Our classification based on genome synteny analysis suggests that the *GPR54* genes previously identified from tilapia,[17] zebrafish,[11,18] gray mullet,[19] cobia,[20] and bullfrog[13] belong to the *GPR54-2* subfamily, while the *KiSS1Rb* gene of zebrafish[18] and goldfish[15] belong to the *GPR54-1* subfamily.

The evolutionary history of *GPR54* is likely similar to that of kisspeptin genes. Duplication of an ancestral *GPR54* gene occurred during early vertebrate evolution, giving rise to the *GPR54-1* and *GPR54-2* genes. Second round gene duplication may have contributed to the generation of subtypes for each *GPR54-1* and *GPR54-2* genes. The amphibian species possesses two forms of *GPR54-1*, one, *GPR54-1a*, is orthologous to human and platypus *GPR54* and the other, *GPR54-1b*, is orthologous to fish *GPR54-1*. Fish/amphibian *GPR54-2* and anole lizard/platypus *GPR54-2* appear to be subtypes of each other, as their *GPR54-2*-neighboring genes are differentially arranged in the genomic region (Fig. 3E).

Ligand selectivity of GPR54s for kisspeptin isoforms

Putative mature forms of each *KiSS* gene can be predicted based on the analysis of the precursor structure.[4] Fish *KiSS1* and *X. tropicalis KiSS1b* may produce a mature pentadecapeptide with N-terminal pyroglutamylation and C-terminal Tyr amidation, while *X. tropicalis KiSS1a* may produce a tetradecapeptide with C-terminal amidation (Fig. 1). Fish and *X. tropicalis KiSS2* genes may produce a mature dodecapeptide with C-terminal Phe amidation (Fig. 1). Recently, we examined the ligand selectivity of these putative mature kisspeptins towards GPR54 isoforms.[4] The predicted KiSS1-derived mature peptides are more potent than KiSS2-derived peptides in activating nonmammalian GPR54-1. For instance, pyroglutamylated KiSS1 pentadecapeptides show the highest potency towards zebrafish GPR54-1 and *X. tropicalis* GPR54-1b, while the *X. tropicalis* KiSS1a tetradecapeptide is most potent in activating *X. tropicalis* GPR54-1a. Interestingly, the shorter KiSS1-derived decapeptide exhibits much lower potency than predicted longer mature peptides for these nonmammalian GPR54-1, although it is well established that the shorter decapeptide exhibits very high potency for mammalian GPR54. Thus, unlike in the mammalian system, shorter KiSS1 decapeptides are not as potent as longer native peptides in activating nonmammalian GPR54-1. Recently, several groups

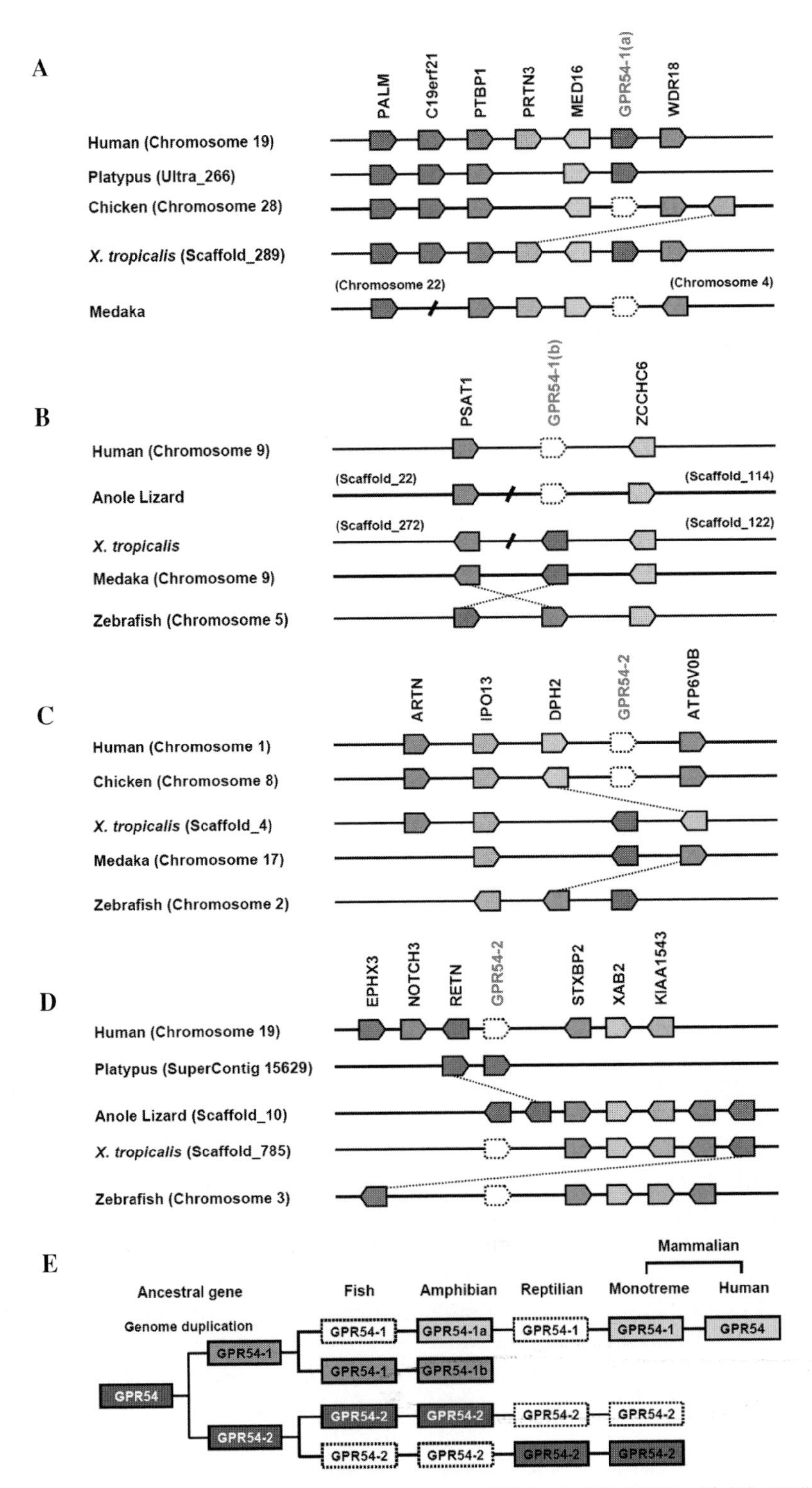

Figure 3. Conserved synteny for the genomic region comprising *GPR54-1a* (A), *GPR54-1b* (B), *GPR54-2* (C and D) genes and proposed evolutionary lineage of kisspeptin genes (E). Gene loci organizations in the genome containing the *GPR54* loci were retrieved from the Ensembl Genome Browser.

have used the shorter mammalian KiSS1 decapeptides in fish systems to explore the ligand selectivity toward fish GPR54s.[15,16,18] However, as the longer KiSS1 peptides are predicted to be more potent than the shorter decapeptides, use of these longer forms should be considered to determine selective KiSS1 effects in nonmammalian systems.

The KiSS2 dodecapeptide exhibits a potency similar to KiSS2 decapeptides in the activation of fish and *X. tropicalis* GPR54-2s. In addition, the potency of the KiSS2 dodecapeptide is not significantly greater than KiSS1-derived peptides for these nonmammalian GPR54-2s, indicating that nonmammalian GPR54-2s may discriminate poorly between KiSS2- and KiSS1-derived peptides. Interestingly, a synthetic *X. tropicalis* KiSS2 dodecapeptide where C-terminal Phe is replaced by Tyr, exhibits the highest potency towards *X. tropicalis* GPR54-2 and bullfrog GPR54-2.[4,13] Thus, this analog may potentially be used as a selective agonist for amphibian GPR54-2s.

Concluding remarks

Based on the current observations, it is likely that fish have two forms of kisspeptin and *GPR54*, amphibians have three forms of kisspeptin and *GPR54*, reptiles have one form of kisspeptin (*KiSS2*) and *GPR54* (*GPR54-2*), avian species do not have any forms of kisspeptin and *GPR54*, mammalian monotreme species have two forms of kisspeptin and *GPR54*, and rodents and human have only one form of kisspeptin (*KiSS1*) and *GPR54*. However, as genome sequencing has not been completed for many vertebrate species discussed in this article, there is possibility for the presence of additional isoforms of kisspeptin and *GPR54* in these species.

The presence of two or three forms of kisspeptin and *GPR54* genes in a single species strongly supports the hypothesis of two round genome duplication followed by degeneration and complementation of the genes.[23] Classification of kisspeptin and *GPR54* isoforms based on genome synteny together with phylogenetic analysis may be helpful in delineating the evolutionary lineages of these genes in vertebrate species. The physiological function of each kisspeptin and *GPR54* isoform, as well as the selective and/or cross-interactions between kisspeptin and *GPR54* isoforms, remains to be further explored. This study may shed new light on the molecular coevolution of the kisspeptin-GPR54 system in vertebrates.

Acknowledgments

This work was supported by a grant (M103KV-01000508K2201-00510) to J.Y.S. from the Brain Research Center of the 21st Century Frontier Research Program and a STAR exchange program to J.Y.S. and H.V.

Conflict of interest

The authors declare no conflict of interest.

References

1. Cho, H.J. *et al.* 2007. Molecular evolution of neuropeptide receptors with regard to maintaining high affinity to their authentic ligands. *Gen. Comp. Endocrinol.* **153:** 98–107.
2. Conlon, J.M. & D. Larhammar. 2005. The evolution of neuroendocrine peptides. *Gen. Comp. Endocrinol.* **142:** 53–59.
3. Lundin, L.G. 1993. Evolution of the vertebrate genome as reflected in paralogous chromosomal regions in man and the house mouse. *Genomics* **16:** 1–19.
4. Lee, Y.R. *et al.* 2009. Molecular evolution of multiple forms of kisspeptins and GPR54 receptors in vertebrates. *Endocrinology* **150:** 2837–2846.
5. Lee, J.H. *et al.* 1996. KiSS-1, a novel human malignant melanoma metastasis-suppressor gene. *J. Natl. Cancer Inst.* **88:** 1731–1737.
6. Lee, J.H. & D.R. Welch. 1997. Suppression of metastasis in human breast carcinoma MDA-MB-435 cells after transfection with the metastasis suppressor gene, KiSS-1. *Cancer Res.* **57:** 2384–2387.
7. Ohtaki, T. *et al.* 2001. Metastasis suppressor gene KiSS-1 encodes peptide ligand of a G-protein-coupled receptor. *Nature* **411:** 613–617.
8. Seminara, S.B. *et al.* 2003. The GPR54 gene as a regulator of puberty. *N. Engl. J. Med.* **349:** 1614–1627.
9. de Roux, N. *et al.* 2003. Hypogonadotropic hypogonadism due to loss of function of the KiSS1-derived peptide receptor GPR54. *Proc. Natl. Acad. Sci. USA* **100:** 10972–10976.
10. d'Anglemont de Tassigny, X. *et al.* 2007. Hypogonadotropic hypogonadism in mice lacking a functional Kiss1 gene. *Proc. Natl. Acad. Sci. USA* **104:** 10714–10719.
11. van Aerle, R. *et al.* 2008. Evidence for the existence of a functional Kiss1/Kiss1 receptor pathway in fish. *Peptides* **29:** 57–64.

12. Kanda, S. *et al.* 2008. Identification of KiSS-1 product kisspeptin and steroid-sensitive sexually dimorphic kisspeptin neurons in medaka (*Oryzias latipes*). *Endocrinology* **149:** 2467–2476.
13. Moon, J.S. *et al.* 2009. Molecular cloning of the bullfrog kisspeptin receptor GPR54 with high sensitivity to Xenopus kisspeptin. *Peptides* **30:** 171–179.
14. Kitahashi, T., S. Ogawa & I.S. Parhar. 2009. Cloning and expression of kiss2 in the zebrafish and medaka. *Endocrinology* **150:** 821–831.
15. Li, S. *et al.* 2009. Structural and functional multiplicity of the kisspeptin/GPR54 system in goldfish (Carassius auratus). *J. Endocrinol.* **201:** 407–418.
16. Felip, A. *et al.* 2009. Evidence for two distinct KiSS genes in non-placental vertebrates that encode kisspeptins with different gonadotropin-releasing activities in fish and mammals. *Mol. Cell. Endocrinol.* **312:** 61–71.
17. Parhar, I.S., S. Ogawa & Y. Sakuma. 2004. Laser-captured single digoxigenin-labeled neurons of gonadotropin-releasing hormone types reveal a novel G protein-coupled receptor (Gpr54) during maturation in cichlid fish. *Endocrinology* **145:** 3613–3618.
18. Biran, J., S. Ben-Dor & B. Levavi-Sivan. 2008. Molecular identification and functional characterization of the kisspeptin/kisspeptin receptor system in lower vertebrates. *Biol. Reprod.* **79:** 776–786.
19. Nocillado, J.N. *et al.* 2007. Temporal expression of G-protein-coupled receptor 54 (GPR54), gonadotropin-releasing hormones (GnRH), and dopamine receptor D2 (drd2) in pubertal female grey mullet, Mugil cephalus. *Gen. Comp. Endocrinol.* **150:** 278–287.
20. Mohamed, J.S. *et al.* 2007. Developmental expression of the G protein-coupled receptor 54 and three GnRH mRNAs in the teleost fish cobia. *J. Mol. Endocrinol.* **38:** 235–244.
21. Kotani, M. *et al.* 2001. The metastasis suppressor gene KiSS-1 encodes kisspeptins, the natural ligands of the orphan G protein-coupled receptor GPR54. *J. Biol. Chem.* **276:** 34631–34636.
22. Gutiérrez-Pascual E. *et al.* 2009. *In vivo* and *in vitro* structure-activity relationships and structural conformation of kisspeptin-10-related peptides. *Mol. Pharmacol.* **76:** 58–67.
23. Force, A. *et al.* 1999. Preservation of duplicate genes by complementary, degenerative mutations. *Genetics* **151:** 1531–1545.

Ann. N.Y. Acad. Sci. ISSN 0077-8923

ANNALS OF THE NEW YORK ACADEMY OF SCIENCES
Issue: *Phylogenetic Aspects of Neuropeptides*

Phylogenetic aspects of gonadotropin-inhibitory hormone and its homologs in vertebrates

Kazuyoshi Tsutsui
Laboratory of Integrative Brain Sciences, Department of Biology, Waseda University, and Center for Medical Life Science of Waseda University, Tokyo, Japan

Address for correspondence: Kazuyoshi Tsutsui, Ph.D., Professor, Laboratory of Integrative Brain Sciences, Department of Biology, Waseda University, Center for Medical Life Science of Waseda University, 2-2 Wakamatsu-cho, Shinjuku-ku, Tokyo 162-8480, Japan. k-tsutsui@waseda.jp

The decapeptide gonadotropin-releasing hormone (GnRH) is the primary factor responsible for the hypothalamic control of gonadotropin secretion in vertebrates, but a hypothalamic neuropeptide inhibiting gonadotropin secretion was, until recently, unknown in vertebrates. In 2000, we discovered a novel hypothalamic dodecapeptide that inhibits gonadotropin release in quail and termed it gonadotropin-inhibitory hormone (GnIH). GnIH acts on the pituitary and GnRH neurons in the hypothalamus via a novel G protein-coupled receptor for GnIH to inhibit gonadal development and maintenance by decreasing gonadotropin release and synthesis. The pineal hormone melatonin is a key factor controlling GnIH neural function. Because GnIH exists and functions in several avian species, GnIH is considered to be a new key neuropeptide controlling avian reproduction. After the discovery of GnIH in birds, the presence of GnIH homologs has been demonstrated in other vertebrates from fish to humans. Interestingly, mammalian GnIH homologs also act to inhibit reproduction by decreasing gonadotropin release in several mammalian species. It is concluded that GnIH and GnIH homologs act to inhibit gonadotropin release in higher vertebrates.

Keywords: gonadotropins; gonadotropin-inhibitory hormone (GnIH); gonadotropin-releasing hormone (GnRH); hypothalamus; pituitary; reproduction

Introduction

Gonadotropin-releasing hormone (GnRH) regulates secretion of both of the gonadotropins, luteinizing hormone (LH) and follicle-stimulating hormone (FSH), and acts as a key neuropeptide for vertebrate reproduction. Since the discovery of GnRH, a hypothalamic decapeptide, in the brain of mammals at the beginning of 1970s,[1,2] several other GnRHs have been identified in the brain of nonmammals.[3–7] It has also been generally accepted that GnRH is the only hypothalamic regulator of the release of pituitary gonadotropins, and that no other neuropeptide has a direct influence on the reproductive axis. Some neurochemicals and peripheral hormones [e.g., γ-aminobutyric acid (GABA), opiates, gonadal sex steroids, and inhibin) can modulate gonadotropin release to a degree, but GnRH was considered to be unusual among hypothalamic neuropeptides in that it appeared to have no hypothalamic antagonist. However, this dogma was challenged by the discovery in 2000 of a vertebrate hypothalamic neuropeptide that inhibits pituitary gonadotropin release.[8]

In a search for novel neuropeptides regulating the release of pituitary hormones, Tsutsui *et al.* identified a novel hypothalamic dodecapeptide (SIKPSAYLPLRFamide) that directly acts on the pituitary to inhibit gonadotropin release in quail and termed it gonadotropin-inhibitory hormone (GnIH).[8] This was the first demonstration of a hypothalamic neuropeptide inhibiting gonadotropin release in any vertebrate and no other hypothalamic neuropeptide has a direct negative influence on pituitary gonadotropin release. From the past 9 years of research, we now know that GnIH exists in several avian species and acts as a new key neuropeptide for the regulation of avian reproduction by decreasing gonadotropin release and synthesis.[8–23]

doi: 10.1111/j.1749-6632.2010.05510.x

Because a gonadotropin-inhibitory system is an intriguing new concept, Tsutsui *et al.* have further identified GnIH homologs in other vertebrates from fish to humans. Interestingly, mammalian GnIH homologs as well as GnIH identified in birds act to inhibit gonadotropin release in several mammalian species, such as rats, hamsters, and sheep.[24–29] Thus, GnIH and GnIH homologs act to inhibit gonadotropin release in higher vertebrates. The discovery of GnIH and GnIH homologs has changed our understanding about regulation of the reproductive axis drastically in the last 9 years, but we are only at the beginning of an exciting new era of research on reproductive neurobiology.

Brief history and discovery of GnIH in birds

GnIH possesses the RFamide (Arg-Phe-NH_2) motif at its C-terminus (i.e., RFamide peptide). The first isolation of RFamide peptide occurred in an invertebrate species almost 30 years ago[30] and 6 years later the first RFamide peptide in vertebrates was discovered in chickens.[31] In vertebrates, the study of RFamide neurobiology has gathered vast momentum in recent years. The molluscan cardioexcitatory neuropeptide Phe-Met-Arg-Phe-NH_2 (FMRFamide) was first found in the ganglia of the venus clam.[30] Immunohistochemical studies using an antiserum against FMRFamide suggested the presence of uncharacterized RFamide peptides in vertebrate nervous systems.[32,33] Some of the FMRFamide-like immunoreactive neurons were seen to project to the hypothalamic region close to the pituitary gland and thus were predicted to play a role in the regulation of pituitary function. Tsutsui *et al.* therefore reasoned that the avian hypothalamus might contain a novel RFamide peptide that regulates anterior pituitary function and developed a strategy to isolate it.[8]

An RFamide peptide was isolated from the brain of the Japanese quail using HPLC and a competitive enzyme-linked immunosorbent assay for the dipeptide Arg-Phe-NH_2.[8] The amino acid sequence of this isolated peptide was found to be Ser(62)-Ile(252)-Lys(233)-Pro(226)-Ser(38)-Ala(194)-Tyr(173)-Leu(148)-Pro(104)-Leu(108)-Arg(45)-Phe(52), with the detected amount (pmol) of each amino acid indicated in parentheses. The isolated native peptide was confirmed as a 12 amino acid sequence (SIKPSAYLPLRFamide) with RFamide at the C-terminus.[8] This neuropeptide had not been previously reported in vertebrates, although the C-terminal LPLRFamide was identical to chicken pentapeptide LPLRFamide peptide,[31] which may be a degraded fragment of the dodecapeptide, as suggested by Dockray and Dimaline.[34] Subsequently, the isolated novel peptide was shown to be located in the quail hypothalamo–hypophysial system and to decrease gonadotropin release from cultured quail anterior pituitaries in a dose-dependent manner.[8] We therefore designated this novel RFamide peptide as gonadotropin-inhibitory hormone (GnIH).[8]

Localization, expression, and mode of action of GnIH in birds

Several studies investigated the precise localization of GnIH in the quail brain by immunohistochemistry.[8,11,12] Clusters of distinct GnIH-immunoreactive (-ir) neurons were found in the paraventricular nucleus (PVN) in the hypothalamus. In addition to the PVN, some scattered small immunoreactive cells were located in the septal area. It is clear that the GnIH-ir neuron population is distinct from vasotocin- or mesotocin-expressing neurons.[12] In contrast, GnIH-ir nerve fibers were widely distributed in the diencephalic and mesencephalic regions. Dense networks of immunoreactive fibers were found in the ventral paleostriatum, septal area, preoptic area (POA), hypothalamus, and optic tectum. The most prominent fibers were seen in the median eminence (ME) of the hypothalamus, and in the dorsal motor nucleus of the vagus in the medulla oblongata.

Further studies investigated GnIH localization in the brain of several species of sparrows, such as song sparrows, house sparrows, Gambel's white-crowned sparrows, etc.[10,14] Dense populations of GnIH-ir neurons were also found exclusively in the PVN of these birds.[10,14] Thus, the presence of GnIH in the PVN appears to be a conserved property among several avian species (for reviews see Refs. 19–22). In addition to the dense population of GnIH-ir neurons within the hypothalamus of all the avian species studied so far, there were extensive networks of branching beaded fibers emanating from those cells. Some of the fibers extended to terminals in the ME, consistent with a role for GnIH in pituitary gonadotropin regulation. Other fibers extended through the brain caudally at least as far as the brain stem and possibly into the spinal cord, consistent with multiple regulatory roles for GnIH.

Interestingly, GnIH fibers were further observed in extremely close proximity to GnRH neurons in the POA in birds.[10,17] Contact also occurred between GnIH and GnRH fibers in the ME.[10] GnIH may influence the GnRH system at the neuron and fiber terminal levels. It is therefore possible that GnIH acts at the level of the hypothalamus to regulate gonadotropin release as well as at the pituitary.[17]

In view of the first immunohistochemical finding indicating that GnIH-ir neurons project to the ME close to the pituitary in quail, Tsutsui *et al.* analyzed the effect of the isolated SIKPSAYLPLRFamide, GnIH, on the release of LH, FSH, and prolactin (PRL) from cultured quail anterior pituitaries.[8] GnIH significantly inhibited LH release, after 100-min incubation. The inhibitory effect on LH release was dose-dependent, and its threshold concentration ranged between 10^{-9} and 10^{-8} M.[8] A similar inhibitory effect of GnIH on FSH release was also detected.[8] However, there was no effect of GnIH on PRL release.[8] Based on these results of this novel RFamide peptide isolated from the quail brain, Tsutsui *et al.* therefore named it GnIH.[8] GnIH was also effective in inhibiting circulating LH *in vivo*. When administered intraperitoneally (i.p.) to quail via osmotic pumps, GnIH significantly reduced plasma LH.[16] GnIH injected simultaneously with GnRH inhibited the GnRH-induced elevation in plasma LH in song sparrows.[14] Furthermore, GnIH injections also decreased levels of LH in breeding free-living Gambel's white-crowned sparrows.[14]

In addition to gonadotropin release, GnIH inhibited gonadotropin biosynthesis.[13,16] Addition of a physiological dose (10^{-7} M) of GnIH to short-term (120 min) cultures of diced pituitary glands from chickens suppressed gonadotropin common α and FSHβ subunit mRNAs.[13] The suppressive effect of GnIH on gonadotropin mRNA was associated with an inhibition of both LH and FSH release.[13] When administered i.p. to quail *in vivo* via osmotic pumps, GnIH significantly reduced gonadotropin common α and LHβ subunit mRNAs, as well as reducing plasma LH.[16] Thus, it is clear that GnIH can act in birds to reduce pituitary gonadotropin release and synthesis.

In addition to the neuroendocrine effects of GnIH, central injections of GnIH rapidly suppress female sexual behavior in Gambel's white-crowned sparrows, as assayed by monitoring copulation-solicitation displays in response to male song.[23] Thus, GnIH is likely to be an important component in the regulation of reproductive behavior of birds.

To investigate the mechanisms that regulate the biosynthesis of GnIH in birds, the precursor polypeptide for GnIH was then examined.[9,14] A cDNA that encoded the GnIH precursor polypeptide was identified in the quail brain by a combination of 3′ and 5′ rapid amplification of cDNA ends (3′/5′ RACE).[9] The deduced GnIH precursor consisted of 173 amino acid residues that encoded one GnIH and two putative GnIH-related peptides (GnIH-RP-1 and GnIH-RP-2) that included -LPXRFamide (X = L or Q) at their C-termini (Table 1). All these peptide sequences were flanked by a glycine C-terminal amidation signal and a single basic amino acid on each end as an endoproteolytic site. Subsequently, Satake *et al.* identified GnIH-RP-2 as a mature peptide (Table 1).[9] Osugi *et al.* also cloned a cDNA that encoded GnIH in the brain of Gambel's white-crowned sparrow.[14] The deduced sparrow GnIH precursor also consisted of 173 amino acid residues, encoding one sparrow GnIH and two sparrow GnIH-related peptides (sparrow GnIH-RP-1 and GnIH-RP-2) that included -LPXRFamide (X = L or Q) at their C-termini (Table 1). The C-terminal structures of GnIH, GnIH-RP-1, and GnIH-RP-2 were all identical in these two species (Table 1).[9,14] A cDNA encoding GnIH and GnIH-RPs was also reported in the chicken from a gene database. These results indicate that the chicken LPLRFamide[31] is a fragment of GnIH and GnIH-RP-1.[8,9,14]

Subsequently, the mechanisms regulating GnIH biosynthesis were investigated in birds. Although many bird species are photoperiodic, a dogma has existed that birds do not use seasonal changes in melatonin secretion to time their reproductive effort, and the evidence for a role for melatonin in birds has remained somewhat confusing.[35,36] Despite the accepted dogma, there is some good evidence that melatonin can be involved in the negative regulation of several seasonal processes, including gonadal activity and gonadotropin secretion.[37–41] In light of these reports and considering GnIH's

Table 1. Amino acid sequences of GnIH and its homologs (LPXRFamide peptides) in vertebrates

Animal	Name	Sequence	Reference
Human	RFRP-1[a]	MPHSFANLPLRFa	25
	RFRP-3	VPNLPQRFa	25
Monkey	RFRP-3	SGRNMEVSLVRQVLNLPQRFa	69
Rat	RFRP-1[a]	SVTFQELKDWGAKKDIKMSPAPANKVPHSAANLPLRFa	65
	RFRP-3	ANMEAGTMSHFPSLPQRFa	68
Hamster	RFRP-1[a]	SPAPANKVPHSAANLPLRFa	24
	RFRP-3[a]	TLSRVPSLPQRFa	24
Bovine	RFRP-1	SLTFEEVKDWAPKIKMNKPVVNKMPPSAANLPLRFa	66
	RFRP-3	AMAHLPLRLGKNREDSLSRWVPNLPQRFa	67
Quail	GnIH	SIKPSAYLPLRFa	8
	GnIH-RP-1[a]	SLNFEEMKDWGSKNFMKVNTPTVNKVPNSVANLPLRFa	9
	GnIH-RP-2	SSIQSLLNLPQRFa	9
Sparrow	GnIH[a]	SIKPFSNLPLRFa	14
	GnIH-RP-1[a]	SLNFEEMEDWGSKDIIKMNPFT ASKMPNSVANLPLRFa	14
	GnIH-RP-2[a]	SPLVKGSSQSLLNLPQRFa	14
Bullfrog	fGRP	SLKPAANLPLRFa	60
	fGRP-RP-1	SIPNLPQRFa	62
	fGRP-RP-2	YLSGKTKVQSMANLPQRFa	62
	fGRP-RP-3	AQYTNHFVHSLDTLPLRFa	62
Goldfish	Goldfish LPXRFa-1[a]	PTHLHANLPLRFa	64
	Goldfish LPXRFa-2[a]	AKSNINLPQRFa	64
	Goldfish LPXRFa-3	SGTGLSATLPQRFa	64

[a]Putative peptides.

inhibitory effects on gonadotropin release and synthesis,[8,13,14,16] we manipulated melatonin levels in quail by removing sources of melatonin and investigating the action of melatonin on GnIH expression in the quail brain.[15] The pineal gland and eyes are the major sources of melatonin in the quail.[42] Pinealectomy combined with orbital enucleation (Px+Ex) decreased the expression of GnIH precursor mRNA and the mature peptide GnIH in the diencephalon, including the PVN and median eminence.[15] Melatonin administration to Px+Ex birds caused a dose-dependent increase in expression of GnIH precursor mRNA and production of mature peptide.[15] The expression of GnIH was photoperiodically controlled and increased under short day (SD) photoperiods,[15] when the nocturnal duration of melatonin secretion increases.[43] Critically, Mel_{1c}, a melatonin receptor subtype was expressed in GnIH-ir neurons in the PVN.[15] Melatonin receptor autoradiography further revealed specific binding of melatonin in the PVN.[15] Melatonin appears to act directly on GnIH neurons via its receptor to induce GnIH expression. Thus, GnIH is capable of transducing photoperiodic information via changes in the melatonin signal, thereby influencing the reproductive axis in birds.

Identification of GnIH receptor is essential to understand the mode of action of GnIH. We therefore identified a novel G protein–coupled receptor (GPCR) for GnIH in the quail.[18] Yin *et al.* first cloned a cDNA encoding a putative GnIH receptor.[18] This putative GnIH receptor is considered from structural and hydrophobic analyses to possess seven transmembrane domains (TMs), and to be a new member of the GPCR superfamily. Furthermore, binding experiments using crude membrane fraction of COS-7 cells transfected with the putative GnIH receptor cDNA demonstrated that this membrane protein specifically binds to GnIH and GnIH-RPs, but not to neuropeptides lacking the C-terminal LPXRFamide (X = L or Q) motif.[18] Taken together, the identified GnIH receptor appears to be a functional receptor in the quail. Subsequently,

GnIH receptor has been identified in the chicken[44,45] and European starling.[17]

To elucidate the mode of action of GnIH on gonadotropin release and synthesis, the expression of GnIH receptor mRNA was further characterized in birds.[18] Southern blotting analysis of reverse-transcriptase-mediated PCR products revealed the expression of GnIH receptor mRNA in the pituitary and several brain regions including the hypothalamus in the quail.[18] In the pituitary, GnIH receptor-ir cells were colocalized with LHβ mRNA-, and FSHβ mRNA-containing cells in quail (unpublished observation) and chickens.[45] These results indicate that GnIH can act directly on gonadotropes in the pituitary via GnIH receptor to inhibit gonadotropin release and synthesis.

Because GnIH receptor was expressed in the hypothalamus, it is also possible that GnIH could be acting at the level of the hypothalamus via GnIH receptor. As already described, *in vivo* treatment with GnIH inhibits GnRH-elicited LH release in sparrows.[14] Recently, Bentley *et al.* and Ubuka *et al.* further found that GnIH neurons project to GnRH neurons as well as to the ME in birds.[11,17] Furthermore, GnIH receptor was expressed in GnRH neurons as well as in the pituitary gland.[17,18,23,45] These findings suggest that GnIH acts on GnRH neurons to inhibit gonadotropin release and synthesis, as well as acting at the pituitary gland. It is unknown whether this dual action occurs over different time frames or in response to different stimuli.

Furthermore, other brain regions, such as the cerebrum and mesencephalon, and the spinal cord also express GnIH receptor, suggesting that GnIH could have multiple regulatory functions in the avian brain.[18,23]

Functional significance of GnIH in birds

We investigated the effect of GnIH on gonadal maintenance in male quail to determine the functional significance of GnIH and its potential role as a key neurohormone in avian reproduction.[16] In mature birds, chronic treatment with GnIH via osmotic pumps for 2 weeks decreased plasma testosterone concentrations as well as gonadotropin synthesis and release.[16] Because LH stimulates the synthesis and release of testosterone in the Leydig cells of birds,[46–48] the decrease in circulating testosterone after GnIH administration is likely to be a result of the decrease in LH synthesis and release. Further, GnIH administration induced testicular apoptosis and decreased spermatogenic activity in mature birds.[16] Apoptotic cell death was detected in spermatocytes, spermatogonia, and Sertoli cells,[16] the same cell types that undergo apoptosis in the testis of starlings during seasonal testicular regression.[49] It is considered that in starlings the decrease in circulating testosterone is the main cause of testicular apoptosis.[49,50] Testosterone is known as a testicular cell-survival factor in rodents.[51–53] It is likely that GnIH decreases testicular testosterone and consequently induces apoptotic cell death in the testis. The decrease in the survival of germ and Sertoli cells would account for the observed reduction in seminiferous tubule diameter. Ubuka *et al.* demonstrated that the expression of GnIH in the hypothalamus increases at the onset of testicular regression in adult quail exposed to SD photoperiods.[15] Therefore, the increase in GnIH action may be one of the main causes of gonadal regression in birds. These results indicate that GnIH can inhibit gonadal maintenance by inhibiting gonadotropin release and synthesis.

We also investigated the inhibitory effect of GnIH on gonadal development using immature male quail. Previous findings indicated that circulating gonadotropin concentrations are negatively correlated with the GnIH content in the hypothalamus during quail development.[11] In other words, GnIH decreases as gonadotropins increase during development. Based on this finding, we hypothesized that the decrease in hypothalamic GnIH content may be involved in the rapid increases in the plasma testosterone concentration and testicular growth observed during sexual development. In accordance with the hypothesis, administration of GnIH to immature male quail suppressed the normal rise in plasma testosterone concentrations.[16] The growth of seminiferous tubules and proliferation of germ cells during development were also suppressed by GnIH administration.[16] Because the rise in circulating testosterone is required for testicular development in quail,[54–57] the suppression of testosterone after GnIH administration may cause the decrease in spermatogenic activity. GnIH further inhibited the transition from juvenile plumage into adult plumage during development.[16] This phenomenon might also be a result of suppressed plasma testosterone; gonadal steroids are known to maintain adult plumage in several birds.[58,59] Thus, treatment with GnIH suppressed normal testicular growth and plasma testosterone concentrations in immature birds.[16] These results indicate that GnIH can also inhibit gonadal development.

Taken together, GnIH appears to inhibit gonadal development and maintenance through the decrease in gonadotropin release and synthesis. Thus, GnIH, a newly discovered hypothalamic neuropeptide, may act as an important neuropeptide controlling avian reproduction (for reviews see Refs. 19–22). GnIH is the first identified hypothalamic neuropeptide inhibiting reproductive function in any vertebrate class.

Phylogenetic aspects of GnIH and its homologs in vertebrates

To give our findings a broader perspective, several studies have been conducted to identify novel hypothalamic neuropeptides closely related to GnIH, namely GnIH homologs, in other vertebrates. To date, GnIH homologs have been documented in a variety of vertebrates, such as mammals including primates and humans (Table 1).

After the identification of GnIH,[8] we further sought to identify GnIH homologs in other vertebrate classes that regulate secretion of hormones by the pituitary gland. We first turned to amphibians. An extract of hypothalami from the bullfrog was subjected to HPLC purification, and RFamide immunoreactivity was measured in the eluted fractions with a dot immunoblot assay.[60] The isolated frog peptide was a 12-amino acid sequence (SLKPAANLPLRFamide) with RFamide at its C-terminus (Table 1).[60] Although this frog RFamide peptide named fGRP had not been previously reported in vertebrates, it was revealed to have a high sequence homology (75%) with quail GnIH (Table 1).[8,60] Collectively, these results show that the novel RFamide peptides, GnIH and fGRP, possess the same C-terminal motif, LPLRFamide (Table 1).[8,60] The fGRP precursor also encoded one fGRP and three fGRP-related peptides (fGRP-RP-1, -2, and -3),[61] which were identified as mature LPXRFamide peptides (Table 1).[62] At the same time as we reported the identification of fGRP,[60] two other research groups independently purified the same peptide as fGRP from the European green frog (R-RFa).[63] Subsequently, from the goldfish brain, a cDNA that encoded three novel fish LPXRFamide peptides (gfLPXRFa-1, -2, and -3) was characterized and gfLPXRFa-3 was identified as a mature peptide (Table 1).[64]

Turning to mammals, cDNAs that encode novel RFamide peptides similar to GnIH have been detected in mammalian brains from a gene database search.[65] The cDNAs of human and bovine peptides encoded three peptides, which were termed RFamide-related peptide-1, -2, and -3 (RFRP-1, -2, and -3). However, RFRP-2 was not an RFamide peptide. Subsequently, bovine RFRP-1 and -3 were purified from bovine hypothalami (Table 1).[66,67] We also identified RFRP-3 in the rat hypothalamus (Table 1)[68] and the human and monkey hypothalamus (Table 1).[25,69] Mammalian RFRP-1 and -3 also contained a C-terminal LPXRFamide (X = L or Q) sequence[25,66–69] like GnIH.

As shown in Table 1, all the identified neuropeptides possessed a LPXRFamide (X = L or Q) motif at their C-termini. We therefore designated these peptides as LPXRFamide peptides, which form a new member of the RFamide peptide family (for reviews see Refs. 70,71). Thus, the presence of GnIH and its homologs is a conserved property in vertebrates.

The identified GnIH homologs also regulate pituitary hormone release (for reviews see Refs. 70–72). As mentioned above, cDNAs that encode GnIH homologs containing a C-terminal LPXRFamide (X = L or Q) motif have been detected in mammalian brains.[24,25,27,65,69] Mammalian cDNAs encode the two biological active GnIH homologs, i.e., RFRP-1 and RFRP-3. Up until now, mammalian RFRP-1 and RFRP-3 have been identified in the bovine (both RFRP-1 and RFRP-3), rat (RFRP-3 only), primate (RFRP-3 only), and human (RFRP-3 only) brain.[25,66–69] Intracerebroventricular (i.c.v.) administration of the deduced human RFRP-1 increased PRL release in the rat.[65] By contrast, i.c.v. injections of RFRP-3 reduced plasma levels of LH in rats.[26,29] When injected i.c.v. or i.p. GnIH also reduced plasma LH levels in ovariectomized Syrian hamsters.[24] GnIH and RFRP-3 are therefore considered to be functional homologs. More recently, Clarke *et al.* also found that peripheral administration of the deduced ovine GnIH homolog RFRP-3 reduces the amplitude of LH pulses in sheep, and reduces LH and FSH release *in vitro*.[27] In addition, both GnIH and RFRP-3 facilitate food intake in chicks[73] and rats,[26,29] respectively. It is considered that GnIH and its mammalian homolog RFRP-3 act to inhibit gonadotropin release in higher vertebrates.

In contrast to higher vertebrates, the frog GnIH homologs, fGRP and fGRP-RP-2, stimulate the release of growth hormone (GH) and/or PRL in amphibians.[11,60] On the other hand, the goldfish GnIH homologs, gfLPXRFa-1, -2, and -3, stimulate the release of gonadotropins and GH but not PRL.[74] Thus, GnIH and its homologs act on the regulation of pituitary function, but hypophysiotropic activities of these LPXRFamide peptides are divergent in vertebrates, particularly in lower vertebrates.

Because gonadotropins and PRL are key hormones controlling reproduction, the discovery of GnIH and its homologous peptides has open a new area of research in reproductive neurobiology from fish to mammals.

Conclusions and future directions

The discovery of GnIH in birds has opened avenues for a new research field in reproductive neurobiology. GnIH homologs were further identified in the hypothalamus of a variety of vertebrates from fish to mammals including humans. All of the identified GnIH homologs also regulate pituitary hormone release. Interestingly, mammalian GnIH homologs also act to inhibit reproduction by decreasing gonadotropin release in several mammalian species. It is concluded that GnIH and GnIH homologs act to inhibit gonadotropin release in higher vertebrates.

Following the discovery of GnIH, a novel neuropeptide, termed kisspeptin, that plays an important role in the development and upregulation of the reproductive system in mammals has been identified.[75–77] Kisspeptin exerts a stimulatory effect on GnRH neurons leading to the release of GnRH and an upregulation of the hypothalamo–pituitary–gonadal (HPG) axis, while GnIH downregulates the HPG axis at the level of the pituitary and/or the level of GnRH neurons. Both GnIH and kisspeptin are new members of the RFamide peptide family. The distinct opposing roles of these two newly discovered RFamide peptides suggest that GnIH and kisspeptin act as key neurohormones controlling reproductive activity (for a review see Ref. 72).

Thus, we now know that GnRH is not the sole hypothalamic neurohormone controlling reproduction in vertebrates. Future studies will shed light onto previously unknown interactions of GnRH with GnIH and kisspeptin.

Acknowledgments

This work was supported by Grants-in-Aid for Scientific Research from the Ministry of Education, Science and Culture, Japan (16086206 and 18107002 to K. T.). I am grateful to the following collaborators, G. E. Bentley, J. C. Wingfield, P. J. Sharp, L. J. Kriegsfeld, T. Ubuka, T. Osugi, V. S. Chowdhury, E. Saigoh, H. Yin, K. Ukena, H. Teranishi, Y. Fujisawa, S. Ishii, H. Minakata, H. Satake, N. L. McGuire, R. Calisi, N. Perfito, S. O'Brien, I. T. Moore, J. P. Jensen, G. J. Kaur, D. W. Wacker, N. A. Ciccone, I. C. Dunn, T. Boswell, S. Kim, Y. C. Huang, J. Reid, J. Jiang, P. Deviche, T. W. Small, R. P. Millar, T. Tachibana, M. Furuse, D. F. Mei, A. Mason, E. M. Gibson, S. A. Humber, S. Jain, W. P. Williams III, S. Zhao, I. J. Clarke, I. P. Sari, Y. Qi, J. T. Smith, H. C. Parkington, J. Iqbal, Q. Li, A. Tilbrook, K. Morgan, A. J. Pawson, M. Murakami, T. Matsuzaki, T. Iwasa, M. Irahara, M. A. Johnson, G. S. Fraley, M. Binns, P. A. Cadigan, H. Lai, S. Kikuyama, H. Vaudry, K. Yamamoto, A. Koda, K. Sawada, E. Iwakoshi-Ukena, I. Hasunuma, M. Amano, H. Kawauchi. I also thank H. Vaudry, G. E. Bentley, and J. Kriegsfeld for their valuable discussion and critical reading of the manuscript. Grant support: Grants-in-Aid for Scientific Research from the Ministry of Education, Science and Culture, Japan (15207007, 16086206, and 18107002).

Conflicts of interest

The author declares no conflicts of interest.

References

1. Matsuo, H. *et al.* 1971. Structure of the porcine LH- and FSH-releasing hormone. I. The proposed amino acid sequence. *Biochem. Biophys. Res. Commun.* **43:** 1334–1339.
2. Burgus, R. *et al.* 1972. Primary structure of the ovine hypothalamic luteinizing hormone-releasing factor (LRF) (LH-hypothalamus-LRF-gas chromatography-mass spectrometry-decapeptide-Edman degradation). *Proc. Natl. Acad. Sci. USA* **69:** 278–282.
3. King, J.A. & R.P. Millar. 1982. Structure of chicken hypothalamic luteinizing hormone-releasing hormone. II. Isolation and characterization. *J. Biol. Chem.* **257:** 10729–10732.

4. Miyamoto, K. *et al.* 1982. Isolation and characterization of chicken hypothalamic luteinizing hormone-releasing hormone. *Biochem. Biophys. Res. Commun.* **107:** 820–827.
5. Miyamoto, K. *et al.* 1984. Identification of the second gonadotropin-releasing hormone in chicken hypothalamus: evidence that gonadotropin secretion is probably controlled by two distinct gonadotropin-releasing hormones in avian species. *Proc. Natl. Acad. Sci. USA* **81:** 3874–3878.
6. Sherwood, N. *et al.* 1983. Characterization of a teleost gonadotropin-releasing hormone. *Proc. Natl. Acad. Sci. USA* **80:** 2794–2798.
7. Sherwood, N.M. *et al.* 1986. Primary structure of gonadotropin-releasing hormone from lamprey brain. *J. Biol. Chem.* **261:** 4812–4819.
8. Tsutsui, K. *et al.* 2000. A novel avian hypothalamic peptide inhibiting gonadotropin release. *Biochem. Biophys. Res. Commun.* **275:** 661–667.
9. Satake, H. *et al.* 2001. Characterization of a cDNA encoding a novel avian hypothalamic neuropeptide exerting an inhibitory effect on gonadotropin release. *Biochem. J.* **354:** 379–385.
10. Bentley, G.E. *et al.* 2003. Gonadotropin-inhibitory peptide in song sparrows (*Melospiza melodia*) in different reproductive conditions, and in house sparrows (*Passer domesticus*) relative to chicken-gonadotropin-releasing hormone. *J. Neuroendocrinol.* **15:** 794–802.
11. Ubuka, T. *et al.* 2003. Developmental changes in gonadotropin-inhibitory hormone in the Japanese quail (*Coturnix japonica*) hypothalamo-hypophysial system. *J. Endocrinol.* **178:** 311–318.
12. Ukena, K., T. Ubuka & K. Tsutsui 2003. Distribution of a novel avian gonadotropin-inhibitory hormone in the quail brain. *Cell Tissue Res.* **312:** 73–79.
13. Ciccone, N.A. *et al.* 2004. Gonadotrophin inhibitory hormone depresses gonadotrophin alpha and follicle-stimulating hormone beta subunit expression in the pituitary of the domestic chicken. *J. Neuroendocrinol.* **16:** 999–1006.
14. Osugi, T. *et al.* 2004. Gonadotropin-inhibitory hormone in Gambel's white-crowned sparrow (*Zonotrichia leucophrys gambelii*): cDNA identification, transcript localization and functional effects in laboratory and field experiments. *J. Endocrinol.* **182:** 33–42.
15. Ubuka, T. *et al.* 2005. Melatonin induces the expression of gonadotropin-inhibitory hormone in the avian brain. *Proc. Natl. Acad. Sci. USA* **102:** 3052–3057.
16. Ubuka, T. *et al.* 2006. Gonadotropin-inhibitory hormone inhibits gonadal development and maintenance by decreasing gonadotropin synthesis and release in male quail. *Endocrinology* **147:** 1187–1194.
17. Ubuka, T. *et al.* 2008. Gonadotropin-inhibitory hormone neurons interact directly with gonadotropin-releasing hormone-I and -II neurons in European starling brain. *Endocrinology* **149:** 268–278.
18. Yin, H. *et al.* 2005. A novel G protein-coupled receptor for gonadotropin-inhibitory hormone in the Japanese quail (*Coturnix japonica*): identification, expression and binding activity. *J. Endocrinol.* **184:** 257–266.
19. Tsutsui, K., G.E. Bentley & N. Ciccone. 2005. Structure, action and functional significance of GnIH. In *Functional Avian Endocrinology*. A. Dawson & P.J. Sharp, Eds.: 73–82. Narosa Publishing House: New Delhi.
20. Tsutsui, K. *et al.* 2006. Mode of action and functional significance of avian gonadotropin-inhibitory hormone (GnIH): a review. *J. Exp. Zoolog. A Comp. Exp. Biol.* **305:** 801–806.
21. Tsutsui, K. *et al.* 2007. Review: the general and comparative biology of gonadotropin-inhibitory hormone (GnIH). *Gen. Comp. Endocrinol.* **153:** 365–370.
22. Tsutsui, K. *et al.* 2007. Review: discovery of gonadotropin-inhibitory hormone in a domesticated bird, and its mode of action and functional significance. *J. Ornithol.* **147:** 53–54.
23. Bentley, G.E. *et al.* 2006. Rapid inhibition of female sexual behavior by gonadotropin-inhibitory hormone (GnIH). *Horm. Behav.* **49:** 550–555.
24. Kriegsfeld, L.J. *et al.* 2006. Identification and characterization of a gonadotropin-inhibitory system in the brains of mammals. *Proc. Natl. Acad. Sci. USA* **103:** 2410–2415.
25. Ubuka, T. *et al.* 2006. Identification of gonadotropin-inhibitory hormone in human brain. *The 36th Annual Meeting for the Society for Neuroscience Abstract*.
26. Johnson, M.A., K. Tsutsui & G.S. Fraley 2007. Rat RFamide-related peptide-3 stimulates GH secretion, inhibits LH secretion, and has variable effects on sex behavior in the adult male rat. *Horm. Behav.* **51:** 171–180.
27. Clarke, I.J. *et al.* 2008. Potent action of RFRP-3 on pituitary gonadotropes indicative of an hypophysiotropic role in the negative regulation of gonadotropin secretion. *Endocrinology* **149:** 5811–5821.
28. Gibson, E.M. *et al.* 2008. Alterations in RFamide-related peptide expression are coordinated with the preovulatory luteinizing hormone surge. *Endocrinology* **149:** 4958–4969.
29. Murakami, M. *et al.* 2008. Hypophysiotropic role of RFamide-related peptide-3 in the inhibition of LH secretion in female rats. *J. Endocrinol.* **199:** 105–112.

30. Price, D.A. & M.J. Greenberg 1977. Structure of a molluscan cardioexcitatory neuropeptide. *Science* **197:** 670–671.
31. Dockray, G.J. *et al.* 1983. A novel active pentapeptide from chicken brain identified by antibodies to FMRFamide. *Nature* **305:** 328–330.
32. Raffa, R.B. 1988. The action of FMRFamide (Phe-Met-Arg-Phe-NH_2) and related peptides on mammals. *Peptides* **9:** 915–922.
33. Rastogi, R.K. *et al.* 2001. FMRFamide in the amphibian brain: a comprehensive survey. *Microsc. Res. Tech.* **54:** 158–172.
34. Dockray, G.J. & R. Dimaline 1985. FMRFamide- and gastrin/CCK-like peptides in birds. *Peptides* **6:** 333–337.
35. Wilson, F.E. 1991. Neither retinal nor pineal photoreceptors mediate photoperiodic control of seasonal reproduction in American tree sparrows (*Spizella-Arborea*). *J. Exp. Zool.* **259:** 117–127.
36. Juss, T.S. *et al.* 1993. Melatonin and photoperiodic time measurement in Japanese quail (*Coturnix coturnix japonica*). *Proc. Biol. Sci.* **254:** 21–28.
37. Ohta, M., C. Kadota & H. Konishi 1989. A role of melatonin in the initial stage of photoperiodism in the Japanese quail. *Biol. Reprod.* **40:** 935–941.
38. Bentley, G.E., T.J. Van't Hof & G.F. Ball. 1999. Seasonal neuroplasticity in the songbird telencephalon: a role for melatonin. *Proc. Natl. Acad. Sci. USA* **96:** 4674–4679.
39. Bentley, G.E. & G.F. Ball. 2000. Photoperiod-dependent and -independent regulation of melatonin receptors in the forebrain of songbirds. *J. Neuroendocrinol.* **12:** 745–752.
40. Guyomarc'h, C. *et al.* 2001. Effect of melatonin supplementation on the sexual development in European quail (*Coturnix coturnix*). *Behav. Processes* **53:** 121–130.
41. Rozenboim, I., T. Aharony & S. Yahav 2002. The effect of melatonin administration on circulating plasma luteinizing hormone concentration in castrated White Leghorn roosters. *Poult. Sci.* **81:** 1354–1359.
42. Underwood, H. *et al.* 1984. Melatonin rhythms in the eyes, pineal bodies, and blood of Japanese quail (*Coturnix coturnix japonica*). *Gen. Comp. Endocrinol.* **56:** 70–81.
43. Cockrem, J.F. & B.K. Follett. 1985. Circadian rhythm of melatonin in the pineal gland of the Japanese quail (*Coturnix coturnix japonica*). *J. Endocrinol.* **107:** 317–324.
44. Maddineni, S.R. *et al.* 2008. Gonadotropin-inhibitory hormone (GnIH) receptor gene is expressed in the chicken ovary: potential role of GnIH in follicular maturation. *Reproduction* **135:** 267–274.
45. Maddineni, S. *et al.* 2008. Gonadotrophin-inhibitory hormone receptor expression in the chicken pituitary gland: potential influence of sexual maturation and ovarian steroids. *J. Neuroendocrinol.* **20:** 1078–1088.
46. Brown, N.L. *et al.* 1975. Chicken gonadotrophins: their effects on the testes of immature and hypophysectomized Japanese quail. *Cell Tissue Res.* **156:** 499–520.
47. Maung, Z.W. & B.K. Follett. 1977. Effects of chicken and ovine luteinizing hormone on androgen release and cyclic AMP production by isolated cells from the quail testis. *Gen. Comp. Endocrinol.* **33:** 242–253.
48. Maung, S.L. & B.K. Follett. 1978. The endocrine control by luteinizing hormone of testosterone secretion from the testis of the Japanese quail. *Gen. Comp. Endocrinol.* **36:** 79–89.
49. Young, K.A., G.F. Ball & R.J. Nelson. 2001. Photoperiod-induced testicular apoptosis in European starlings (*Sturnus vulgaris*). *Biol. Reprod.* **64:** 706–713.
50. Young, K.A. & R.J. Nelson. 2001. Mediation of seasonal testicular regression by apoptosis. *Reproduction* **122:** 677–685.
51. Tapanainen, J.S. *et al.* 1993. Hormonal control of apoptotic cell death in the testis: gonadotropins and androgens as testicular cell survival factors. *Mol. Endocrinol.* **7:** 643–650.
52. Woolveridge, I. *et al.* 1999. Apoptosis in the rat spermatogenic epithelium following androgen withdrawal: changes in apoptosis-related genes. *Biol. Reprod.* **60:** 461–470.
53. Nandi, S., P.P. Banerjee & B.R. Zirkin. 1999. Germ cell apoptosis in the testes of Sprague Dawley rats following testosterone withdrawal by ethane 1,2-dimethanesulfonate administration: relationship to Fas? *Biol. Reprod.* **61:** 70–75.
54. Tsutsui, K. & S. Ishii. 1980. Hormonal regulations of follicle-stimulating hormone receptors in the tests of Japanese quail. *J. Endocrinol.* **85:** 511–518.
55. Ottinger, M.A. & M.R. Bakst. 1981. Peripheral androgen concentrations and testicular morphology in embryonic and young male Japanese quail. *Gen. Comp. Endocrinol.* **43:** 170–177.
56. Brown, N.L. & B.K. Follett. 1977. Effects of androgens on the testes of intact and hypophysectomized Japanese quail. *Gen. Comp. Endocrinol.* **33:** 267–277.
57. Tsutsui, K. & S. Ishii. 1978. Effects of follicle-stimulating hormone and testosterone on receptors of follicle-stimulating hormone in the testis of the immature Japanese quail. *Gen. Comp. Endocrinol.* **36:** 297–305.
58. Tanabe, Y., K. Himeno & H. Nozaki. 1957. Thyroid and ovarian function in relation to molting in the hen. *Endocrinology* **61:** 661–666.

59. Payne, R.B. 1972. Mechanisms and control of molt. *Avian Biol.* **2:** 103–155.
60. Koda, A. *et al.* 2002. A novel amphibian hypothalamic neuropeptide: isolation, localization, and biological activity. *Endocrinology* **143:** 411–419.
61. Sawada, K. *et al.* 2002. Identification of a cDNA encoding a novel amphibian growth hormone-releasing peptide and localization of its transcript. *J. Endocrinol.* **174:** 395–402.
62. Ukena, K. *et al.* 2003. Novel neuropeptides related to frog growth hormone-releasing peptide: isolation, sequence, and functional analysis. *Endocrinology* **144:** 3879–3884.
63. Chartrel, N. *et al.* 2002. Isolation, characterization, and distribution of a novel neuropeptide, Rana RFamide (R-RFa), in the brain of the European green frog *Rana esculenta*. *J. Comp. Neurol.* **448:** 111–127.
64. Sawada, K. *et al.* 2002. Novel fish hypothalamic neuropeptide: cloning of a cDNA encoding the precursor polypeptide and identification and localization of the mature peptide. *Eur. J. Biochem.* **269:** 6000–6008.
65. Hinuma, S. *et al.* 2000. New neuropeptides containing carboxy-terminal RFamide and their receptor in mammals. *Nat. Cell Biol.* **2:** 703–708.
66. Fukusumi, S. *et al.* 2001. Characteristics and distribution of endogenous RFamide-related peptide-1. *Biochim. Biophys. Acta.* **1540:** 221–232.
67. Yoshida, H. *et al.* 2003. Molecular properties of endogenous RFamide-related peptide-3 and its interaction with receptors. *Biochim. Biophys. Acta.* **1593:** 151–157.
68. Ukena, K. *et al.* 2002. A novel rat hypothalamic RFamide-related peptide identified by immunoaffinity chromatography and mass spectrometry. *FEBS Lett.* **512:** 255–258.
69. Ubuka, T. *et al.* 2009. Gonadotropin-inhibitory hormone distribution in rhesus macaque brain. *J. Comp. Neurol.* in press.
70. Ukena, K. & K. Tsutsui. 2005. Review: a new member of the hypothalamic RF-amide peptide family, LPXRF-amide peptides: structure, localization, and function. *Mass Spectrom. Rev.* **24:** 469–486.
71. Tsutsui, K. & K. Ukena. 2006. Review: hypothalamic LPXRF-amide peptides in vertebrates: identification, localization and hypophysiotropic activity. *Peptides* **27:** 1121–1129.
72. Tsutsui, K. 2009. Review: a new key neurohormone controlling reproduction, gonadotropin-inhibitory hormone (GnIH): biosynthesis, mode of action and functional significance. *Prog. Neurobiol.* **88:** 76–88.
73. Tachibana, T. *et al.* 2005. Gonadotropin-inhibiting hormone stimulates feeding behavior in chicks. *Brain Res.* **1050:** 94–100.
74. Amano, M. *et al.* 2006. Novel fish hypothalamic neuropeptides stimulate the release of gonadotrophins and growth hormone from the pituitary of sockeye salmon. *J. Endocrinol.* **188:** 417–423.
75. de Roux, N. *et al.* 2003. Hypogonadotropic hypogonadism due to loss of function of the KiSS1-derived peptide receptor GPR54. *Proc. Natl. Acad. Sci. USA* **100:** 10972–10976.
76. Funes, S. *et al.* 2003. The KiSS-1 receptor GPR54 is essential for the development of the murine reproductive system. *Biochem. Biophys. Res. Commun.* **312:** 1357–1363.
77. Seminara, S.B. *et al.* 2003. The GPR54 gene as a regulator of puberty. *N. Engl. J. Med.* **349:** 1614–1627.

Ann. N.Y. Acad. Sci. ISSN 0077-8923

ANNALS OF THE NEW YORK ACADEMY OF SCIENCES
Issue: *Phylogenetic Aspects of Neuropeptides*

Evolution of the opioid/ORL-1 receptor gene family

Jazalle McClendon, Stephanie Lecaude, Anthony R. Dores, and Robert M. Dores

University of Denver, Department of Biological Sciences, Denver, Colorado

Address for correspondence: Robert M. Dores, University of Denver, Department of Biological Sciences, 2190 E. Iliff, Olin Hall 102, Denver, CO. rdores@du.edu

In gnathostomes the kappa, mu, delta, ORL-1 receptor genes constitute the opioid/ORL-1 receptor gene family. These genes are most likely the result of two (2R) genome duplication events that occurred during the radiation of the chordates. *In stilico* analysis of the genome of the lamprey, *Petromyzon marius*, revealed the partial sequences of four genes that may be the result of a lineage specific genome duplication event in the lamprey lineage. The sequencing of cDNAs from the lamprey CNS supports the assumption that these putative lamprey opioid-like receptor genes are expressed by lamprey neurons. Analysis of gnathostome ORL-1 receptor sequences support the hypothesis that the ORL-1 gene has undergone a transition from an opioid receptor that could bind several types of opioid ligands to a receptor in mammals that can only be activated by the FGGF form of the orphanin ligand.

Keywords: opioid receptors; ORL-1 receptor; evolution; lamprey

Introduction

The enigma as to why an alkaloid by-product, like morphine, would promote an analgesic sensation in the central nervous system (CNS) of humans was resolved by the twin discoveries that neurons in the CNS of vertebrates (a) express opiate-like (opioid) receptors,[1] and (b) synthesize endogenous opiate-like (opioid) neuropeptides.[2] Following these seminal discoveries, pharmacological studies established that there were three types of opioid receptors (mu, delta, and kappa).[3] Later studies would reveal that each pharmacological type of opioid receptor was encoded on a distinct gene.[4–7] Analyses of the amino acid sequences of these receptors indicated that opioid receptors are: (a) G protein–coupled receptors (GPCRs), (b) members of the Rhodopsin Class of GPCRs (Subfamily A4), and (c) are mostly likely derived from a common ancestral opioid-like receptor gene.[8,9]

Earlier studies had already established that neurons in the CNS of vertebrates synthesized several distinct, yet related, opioid receptor ligands that were identified as the enkephalins, the dynorphins, and β-endorphin.[10] The unifying features of these polypeptides are that: (a) all of these neuropeptides have the opioid core sequence motif, YGGF, (b) the opioid ligands are synthesized on larger precursor proteins, and (c) these precursor proteins (Proenkephalin, Prodynorphin, and Proopiomelanocortin) are encoded on distinct genes that are believed to have evolved from a common ancestral gene (for reviews see Refs. 11,12).

In the CNS of mammals, the opioid ligand-coding genes are expressed in distinct neurons,[13] and the result is an elaborate network of enkephalinergic, dynorphinergic, and endorphinergic circuits that release inhibitory neuropeptides that bind to opioid receptors on downstream target neurons and influence a number of processes in the CNS.[14,15] These networks are not restricted to mammals but have been mapped in representatives from every major class of the gnathostome vertebrates.[16–19] Hence, the radiation of opioidinergic networks during the evolution of the chordates has been the result of the co-evolution of the opioid ligand-coding genes and the opioid receptor-coding genes. The radiation of these ligand-coding genes and receptor-coding genes in Phylum Chordata can best be explained when viewed from the perspective of the 2R hypothesis and the role that genome duplication events have played during the evolution of the chordates.

doi: 10.1111/j.1749-6632.2010.05515.x

The 2R hypothesis and opioid/ORL-1 receptor phylogeny

In the 2R hypothesis,[20,21] an ancestral protochordate lineage underwent a genome duplication event, and that lineage is proposed to be at the base of the radiation that gave rise to the jawless vertebrates (Agnatha) represented today by the lampreys and hagfish. As a result where there had been one copy of a gene in the ancestral protochordate lineage, there were now two copies of that gene in the ancestral agnathans. These duplicated genes could retain the same function as the ancestral gene (subfunctionalization), evolve toward a new function (neofunctionalization), or become pseudogenes.[22] During the radiation of the agnathans, a second genome duplication event is proposed to have occurred, and this proposed ancestral agnathan lineage was at the base of the radiation that gave rise to the jawed vertebrates (Gnathostomata). The predicted outcome of these two genome duplication events is the potential for four copies of the original ancestral protochordate gene in extant gnathostomes. Given that three opioid ligand-coding genes and three opioid receptor-coding genes had been characterized between 1979 and 1994, the 2R hypothesis predicted that a fourth opioid-related ligand-coding gene and a fourth opioid-related receptor-coding gene could be present in the genomes of extant gnathostomes. In fact by 1995 the fourth gene in each family had been discovered. The ligand-coding gene, Proorphanin, was first characterized from mammals[23,24] and later from a bony fish, the white sturgeon.[25] The unique feature of the mammalian Proorphanin gene was not just the presence of a single end-product (FGGFTGARKSARKLANQ) that was referred to as orphanin or nociceptin, but rather the fact that the N-terminal motif for this ligand was FGGF rather than the YGGF motif present in classical opioid ligands like the enkephalins, dynorphins, and β-endorphin.[26] It quickly became apparent that mammalian orphanin could not bind and activate the classical opioid receptors (i.e., mu, delta, or kappa), but was the ligand for an orphan GPCR that was identified as ORL-1.[27] The unique feature of the mammalian ORL-1 gene was not that this orphanin receptor had very clear sequence identity with the classic opioid receptors, but rather that the mammalian ORL-1 receptor could not be activated by any of the classic opioid ligands, but could only be activated by mammalian orphanin.[27] This strict one-to-one ligand selectivity, must have evolved during the divergence of the ORL-1 gene from the other opioid-receptor genes following the second genome duplication event. If so then which classical opioid receptor (i.e., mu, delta, and kappa) is the sister gene to the ORL-1 gene? At what point in vertebrate evolution did the activation of the ORL-1 receptor become solely dependent on the orphanin ligand?

With respect to the first question, it appears that the kappa opioid receptor gene and the ORL-1 receptor gene could have evolved from a common ancestral gene. As presented in Figure 1, the rat kappa receptor and the rat ORL-1 receptor can be aligned by the insertion of three gaps. The overall sequence identity of the two receptors at the amino acid level is only 42%. However, in regions of the receptor known to be important for ligand binding (TM3, TM5, TM6, and TM7) the sequence identify is 70%. Note that there are amino acid positions in these transmembrane spanning domains (marked with +; Fig. 1) that have been shown to be critical for dynorphin binding to the rat kappa receptor and orphanin binding to the ORL-1 receptor.[28] Since these amino acid positions are identical in both receptors, ligand selectivity can not be attributed to these sites. However, Meng *et al.*[29] found that by performing site directed mutagenesis at two positions in the rat ORL-1 sequence, it was possible to alter the ligand selectivity of the ORL-1 receptor. After replacing the VQV motif at positions 291–293 (Fig. 1; [a]) in the rat ORL-1 receptor with the IHI motif located at the same positions in the rat kappa receptor, and by replacing the threonine (T) residue at position 317 in the rat ORL-1 receptor with an isoleucine (I) residue present at this position in the rat kappa receptor, a chimeric rat ORL-1 receptor was formed that could bind dynorphin ligands, but not mu or delta agonists, with subnanomolar affinity.[29] Hence, from a functional standpoint these amino acid positions appear to determine whether an opioid receptor is exclusively activated by mammalian FGGF form of orphanin (wild-type rat ORL-1 receptor), or is a "mixed" ligand opioid receptor (rat chimeric ORL-1 receptor) that could be activated by either kappa ligands or mammalian orphanin. The prediction would be that the ancestral condition for the ORL-1 receptor was as a "mixed" ligand receptor, which evolved the exclusive selectivity for the

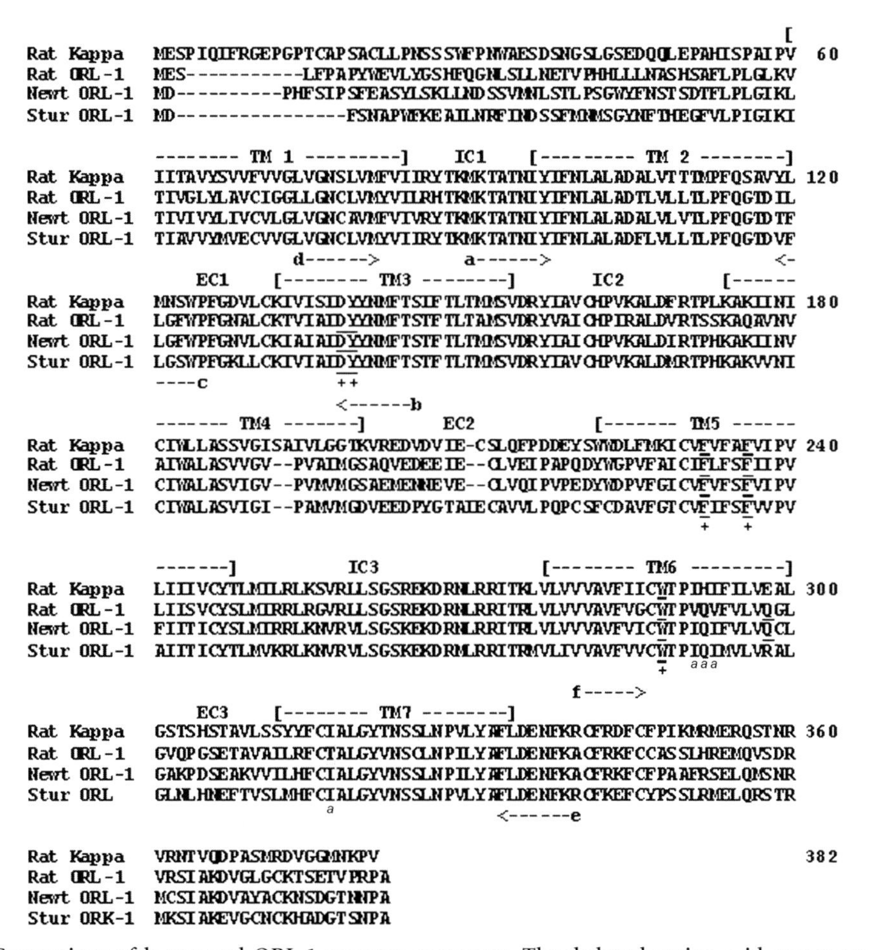

Figure 1. Comparison of kappa and ORL-1 receptor sequences. The deduced amino acid sequences for the rat kappa receptor,[43] the rat ORL-1 receptor,[29] the newt, *Taricha granulosa* ORL-1 receptor,[44] and the white sturgeon (Stur), *Acipenser transmontanus*, ORL-1 (accession #: GU228526) were manually aligned by inserting three gaps. Amino acids considered essential for both kappa agonist binding and orphanin binding[28] are underlined and marked with a (+). Amino acid positions that were altered in rat ORL-1 are shaded in gray. Amino acid positions in the rat kappa receptor, the newt ORL-1 receptor, and the sturgeon ORL-1 receptor that are essential for kappa agonist binding are marked with a ([a]).TM, transmembrane; IC, intracellular loop; EC, extracellular loop. For the cloning of sturgeon ORL-1, mRNA was isolated from a whole brain extract and converted to cDNA as described previously.[45] The full length ORL-1 cDNA was characterized by a series of RT-PCR and RACE protocols. Initially the degenerate primer a (5′GAARACIGCIACIAAYATHTA3′) and the degenerate primer b (D′GTRAACATRTTRTARTARTC3′)[46] were used to generate an amplicon that corresponded to positions 89–145 (Fig. 1). Eleven identical clones were sequenced. Next 5′RACE was performed[45] using the homologous reverse primer c (5′GCCAGGACCCAAGGAAAA3′) to generate an amplicon that included the 5′UTA of the ORL-1 cDNA and positions 1–125. Five identical clones were sequenced. This was followed by RT-PCR using primer d (5′GGGAAACTGCCTGGTCAT3′) and primer e (′TTRAARTTYTCRCIARRAA3′) to generate an amplicon that corresponded to positions 74–340. Five identical clones were sequenced. Finally, 3′RACE was performed[45] using the homologous forward primer f (5′TGGCGGTCTTCGTTGTCT3′) to generate an amplicon that corresponded to positions 280–383 (Fig. 3) and the 3′UTA. Six identical clones were sequenced. Subcloning and sequencing of each amplicon was done as described previously.[45]

mammalian FGGF form of orphanin at some later time in vertebrate evolution.

Support for this hypothesis would come from ligand binding studies done on the ORL-1 receptor cloned from the CNS of the amphibian *Taricha granulosa*.[30] The amphibian lineage predates the mammalian lineage by at least 100 million years.[31] As shown in Figure 1, the *T. granulosa* ORL-1 has an IQI motif at position 291–293 and a I reside at position 317. This receptor can bind either dynorphin-related ligands with nanomolar affinity or the mammalian FGGF form of orphanin with subnanomolar affinity.[30] However, *T. granulosa* ORL-1 had negligible affinity for either mu or delta agonists. Also presented in Figure 1 is the deduced amino acid sequence for the sturgeon ORL-1 receptor. This ORL-1 sequence also has the IQI motif at position 291–293 and the I residue at position 317. The bony fish lineage predates the amphibian lineage by at least 50 million years,[31] hence these observations support the hypothesis that the ancestral ORL-1 receptor was most likely a "mixed" opioid ligand (kappa/orphanin) receptor. Additional support for this hypothesis comes from the observation that the orphanin-related ligand in the sturgeon Proorphanin precursor has the amino acid sequence, YGGFIGIRKSARKWNP.[25] Binding studies indicated that this ligand could bind to either the rat kappa receptor or the rat ORL-1 receptor with nanomolar affinity, but was not an agonist for either the rat mu or delta receptors.[25] Based on these observations, the hypothesis that emerges is that in the ancestral gnathostomes the Proorphanin gene encoded an orphanin sequence with a YGGF N-terminal motif; a ligand that could bind to either the ancestral kappa receptor or the ancestral ORL-1 receptor. Conversely, the ancestral ORL-1 receptor could bind either kappa or orphanin ligands. It would then be reasonable to propose that at some point in the evolution of the ancestral mammalian lineage, the orphanin sequence in Proorphanin mutated to an N-terminal FGGF sequence, and the ancestral ORL-1 gene had the mutations in TM6 and TM7 that resulted in a receptor that could only be activated by the FGGF form of the orphanin ligand. If these predictions are correct then the cartilaginous fish, ray-finned fish, lobe finned fish, and amphibian Proorphanin and ORL-1 sequences should have the amino acid features found in the sturgeon Proorphanin and ORL-1 genes. Furthermore, the distinct amino acid motifs in the mammalian Proorphanin and ORL-1 sequences should have been the result of mutations that occurred during the radiation of the reptilian lineages that give rise to extant reptiles, birds, and mammals.[31] At present the studies to confirm or refute these predictions have not been done.

Opioid/ORL-1 receptor evolution: studies on the lamprey genome

When modeling the evolution of the opioid/ORL-1 receptors, the scheme that would be most consistent with the 2R hypothesis is shown in Figure 2(A). In this scheme the ancestral protochordate opioid receptor gene would have been duplicated to give rise to a hypothetical kappa/ORL-1-like gene and a mu/delta-like gene in agnathan vertebrates. The second genome duplication event would have yielded the distinct ORL-1, kappa, mu, and delta genes found in extant gnathostomes. In support of this hypothesis, synteny analyses indicate that the four members of the opioid receptor gene family are located on different chromosomes in nearly all species of gnathostomes that have been studied.[32] Finally, this model would predict that the ORL-1 and kappa receptors should form a clade and the mu and delta receptors should form a clade.

A way to test the validity of this model would be to perform maximum parsimony analysis on the opioid receptors from individual species. In Figures 2(B) and (C) these relationships were evaluated for the four opioid receptors that have been characterized from the zebrafish and human genomes. While the predicted mu/delta clade formed with a robust bootstrap value in both analyses, the kappa receptor and the ORL-1 receptor did not form a clade in either analysis. Instead the kappa receptor was the sister sequence to the mu/delta clade and the ORL-1 receptor was the sister sequence to a clade composed of the three classical opioid receptors. These results indicate that the ORL-1 receptor is the most divergent of the four members of the Opioid/ORL-1 receptor gene family. In addition, these analyses provide examples of how subfunctionalization (the overlapping of ligand selectivity and perhaps the functional roles of the ancestral gnathostome opioid receptors) can result in neofunctionalization (the unique ligand selectivity and functional role of the ORL-1 receptor in mammals).[33] While

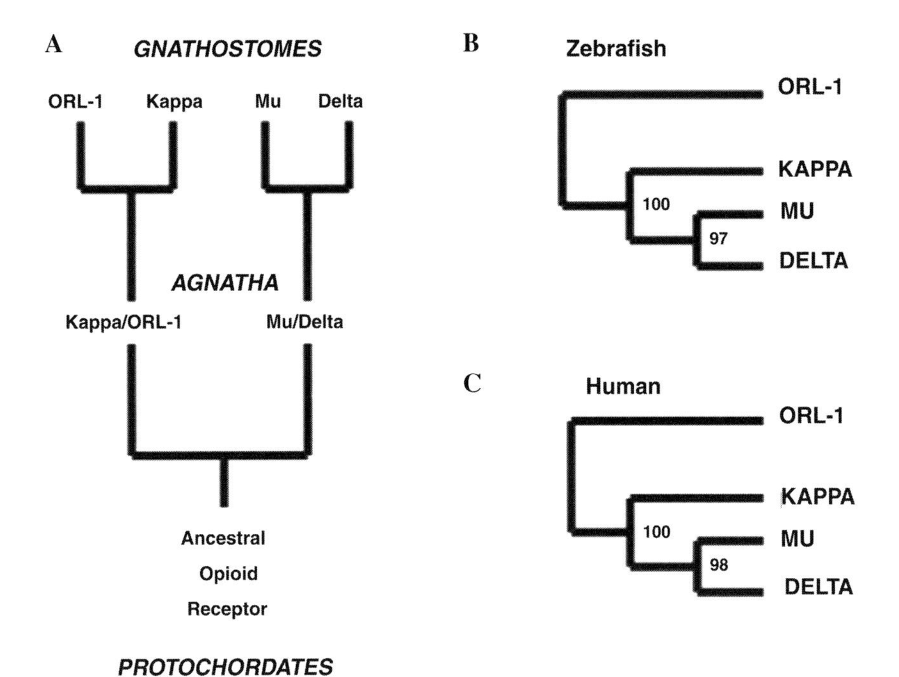

Figure 2. Modeling the evolution of opioid/ORL-1 receptors. (A) Hypothetical scheme for the evolution of the opioid/ORL-1 Receptor gene family based on the 2R hypothesis. (B) The amino acid sequences for the zebrafish mu (accession # NP_571782.1), delta (accession # NP_571333.1), kappa (accession # NP_878306), and ORL (accession # AAN46774.1) receptors were manually aligned and analyzed using the maximum parsimony algorithm. The exhaustive search mode was used, and the tree was rooted using the midpoint rooting algorithm.[47] The bootstrap values are shown at the midpoint of each branch based on 1000 replicates. (C) The amino acid sequences for the human mu (accession # AAC51877.1), delta (accession # AAA18789.2), kappa (accession # AAC50158.1), and ORL (accession # AF115266.1) receptors were manually aligned and analyzed using the maximum parsimony algorithm. The exhaustive search mode was used, and the tree was rooted using the midpoint rooting algorithm.[47] The bootstrap values are shown at the midpoint of each branch based on 1000 replicates.

it would be reasonable to assume that the ORL-1 and kappa receptor genes were the result of the duplication of an ancestral agnathan ORL-1/kappa receptor gene, it would appear that the ORL-1 receptor gene has accumulated a greater number of mutations than the kappa receptor since that duplication event. While the kappa receptor remained a "mixed" ligand receptor that has a higher affinity for dynorphin-related ligands, than for mu or delta ligands,[3] the ORL-1 receptor has gone through two evolutionary phases. In the first phase, ORL-1 was a "mixed" ligand receptor. During the second phase ORL-1 transitioned to a receptor that could only be activated by an orphanin ligand with the FGGF N-terminal motif. It is conceivable that during phase 1 the ORL-1 gene in some lineages could have become a pseudogene. To date ORL-1 genes have been detected in bony fish (Fig. 1) but have not been cloned from the CNS of any cartilaginous fish or detected in the genome of the elephant shark, *Callorhinchus milii* (http://eshark.imcb.a-star.edu.sg/). Since the elephant shark genome project is not complete, there still is the possibility that an ORL-1 homolog could be detected. Assuming that the ORL-1 gene may have been secondarily lost in the cartilaginous fish lineage, it still seems reasonable to conclude that the presence of four distinct opioid-related receptor-coding genes paralleled the emergence of the gnathostomes. Furthermore, based on the model presented in Figure 2(A), opioid-like receptor genes could also be present in the genomes of extant protochordates and agnathan vertebrates.

In this regard, blast analyses of the tunicate, *Ciona intestinalis*, genome (http://www.ensembl.org/Multi/blastview) or the amphioxus, *Branchiostoma floridae* genome (http://genome.jgi-psf.org/

Brafl1.home.html) have not detected any opioid-like receptor (OLR) gene sequences. Because these genome projects are fairly complete, these outcomes may indicate that OLR gene sequences have been secondarily lost in these species. However, a more likely alternative is that the tunicate and amphioxus lineages may not be derived from the protochordate lineage that evolved the first opioid-like GPCR. It is noteworthy that none of the opioid ligand-coding genes have been detected in either the tunicate or amphioxus genomes, and neither opioid receptor nor opioid ligand coding genes have been detected in an echinoderm genome or the *Caenorhabditis elegans* genome. Collectively, these negative data would indicate that the opioid ligand and receptor genes may be "recent" additions to the collective metazoan genome emerging at some point in the Cambrian in a lineage of protochordates that underwent the first genome duplication.

While analyses of extant protochordate genomes have yielded negative results, an analysis of the genome of the marine lamprey, *Petromyzon marinus*, has been more rewarding (http://pre.emsemble.org/Petromyzon.marinus/Info/Index). The amino acid sequence of the zebrafish kappa receptor (accession number: NP_878306) was used to probe the lamprey database using the TBLASTN protocol. This approach detected several contigs with DNA sequences that corresponded to OLR-like genes. The nucleotide sequence for each putative lamprey OLR gene was used as a template to design primers for PCR to determine if these ORL sequences are actually expressed in the lamprey CNS. To this end mRNA was isolated from a lamprey CNS extract and converted to cDNA. The PCR primer sets (Fig. 3) detected four cDNAs that encoded the partial sequences of what appear to be four distinct, yet related lamprey ORL genes. The deduced amino acid sequences for these lamprey contigs are shown in Figure 3. The four sequences were labeled as lamprey opioid-like receptor (OLR) 1a, 1b, 2a, and 2b. OLR 1a is 283 amino acid in length, OLR 2a is 197 amino acids in length, ORL 1b is 113 amino acids in length, and ORL 2b is 112 amino acids in length. The presence of four putative opioid-like receptor genes in the lamprey genome would appear to run counter to the 2R hypothesis, however it is possible that OLR 1b and OLR 2b may be the result of a gene or genome duplication that yielded copies of the lamprey OLR 1a and OLR 2a, respectively. Evidence for a lamprey lineage-specific genome duplication is suggested by the characterization of two distinct Proopiomelanocortin genes in the marine lamprey genome.[34,35] Based on these considerations, this analysis will only focus on OLR 1a and OLR 2a genes.

Overlapping sequences of the lamprey OLR 1a gene were found on contig 51033.1 and contig 50736.2 (Fig. 3). This partial sequence lacks the N-terminal region and TM1, but includes part of the Intracellular Loop 1 and all of TM2, TM3, TM4, TM5, TM6, TM7, and the C-terminal region of the putative receptor. As shown in Figure 3 the Lamprey OLR sequences were aligned to the zebrafish kappa receptor amino acid sequence. Shaded positions indicate residues that are identical in the zebrafish kappa sequence and the lamprey OLR sequences. Based on this alignment, the lamprey ORL 1a sequence has 60% primary sequence identify to the corresponding positions in the zebrafish kappa receptor. However, within the transmembrane regions the sequence identity is 79%. The residues identified in Figure 1 that are considered critical for opioid ligand binding (Fig. 3; +) are all present in lamprey OLR 1 and in the sequence of lamprey ORL 2a. The later deduced amino acid sequence is from contig 28623.2. Lamprey ORL 2a extends from Intracellular Loop 1 to TM6. The sequence identity between lamprey ORL 2a and the zebrafish kappa receptor in the transmembrane spanning regions was 76%. Thus from a primary sequence perspective it seems reasonable to assume that lamprey ORL 1a and lamprey ORL 2a are opioid-like GPCRs. In the absence of binding studies it is not possible to determine which of the lamprey OLRs correspond to the ancestral agnathan opioid receptors predicted in Figure 2(A). However, by performing maximum parsimony analysis it might be possible to gain some insights into the identities of lamprey OLR 1a and OLR 2a.

The amino acid sequences for Lamprey OLR 1a and OLR 2a were aligned to the mu, delta, kappa, and ORL-1 sequences for human, amphibian (newt and frog sequences), and zebrafish (data not shown). Maximum parsimony analysis (Fig. 4) revealed the same relationships for the gnathostome opioid receptors that were observed when the opioid receptor sequences from a single species were analyzed (Figs. 2A and B). The kappa, mu, and delta sequences formed distinct clades with

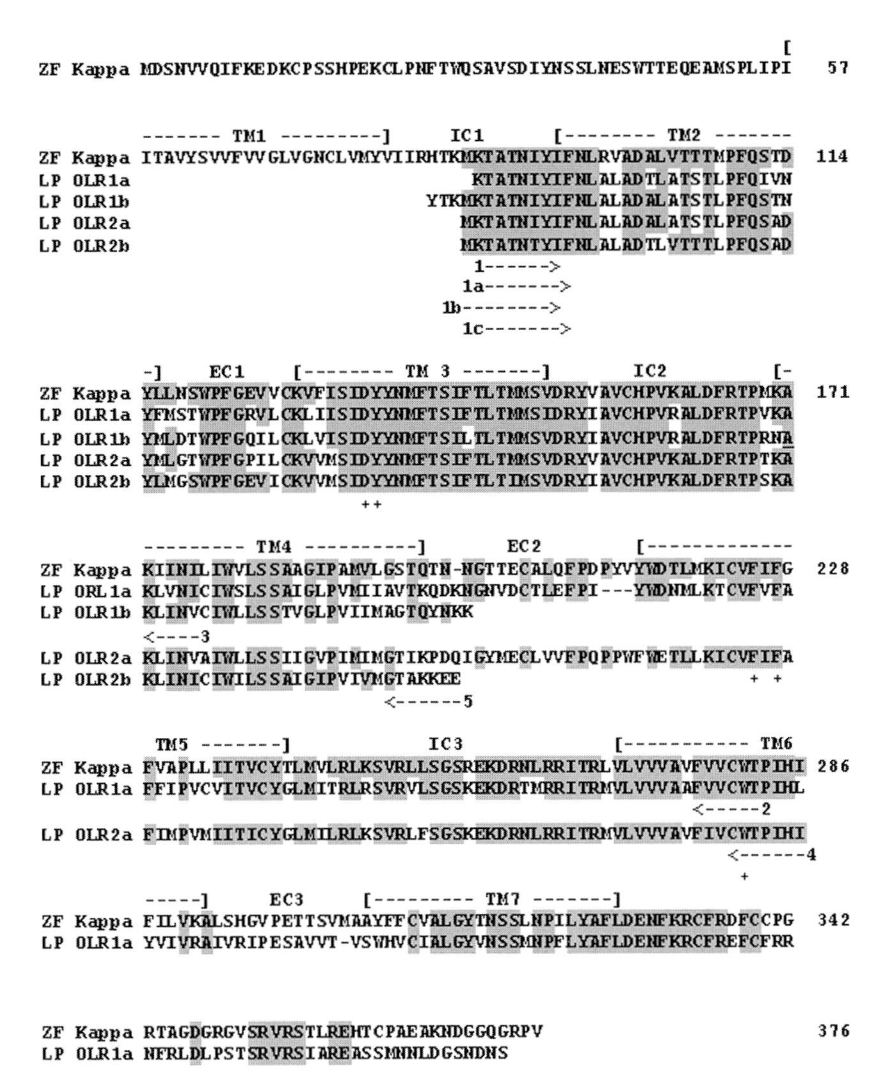

Figure 3. Sequence of lamprey opioid-like receptors. Four opioid-like receptors (OLRs) were found in the *Petromyzon marinus* genome database (http://pre.emsemble.org/Petromyzon.marinus/Info/Index). The deduced amino acid sequence for OLR 1a was found on two overlapping contigs (51033.1 and 50736.2). The deduced amino acids sequences for OLR 1b, OLR 2a, and OLR 2b were found on contigs 55613.1, 28623.2, and 7520.3, respectively. To demonstrate that these four putative OLR genes are express, mRNA was isolated from the brain of an adult post-spawning *P. marinus* and converted to cDNA as described previously.[45] RT-PCR was done using the following primer sets: OLR 1a, primer 1 (5′-AAAACCGCCACCAACAT-3′) and primer 2 (5′-CCAGCACACCACGAAG-3′); OLR 1b, primer 1a (5′-ACCGCTACCAACATCTAC-3′) and primer 3 (5′-ACGTTGATGAGCTTGGC-3′); OLR 2a, primer 1b (5′-ARACSGCYACMAACAYBTAY-3′) and primer 4 (5′-GATAGACGAGAGCAGCC-3′); and OLR 2b, primer 1c (5′-GAAGACGGCTACCAACACG-3′) and primer 5 (5′-TTCTTGGCCGTGCCC-3′). Subcloning and sequencing of each amplicon was done as described previously.[45]

the ORL-1 sequences as the sister sequences to the classic opioid receptor clades. Once again the divergent nature of the ORL-1 receptors relative to the classic opioid receptors is apparent. Interestingly, the lamprey ORL 1a sequence grouped with the kappa opioid receptors. This outcome would appear to indicate that the lamprey ORL-1 might represent the predicted agnathan kappa/ORL-1 receptor. However, this outcome is perplexing because dynorphin-related ligands have not been

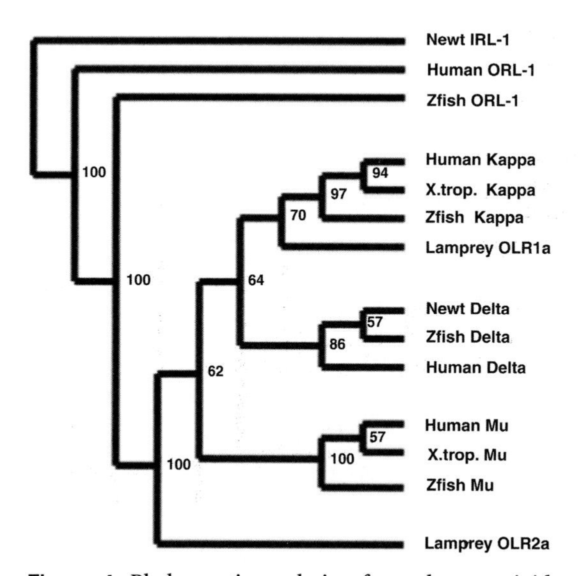

Figure 4. Phylogenetic analysis of vertebrate opioid receptor sequences. The amino acids sequences of Lamprey OLR 1a and OLR 2a, and the following receptors: zebrafish mu, delta, kappa, and ORL-1, human mu, delta, kappa, and ORL-1, *Taricha granulosa* delta (accession # AY751785.1) and ORL-130 (accession # AY728087.1), and *Xenopus tropicalis* mu (accession # ENSXETP00000018406) and kappa (ENSXETP00000017733) were manually aligned and analyzed using the Maximum Parsimony algorithm. The exhaustive search mode was used, and the tree was rooted using the midpoint rooting algorithm.[47] The bootstrap values are shown at the midpoint of each branch based on 1000 replicates. Newt, *Taricha granulosa*; X. trop., *Xenopus tropicalis*; Zfish, zebrafish.

isolated from the CNS of the lamprey. By contrast the lamprey OLR 2a sequence did not group with any of the classical opioid receptor clades, but was the sister sequence to the classic opioid receptors. Could Lamprey OLR 2a correspond to the predicted ancestral agnathan mu/delta opioid receptor? This is possible since both enkephalins and β-endorphin have been detected in the pituitary of lampreys.[36–39] However, these phylogenetic relations for lamprey OLR 1a and OLR 2a have to be viewed as possibilities rather than realities until the full length sequences of these putative receptors have been characterized, the full complement of opioid ligands in the lamprey CNS have been identified, and the appropriate binding and activation studies on lamprey OLRs expressed in heterologous cell lines have been completed.

Conclusions

The opioid/ORL-1 receptor gene family provides a useful model for testing the 2R hypothesis, and also provides insights into how the evolution of a gene family can deviate from the predicted hypothetical 2R scheme during the radiations that occur after a genome duplication event. Based on recent analyses of metazoan genome databases[32] and this study, the opioid receptor-coding genes and the opioid ligand-coding genes appear to be relatively recent additions to the metazoan genome appearing at some point during the Cambrian. In this scenario some protochordate lineage evolved the ancestral opioid receptor-coding gene and the opioid ligand-coding gene and then underwent a genome duplication event that led to the evolution of the jawless vertebrates (Agnatha) that are represented by the extant lampreys. Whether jawless vertebrates evolved several times from different protochordate lineages in this post-Cambrian period or just once is not resolved.[31] What does seem certain is that the protochordate lineage that evolved the ancestral opioid receptor and ligand coding genes is extinct. Hence, modeling studies based on a phylogenetic reconstruction algorithm will be needed to predict what these genes might have looked like.[40]

In this study, partial sequences for proposed opioid-like receptor (OLRs) genes are presented in Figure 3. The presence of four of these genes instead of the predicted two agnathan opioid receptor-coding genes is initially perplexing, unless a lineage specific genome duplication event has occurred within the lamprey lineage. The presence of the two distinct Proopiomelanocortin genes would support this hypothesis.[32,33] Hence, it appears that there are two candidates in the lamprey genome that represent contemporary forms of the predicted ancestral ORL-1/kappa gene and the predicted ancestral mu/delta gene. Obtaining the full length sequences for these receptors, and performing functional expression studies will be critical in understanding the evolutionary position of these putative opioid-like receptor. However, there is the long overdue need for a parallel study to characterize the opioid ligands synthesized by neurons in the lamprey CNS.

It will not be surprising if the functional expression of the lamprey opioid receptors indicates that these receptors can bind and be activated by the full array of lamprey opioid ligands. The lack of a one-to-one ligand to receptor specificity may be the ancestral condition for the opioid/ORL-1 receptors. Ligand/receptor selectivity studies have been done on several gnathostome opioid/ORL-1 receptor genes and some trends are apparent. Binding and activation studies indicate that all of the classic opioid ligands can bind to and activate the mammalian mu and delta receptors.[26] The positively charged dynorphin ligands, the enkephalin ligands and to a lesser extent β-endorphin can bind to the mammalian kappa receptor, but the dynorphins preferentially bind to this receptor.[26] However, the mammalian ORL-1 receptor can only be activated by the mammalian FGGF form of the orphanin ligand.[26] Establishing when the gnathostome ORL-1 receptor made the transition from a "mixed" ligand receptor to a receptor that would only bind the FGGF form of orphanin remains to be determined. It is very possible that this strict ligand to receptor selectivity is a unique feature of the mammals.

Finally, the mapping of opioidinergic networks in the mammalian CNS[13] revealed that there are terminal field in the brain where multiple opioid ligands are released (i.e., enkephalinergic synapses and dynorphinergic synapses). How, for example, is activation of the kappa receptor affected when α-neoendorphin, dynorphin A, and dynorphin B are all interacting with the receptor in equimolar concentrations? For that matter in the CNS of the lungfish where Prodynorphin encodes not only dynorphin-related end-products, but also enkephalin-related end-products how do the opioid receptors in those synapses respond to an array of distinct opioid ligands?[41,42] Although opioid receptors and opioid neuropeptides were first discovered over 35 years ago,[1,2] there are still several unresolved issues that are fundamental for understanding the role of opioidinergic networks in the evolution of the chordates.

Acknowledgments

This study was support by NSF IOB 0516958 (R.M.D.).

Conflicts of interest

The authors declare no conflicts of interest.

References

1. Pert, C.B. & Snyder, S.H. 1973. Opiate receptor: demonstration in nervous tissue. *Science* **179:** 1011–1014.
2. Hughes, J. *et al.* 1975. Identification of two related pentapeptides from the brain with potent opiate agonist activity. *Nature* **258:** 577–580.
3. Pechnick, R. *et al.* 1985. Identification of multiple opiate receptors through neuroendocrine responses. II. Antagonism of mu, kappa and sigma agonists by naloxone and WIN 44,441–3. *J. Pharmacol. Exp. Ther.* **232:** 170–177.
4. Evans, C.J. *et al.* 1992. Cloning of a delta opioid receptor by functional expression. *Science* **258:** 1952–1955.
5. Kieffer, B.L. *et al.* 1992. The delta-opioid receptor: isolation of a cDNA by expression cloning and pharmacological characterization. *Proc. Natl. Acad. Sci. USA* **89:** 12048–12052.
6. Chen, Y. *et al.* 1993a. Molecular cloning and functional expression of a mu-opioid receptor from rat brain. *Mol. Pharmacol.* **44:** 8–12.
7. Chen, Y. *et al.* 1993b. Molecular cloning of a rat κ opioid receptor reveals sequence similarities to the μ and δ opioid receptors. *Biochem. J.* **295:** 625–628.
8. Foord, S.M. *et al.* 2005. International Union of Pharmacology. XLVI. G protein-coupled receptor list. *Pharmacol. Rev.* **57:** 279–288.
9. Bjarnadottir, T.K. *et al.* 2006. Comprehensive repertoire and phylogenetic analysis of the G protein-coupled receptors in human and mouse. *Genomics* **88:** 263–273.
10. Akil, H. *et al.* 1984. Endogenous opioids: biology and function. *Annu. Rev. Neurosci.* **7:** 223–255.
11. Khalap, A. *et al.* 2005. Trends in the Evolution of the Proenkephalin and Prodynorphin Genes in Gnathostomes. *Ann. New York Acad. Sci.* **1040:** 22–37.
12. Dores, R.M. & Lecaude, S. 2005. Trends in the evolution of the proopiomelanocortin gene. *Gen. Comp. Endocrinol.* **142:** 81–93.
13. Khachaturian, H. *et al.* 1985. Anatomy of the CNS opioid systems. *Trends Neurosci.* **8:** 111–119.
14. Mansour, A. *et al.* 1995. The cloned Mu, Delta and Kappa receptors and their endogenous ligands: evidence for two opioid peptide recognition cores. *Brain Res.* **700:** 89–98.
15. Kieffer, B.L. 1995. Recent advances in molecular recognition and signal transduction of active peptides: receptors for opioid peptides. *Cell. Mol. Neurobiol.* **15:** 615–635.
16. Reiner, A. & Northcutt, R.G. 1987. An immunohistochemical study of the telencephalon of the African lungfish, *Protopterus annectens*. *J. Comp. Neurol.* **256:** 463–481.
17. Northcutt, R.G. *et al.* 1988 Immunohistochemical study of the teleoncephalon of the spiny dogfish, *Squalus acanthias*. *J. Comp. Neurol.* **277:** 250–267.
18. Reiner, A. & Northcutt, R.G. 1992. An immunohistochemical study of the telecephalon of the Senegal

bichir (*Polypterus senegalus*). *J. Comp. Neurol.* **319:** 359–386.

19. Vallarino, M. *et al.* 1998. Immunocytochemical localization of enkephalins in the brain of the African lungfish, *Protopterus annectens*, provides evidence for differential distribution of met-enkephalin and leu-enkephalin. *J. Comp. Neurol.* **396:** 275–287.
20. Ohno, S. *et al.* 1968. Evolution from fish to mammals by gene duplication. *Hereditas* **59:** 169–187.
21. Holland, P.W. *et al.* 1994. Gene duplications and the origins of vertebrate development. *Development* (Suppl.) 125–133.
22. Force, A. *et al.* 1999. Preservation of duplicate genes by complementary, degenerative mutations. *Genetics* **151:** 1531–1545.
23. Meunier, J. C. *et al.* 1995. Isolation and structure of the endogenous agonist of opioid receptor-like Orl1 receptor. *Nature* **377:** 532–535.
24. Reinscheid, R. K. *et al.* 1995. Orphanin Fq: a neuropeptide that activates an opioid-like G protein-coupled receptor. *Science* **270:** 792–794.
25. Danielson, P.B. *et al.* 2001. Sturgeon orphanin: a molecular "Fossil" that bridges the gap between the opioids and orphanin FQ/N. *J. Biol. Chem.* **276:** 22114–22119.
26. Satoh, M. & Minami, M. 1995. Molecular pharmacology of the opioid receptors. *Pharmacol. Ther.* **68:** 343–364.
27. Mollereau, C. *et al.* 1994. ORL1, a novel member of the opioid receptor family. Cloning, functional expression and localization. *FEBS Lett.* **341:** 33–38.
28. Mouledous, L. *et al.* 2000. Functional inactivation of the nociceptin receptor by transmembrane segment VI: evidence for a site-directed mutagenesis study of the ORL-1 receptor transmembrane-binding domain. *Mol. Pharmacol.* **52:** 495–502.
29. Meng, F. *et al.* 1996. Moving from the Orphanin Fq receptor to an opioid receptor using four point mutations. *J. Biol. Chem.* **271:** 32016–32020.
30. Walthers, E.A. *et al.* 2005. Cloning, pharmacological characterization and tissue distribution of an ORL-1 opioid receptor fro an amphibian the rough-skinned newt *Taricha granulosa. Gen. Comp. Endocrinol.* **34:** 247–256.
31. Carroll, R.L. 1988. *Vertebrate Paelontology and Evolution.* W.H. Freeman and Co. New York.
32. Dreborg, S. *et al.* 2008. Evolution of vertebrate opioid receptors. *Proc. Natl. Acad. Sci. USA* **105:** 15487–15492.
33. Meunier, J.C. 2003. Utilizing functional genomics to identify new pain treatments: the example of nociceptin. *Am. J. Pharmacogenomics* **3:** 117–130.
34. Heinig, J.A. *et al.* 1995. The appearance of proopiomelanocortin early in vertebrate evolution; cloning and sequencing of POMC from a lamprey pituitary cDNA library. *Gen. Comp. Endocrinol.* **99:** 137–144.
35. Takahashi, A. *et al.* 1995. Melanotropin and corit-cotropin are encoded on two distinct genes in the lamprey, the earliest evolved extant and sequencing of POMC vertebrate. *Biochem. Biophys. Res. Commun.* **213:** 490–496.
36. Dores, R.M. *et al.* 1984. Immunohistochemical localization of enkephalin- and ACTH-related substances in the pituitary of the lamprey. *Cell Tissue Res.* **235:** 107–115.
37. Dores, R.M. & Gorbman, A. 1990. Detection of Met-enkephalin and Leu-enkephalin in the brain of the hagfish, *Eptatretus stouti*, and the lamprey, *Petromyzon marinus. Gen. Comp. Endocrinol.* **77:** 489–500.
38. Dores, R.M. & McDonald, L.K. 1992. Detection of Met-enkephalin in the pars intermedia of the lampreys, *Ichthyomyzon marinus* and *Petromyzon marinus. Gen. Comp. Endocrinol.* **88:** 292–297.
39. Takahasi, A. *et al.* 2006. Posttranslational processing of proopiomelanocortin family molecules in sea lamprey based on mass spectrometric and chemical analyses. *Gen. Comp. Endocrinol.* **148:** 79–84.
40. Bridgham, J.T. *et al.* 2006. Evolution of hormone-receptor complexity by molecular exploitation. *Science* **312:** 97–101.
41. Sollars, C. *et al.* 2000. Deciphering the rigin of Met-enkephalin and Leu-enkephalin in lobe-finned fish: cloning of Australian Lungfish Proenkephalin. *Brain Res.* **874:** 131–136.
42. Dores, R.M. *et al.* 2000. In the African lungfish Met-enkephalin and Leu-enkephalin are derived from separate genes: cloning of a Proenkephalin cDNA. *Neuroendocrinology* **72:** 224–230.
43. Minami, M. *et al.* 1993. Cloning and expression of a cDNA for the rat kappa-opioid receptor. *FEBS Lett.* **329:** 291–295.
44. Walthers, E.A. *et al.* 2005. Cloning, pharmacological characterization and tissue distribution of an ORL1 opioid receptor from an amphibian, the rough-skinned newt *Taricha granulosa. J. Mol. Endocrinol.* **34:** 247–256.
45. Lecaude, S. *et al.* 1999. Organization of proenkephalin in amphibians: cloning a proenkephalin cDNA from the brain of the anuran amphibian, *Spea multiplicatus. Peptides* **21:** 339–344.
46. Li, X. *et al.* 1996. Multiple opioid receptor-like genes are identified in diverse vertebrate phyla. *FEBS Lett.* **397:** 25–29.
47. Hess, P.N. & DeMorae Russo, C.A. 2007. An empirical test of the midpoint rooting method. *Biol. J. Linn. Soc.* **92:** 669–674.

Ann. N.Y. Acad. Sci. ISSN 0077-8923

ANNALS OF THE NEW YORK ACADEMY OF SCIENCES
Issue: *Phylogenetic Aspects of Neuropeptides*

Sexual differentiation of kisspeptin neurons responsible for sex difference in gonadotropin release in rats

Hiroko Tsukamura, Tamami Homma, Junko Tomikawa, Yoshihisa Uenoyama, and Kei-ichiro Maeda

Graduate School of Bioagricultural Sciences, Nagoya University, Nagoya, Japan

Address for correspondence: Hiroko Tsukamura, Graduate School of Bioagricultural Sciences, Nagoya University, Nagoya 464-8601, Japan. htsukamura@nagoya-u.jp

The brain mechanism regulating GnRH/luteinizing hormone (LH) release is sexually differentiated in rodents. Estrogen induces a GnRH/LH surge in females but not in males. Kisspeptin neurons in the anteroventral periventricular nucleus (AVPV) have been reported to be sexually dimorphic and suggested to be involved in the GnRH/LH surge generation. Neonatal testicular androgen may cause the reduction of AVPV kisspeptin expression and a lack of LH surge in male rats. Thus, it is plausible that perinatal testicular androgen causes defeminization of the AVPV kisspeptin system, resulting in the loss of the surge system in male rats.

Keywords: GPR54; Kiss1r; Kiss1; sex difference; LH; GnRH

Introduction

At the beginning of the 20th century, Marshall and Jolly[1] first suggested the sex difference in gonadotropin secretion in their excellent study on the transplantation of ovaries to adult gonadectomized male and female rats. They found that ovaries transplanted in the female show ovulation, whereas those transplanted in the male do not. The first evidence for the testis-derived humoral factors being the cause of the sex difference in gonadotropin release was provided by Pheiffer,[2] who found that while ovaries transplanted to the anterior eye chamber of neonatally castrated male rats show ovulation, those transplanted to female rats bearing neonatal testis transplantation do not. Takasugi demonstrated that neonatal treatment of female rats with androgen or estrogen causes anovulation and persistent estrus, suggesting that neonatal steroids cause defeminization of the ovulatory system.[3] Barraclough and Gorski[4] provided further evidence for the action site of neonatal steroids to cause defeminized brain by showing that electrical stimulation of the preoptic area (POA) does not stimulate ovulation in neonatally androgenized female rats. They concluded that the POA is an action site for neonatal androgen to defeminize cyclic LH secretion. Gorski *et al.*[5] finally found that the POA shows a sexual dimorphism, which is determined by the neonatal steroidal conditions. Since then, a number of sexual dimorphic nuclei have been found in various brain regions.[6]

The anteroventral periventricular nucleus (AVPV) was first found to be sexually dimorphic in rodents by Bleier *et al.*,[7] who reported that the preoptic anterior hypothalamic area including the AVPV was more densely cellular in the female than in the male in four animal species examined. Simmerly *et al.* first identified the cell type showing a sex difference in the rat AVPV. They found that AVPV tyrosine hydroxylase (TH)-positive neurons have a greater number of the cells in females than in males[8] and later showed that this clear sex difference is caused by steroid-induced apoptotic cell death in the neonatal period.[9] Because the AVPV has been suggested to be a center for GnRH/LH surge,[10] the nucleus became highlighted as a target of estrogen positive feedback action to induce GnRH/LH surges, which are sexually differentiated in rodents.[7,8]

Kisspeptin, the product of *Kiss1* gene, which was discovered as a metastasis suppressor gene,[11] was first identified as an endogenous ligand of

doi: 10.1111/j.1749-6632.2010.05645.x
 © 2010 New York Academy of Sciences.

GPR54.[12–14] Accumulating evidence suggests that kisspeptin-GPR54 signaling plays a critical role in regulating HPG axis, governing the onset of puberty and ovarian cyclicity, via stimulation of GnRH/gonadotropin release in mammals.[15–26] Rodents have been reported to have two kisspeptin neuronal populations: the AVPV and hypothalamic arcuated nucleus (ARC).[27–30] Of these two populations, AVPV kisspeptin neurons show an apparent sexual dimorphism. Males have the ARC kisspeptin neuronal population but show few kisspeptin neurons in the AVPV. Interestingly, estrogen stimulates *Kiss1* and kisspeptin expressions in the AVPV in female rodents, and estrogen receptor α (ERα), a receptor mediating the estrogen positive feedback effect,[31,32] is coexpressed in the kisspeptin neuron.[27,33–35] These facts suggest that AVPV kisspeptin neuron is a target of estrogen positive feedback action to induce GnRH/LH surge. Previous studies demonstrated that neonatal androgen or estrogen in female rats causes a male-like pattern of reduced AVPV *Kiss1* expression.[29,30,36,37] These results suggest that the AVPV neurons are also a target of the estrogen organizational effects in the developing brain to cause reduced AVPV kisspeptin population and then inhibition of GnRH/LH surge mechanism in adult male rats.

Kauffman has extensively reviewed the issue of sexual differentiation of kisspeptin neurons, which might cause the sex difference of the GnRH/LH surge system.[38,39] This review focuses on the physiological significance of neonatal testicular androgen to defeminize AVPV kisspeptin neurons and then the GnRH/LH surge generation system.

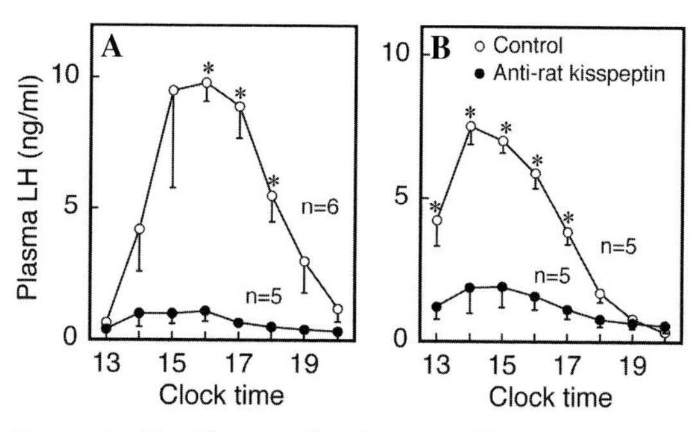

Figure 1. **Significance of endogenous kisspeptin in generating LH surge. Effects of central infusion of anti-rat kisspeptin monoclonal antibody (Takeda #156) on proestrous (A) and estrogen-induced (B) LH surge were investigated in female rats. Anti-rat kisspeptin antibody (4.68 mg/mL) was infused into the preoptic area (POA) at 1 mL/h with a microinfusion pump from 10:00 to 18:00 h. The same volume of vehicle (phosphate-buffered saline) (A) or normal mouse IgG (B) was infused into the POA of control rats. Blood samples were collected every hour at 13:00 to 20:00 through an indwelling atrial cannula from freely moving conscious rats. Values are means ± SEM. $^*P < 0.05$, significant difference between the antibody-infused rats and control rats (one-way ANOVA with repeated measures). The graphs are modified from our original papers.[18,27]**

Results/Discussion

Involvement of endogenous kisspeptin in regulating LH surge

Previous studies have reported a potent stimulatory effect of kisspeptin on gonadotropin release[20,40,41] since its discovery in 2001,[12–14] providing evidence for this peptide as a regulator of HPG axis. There also have been previous reports providing direct evidence indicating the physiological significance of kisspeptin-GPR54 signaling in the control of gonadotropin release and then gonadal functions. Knockout (KO) studies revealed that *Kiss1*- or *Gpr54*-deficient mice do not show normal reproductive functions.[25,42] A GPR54 antagonist recently developed by a group led by Miller, showed an inhibition of gonadotropin release in various animal models.[43] Another strategy is to trap endogenous local kisspeptin by infusing a specific kisspeptin antibody into the POA, where GnRH neuronal cell bodies are located.[44] Specific antibody infusion blocked spontaneous LH surge as well as the estrogen-induced surge (Fig. 1).[18,27] All of these reports suggest that endogenous kisspeptin released into the POA mediates an estrogen positive feedback effect on GnRH/LH surges in the rat, probably via a direct stimulation of GnRH neuronal activity. The notion is consistent with the fact that GPR54 mRNA is expressed in GnRH neuons,[17,45] and kisspeptin directly triggers the activity of GnRH neurons.[46]

The role of AVPV kisspeptin neurons in surge mode of GnRH/LH release is also supported by some other observations, such as the increase in AVPV kisspeptin expressions at proestrous afternoon and coexpression of ERα in the AVPV kisspeptin neurons.[27,33–35] Interestingly, Dungan *et al.* reported that *GPR54* KO mice retain the ability to exhibit positive feedback effects of estrogen on GnRH/LH secretion.[47] This discrepancy could indicate the presence of other receptor(s) for kisspeptin besides GPR54 for the GnRH surge-generating system. Further studies are required to clarify the discrepancy.

Role of testicular steroids to defeminize GnRH/LH surge-generating system and AVPV kisspeptin neurons

In the rat, the LH surge-generating system is abolished by excessive estrogen exposure during the neonatal period, with estrogens being locally produced by aromatization of testis-derived testosterone. Neonatal estrogen treatment causes persistent estrus and failure of LH surge and ovulation in female rats,[48,49] and treatment with unaromatizable androgen in neonatal female rats did not affect estrous cyclicity at adulthood.[50] In this context, sex steroids act on key components of the brain pathway during perinatal development to organize neural systems that show sexual dimorphism. This steroidal effect in the developing brain has been well established as the "organizational effect" as extensively reviewed by several investigators.[6,51,52]

Kauffman *et al.* first showed that neonatally androgenized female rats display a male-like pattern of adulthood *Kiss1* expression in the AVPV.[30] Down-regulation of kisspeptin expression in the AVPV was shown in α-fetoprotein KO female mice compared with wild-type females,[53] providing potential evidence that testicular androgen in the developing brain mediates the sexual dimorphism of AVPV kisspeptin neurons. Further evidence for neonatal testicular androgen as a determinant for defeminized AVPV kisspeptin and surge-generating system was obtained from neonatal castration after birth. Castration on the day of birth rescues the GnRH/LH surge-generating system, as well as AVPV kisspeptin neurons in genetically male rats. Neonatally castrated male rats showed a surge-like LH increase as shown in female rats when they are treated with preovulatory level of estradiol at adulthood (Fig. 2).[29,54] Thus, genetically male rats are potentially equipped with a GnRH/LH surge-generating system that is either abolished or inactivated by the neonatal estrogen converted from testicular androgen at the perinatal period.

As mentioned already, *Kiss1* and kisspeptin expressions in the AVPV were largely affected by neonatal steroid manipulations.[30] Castration in genetically male neonate rats immediately after birth rescued LH surge and AVPV kisspeptin neurons (Fig. 2).[29] To the contrary, neonatal estrogen reduced the number of AVPV kisspeptin neurons and abolished LH surge generation in female rats. Collectively, these facts suggest that neonatal testicular androgen, but not genetic factors, causes the reduction in AVPV kisspeptin expression, which may result in the absence of the LH surge in male rats (Fig. 3). The steroid action in the early postnatal rats would be an example of an organizational effect of the sex steroid by which AVPV kisspeptin signaling irreversibly attenuated later in life.

Contrary to the AVPV, no apparent sex difference was found in the ARC kisspeptin neurons in gonadectomized adult rats. Indeed, neonatal steroidal manipulation causes no obvious change in the number of ARC kisspeptin neurons in males and female rats (Fig. 2).[29,30] Thus, the ARC kisspeptin expression is largely dependent on the steroidal milieu at adulthood, and neonatal steroid may not have a large effect on the expressions. These facts suggest that the activational effect (negative feedback effect in this case) of estrogen at adulthood may not be influenced by the neonatal steroidal milieu.

Interestingly, steroidal regulation of kisspeptin signaling is different between the different kisspeptin populations and between different stages of life, including early postnatal and pubertal development as well as adulthood. A recent study by Kauffman *et al.* suggested that the mechanisms controlling pubertal activation of reproduction show a sex difference in mice, because a gonadal hormone-independent restraint on ARC kisspeptin neurons was shown in prepubertal male mice, but not in females.[55] In addition, postnatal estrogen is required for the prepubertal development of the rostral periventricular area of the third ventricle (RP3V) kisspeptin neurons in mice.[28]

Lack of sex difference in kisspeptin receptor (GPR54) expression in the rat brain

As reviewed in a number of articles, lack of GPR54 causes reproductive deficiency, implying the physiological significance of the receptor in reproduction.[15,25] Nevertheless, sexual dimorphism of the receptor expression has not been reported. Neither neonatal nor adult steroidal manipulation induces an apparent change in *Gpr54* gene expressions in discrete brain areas, such as the AVPV, ARC-median

estrogen organizational effects that defeminize the GnRH/LH surge-generating system.

Acknowledgments

The authors are grateful to Drs. Orikasa and Sakuma for the neonatal castration technique, and the National Hormone and Pituitary Program for the LH assay kit. We also thank Drs. S. Ohkura and N. Inoue for valuable suggestions, and Ms. Y. Iwata, M. Maeda, and Y. Inamoto for their technical assistance. The RIA was performed at the Nagoya University Radioisotope Center. This research was supported by grant nos. 18658106 and 19380157 to H.T. and grant no. 18208025 to K.-I.M. from the Japan Society for the Promotion of Science's Grants-in-Aid and by the Program for Promotion of Basic Research Activities for Innovative Biosciences of Japan.

Conflicts of interest

The authors declare no conflict of interest

References

1. Marshall, F.H.L. & W.A. Jolly. 1907. Results of removal and transplantation of ovaries. *Trans. Roy. Soc. Edinburgh* **45:** 589–600.
2. Pfeiffer, C.A. 1936. Sexual differences of the hypophyses and their determination by the gonads. *Am. J. Anatomy* **58:** 195–226.
3. Takasugi, N. 1952. Einflusse von androgen und oestrogen auf ovarien der neugeboren und reifen, weiblichen ratten. *Annot. Zool. Japon* **25:** 120–127.
4. Barraclough, C.A. & R.A. Gorski. 1961. Evidence that the hypothalamus is responsible for androgen-induced sterility in the female rat. *Endocrinology* **68:** 68–79.
5. Gorski, R.A., J.H. Gordon, J.E. Shryne, *et al.* 1978. Evidence for a morphological sex difference within the medial preoptic area of the rat brain. *Brain Res.* **148:** 333–346.
6. Cooke, B., C.D. Hegstrom, L.S. Villeneuve, *et al.* 1998. Sexual differentiation of the vertebrate brain: principles and mechanisms. *Front. Neuroendocrinol.* **19:** 323–362.
7. Bleier, R., W. Byne & I. Siggelkow. 1982. Cytoarchitectonic sexual dimorphisms of the medial preoptic and anterior hypothalamic areas in guinea pig, rat, hamster, and mouse. *J. Comp. Neurol.* **212:** 118–130.
8. Simerly, R.B., L.W. Swanson, R.J. Handa, *et al.* 1985. Influence of perinatal androgen on the sexually dimorphic distribution of tyrosine hydroxylase-immunoreactive cells and fibers in the anteroventral periventricular nucleus of the rat. *Neuroendocrinology* **40:** 501–510.
9. Waters, E.M. & R.B. Simerly. 2009. Estrogen induces caspase-dependent cell death during hypothalamic development. *J. Neurosci.* **29:** 9714–9718.
10. Wiegand, S.J. & E. Terasawa. 1982. Discrete lesions reveal functional heterogeneity of suprachiasmatic structures in regulation of gonadotropin secretion in the female rat. *Neuroendocrinology* **34:** 395–404.
11. Lee, J.H., M.E. Miele, D.J. Hicks, *et al.* 1996. KiSS-1, a novel human malignant melanoma metastasis-suppressor gene. *J. Natl. Cancer Inst.* **88:** 1731–1737.
12. Kotani, M., M. Detheux, A. Vandenbogaerde, *et al.* 2001. The metastasis suppressor gene KiSS-1 encodes kisspeptins, the natural ligands of the orphan G protein-coupled receptor GPR54. *J. Biol. Chem.* **276:** 34631–34636.
13. Muir, A.I., L. Chamberlain, N.A. Elshourbagy, *et al.* 2001. AXOR12, a novel human G protein-coupled receptor, activated by the peptide KiSS-1. *J. Biol. Chem.* **276:** 28969–28975.
14. Ohtaki, T., Y. Shintani, S. Honda, *et al.* 2001. Metastasis suppressor gene KiSS-1 encodes peptide ligand of a G-protein-coupled receptor. *Nature* **411:** 613–617.
15. de Roux, N., E. Genin, J.C. Carel, *et al.* 2003. Hypogonadotropic hypogonadism due to loss of function of the KiSS1-derived peptide receptor GPR54. *Proc. Natl. Acad. Sci. USA* **100:** 10972–10976.
16. Gottsch, M.L., M.J. Cunningham, J.T. Smith, *et al.* 2004. A role for kisspeptins in the regulation of gonadotropin secretion in the mouse. *Endocrinology* **145:** 4073–4077.
17. Irwig, M.S., G.S. Fraley, J.T. Smith, *et al.* 2004. Kisspeptin activation of gonadotropin releasing hormone neurons and regulation of KiSS-1 mRNA in the male rat. *Neuroendocrinology* **80:** 264–272.
18. Kinoshita, M., H. Tsukamura, S. Adachi, *et al.* 2005. Involvement of central metastin in the regulation of preovulatory luteinizing hormone surge and estrous cyclicity in female rats. *Endocrinology* **146:** 4431–4436.
19. Maeda, K., S. Adachi, K. Inoue, *et al.* 2007. Metastin/kisspeptin and control of estrous cycle in rats. *Rev. Endocr. Metab. Disord.* **8:** 21–29.
20. Matsui, H., Y. Takatsu, S. Kumano, *et al.* 2004. Peripheral administration of metastin induces marked gonadotropin release and ovulation in the rat. *Biochem. Biophys. Res. Commun.* **320:** 383–388.
21. Navarro, V.M., J.M. Castellano, R. Fernandez-Fernandez, *et al.* 2005. Effects of KiSS-1 peptide, the natural ligand of GPR54, on follicle-stimulating hormone secretion in the rat. *Endocrinology* **146:** 1689–1697.

22. Novaira, H.J., Y. Ng, A. Wolfe, *et al.* 2009. Kisspeptin increases GnRH mRNA expression and secretion in GnRH secreting neuronal cell lines. *Mol. Cell Endocrinol.* **311:** 126–134.
23. Ohkura, S., Y. Uenoyama, S. Yamada, *et al.* 2009. Physiological role of metastin/kisspeptin in regulating gonadotropin-releasing hormone (GnRH) secretion in female rats. *Peptides* **30:** 49–56.
24. Pheng, V., Y. Uenoyama, T. Homma, *et al.* 2009. Potencies of centrally- or peripherally-injected full-length kisspeptin or its C-terminal decapeptide on LH release in intact male rats. *J. Reprod. Dev.* **55:** 378–382.
25. Seminara, S.B., S. Messager, E.E. Chatzidaki, *et al.* 2003. The GPR54 gene as a regulator of puberty. *N. Engl. J. Med.* **349:** 1614–1627.
26. Takase, K., Y. Uenoyama, N. Inoue, *et al.* 2009. Possible role of oestrogen in pubertal increase of Kiss1/kisspeptin expression in discrete hypothalamic areas of female rats. *J. Neuroendocrinol.* **21:** 527–537.
27. Adachi, S., S. Yamada, Y. Takatsu, *et al.* 2007. Involvement of anteroventral periventricular metastin/kisspeptin neurons in estrogen positive feedback action on luteinizing hormone release in female rats. *J. Reprod. Dev.* **53:** 367–378.
28. Clarkson, J. & A.E. Herbison. 2006. Postnatal development of kisspeptin neurons in mouse hypothalamus; sexual dimorphism and projections to gonadotropin-releasing hormone neurons. *Endocrinology* **147:** 5817–5825.
29. Homma, T., M. Sakakibara, S. Yamada, *et al.* 2009. Significance of neonatal testicular sex steroids to defeminize anteroventral periventricular kisspeptin neurons and the GnRH/LH surge system in male rats. *Biol. Reprod.* **81:** 1216–1225.
30. Kauffman, A.S., M.L. Gottsch, J. Roa, *et al.* 2007. Sexual differentiation of Kiss1 gene expression in the brain of the rat. *Endocrinology* **148:** 1774–1783.
31. Couse, J.F., M.M. Yates, V.R. Walker, *et al.* 2003. Characterization of the hypothalamic-pituitary-gonadal axis in estrogen receptor (ER) null mice reveals hypergonadism and endocrine sex reversal in females lacking ERalpha but not ERbeta. *Mol. Endocrinol.* **17:** 1039–1053.
32. Wintermantel, T.M., R.E. Campbell, R. Porteous, *et al.* 2006. Definition of estrogen receptor pathway critical for estrogen positive feedback to gonadotropin-releasing hormone neurons and fertility. *Neuron* **52:** 271–280.
33. Clarkson, J., X. d'Anglemont de Tassigny, A.S. Moreno, *et al.* 2008. Kisspeptin-GPR54 signaling is essential for preovulatory gonadotropin-releasing hormone neuron activation and the luteinizing hormone surge. *J. Neurosci.* **28:** 8691–8697.
34. Franceschini, I., D. Lomet, M. Cateau, *et al.* 2006. Kisspeptin immunoreactive cells of the ovine preoptic area and arcuate nucleus co-express estrogen receptor alpha. *Neurosci. Lett.* **401:** 225–230.
35. Smith, J.T., M.J. Cunningham, E.F. Rissman, *et al.* 2005. Regulation of Kiss1 gene expression in the brain of the female mouse. *Endocrinology* **146:** 3686–3692.
36. Bateman, H.L. & H.B. Patisaul. 2008. Disrupted female reproductive physiology following neonatal exposure to phytoestrogens or estrogen specific ligands is associated with decreased GnRH activation and kisspeptin fiber density in the hypothalamus. *Neurotoxicology* **29:** 988–997.
37. Navarro, V.M., M.A. Sanchez-Garrido, J.M. Castellano, *et al.* 2009. Persistent impairment of hypothalamic KiSS-1 system after exposures to estrogenic compounds at critical periods of brain sex differentiation. *Endocrinology* **150:** 2359–2367.
38. Kauffman, A.S. 2009. Sexual differentiation and the Kiss1 system: hormonal and developmental considerations. *Peptides* **30:** 83–93.
39. Kauffman, A.S., D.K. Clifton & R.A. Steiner. 2007. Emerging ideas about kisspeptin- GPR54 signaling in the neuroendocrine regulation of reproduction. *Trends Neurosci.* **30:** 504–511.
40. Navarro, V.M., J.M. Castellano, R. Fernandez-Fernandez, *et al.* 2004. Developmental and hormonally regulated messenger ribonucleic acid expression of KiSS-1 and its putative receptor, GPR54, in rat hypothalamus and potent luteinizing hormone-releasing activity of KiSS-1 peptide. *Endocrinology* **145:** 4565–4574.
41. Thompson, E.L., M. Patterson, K.G. Murphy, *et al.* 2004. Central and peripheral administration of kisspeptin-10 stimulates the hypothalamic-pituitary-gonadal axis. *J. Neuroendocrinol.* **16:** 850–858.
42. d'Anglemont de Tassigny, X., L.A. Fagg, J.P. Dixon, *et al.* 2007. Hypogonadotropic hypogonadism in mice lacking a functional Kiss1 gene. *Proc. Natl. Acad. Sci. USA* **104:** 10714–10719.
43. Roseweir, A.K., A.S. Kauffman, J.T. Smith, *et al.* 2009. Discovery of potent kisspeptin antagonists delineate physiological mechanisms of gonadotropin regulation. *J. Neurosci.* **29:** 3920–3929.
44. Kawano, H. & S. Daikoku. 1981. Immunohistochemical demonstration of LHRH neurons and their pathways in the rat hypothalamus. *Neuroendocrinology* **32:** 179–186.
45. Messager, S., E.E. Chatzidaki, D. Ma, *et al.* 2005. Kisspeptin directly stimulates gonadotropin-releasing

hormone release via G protein-coupled receptor 54. *Proc. Natl. Acad. Sci. USA* **102:** 1761–1766.
46. Han, S.K., M.L. Gottsch, K.J. Lee, *et al.* 2005. Activation of gonadotropin-releasing hormone neurons by kisspeptin as a neuroendocrine switch for the onset of puberty. *J. Neurosci.* **25:** 11349–11356.
47. Dungan, H.M., M.L. Gottsch, H. Zeng, *et al.* 2007. The role of kisspeptin-GPR54 signaling in the tonic regulation and surge release of gonadotropin-releasing hormone/luteinizing hormone. *J. Neurosci.* **27:** 12088–12095.
48. Aihara, M. & S. Hayashi. 1989. Induction of persistent diestrus followed by persistent estrus is indicative of delayed maturation of tonic gonadotropin-releasing systems in rats. *Biol. Reprod.* **40:** 96–101.
49. Gorski, R.A. 1963. Modification of ovulatory mechanisms by postnatal administration of estrogen to the rat. *Am. J. Physiol.* **205:** 842–844.
50. William, G.L. & E.W. Richard. 1970. Dihydrotestosterone, androstenedione, testosterone: comparative effectiveness in masculinizing and defeminizing reproductive systems in male and female rats. *Horm. Behav.* **1:** 265–281.
51. McCarthy, M.M. 2008. Estradiol and the developing brain. *Physiol. Rev.* **88:** 91–124.
52. Simerly, R.B. 2002. Wired for reproduction: organization and development of sexually dimorphic circuits in the mammalian forebrain. *Annu. Rev. Neurosci.* **25:** 507–536.
53. Gonzalez-Martinez, D., C. De Mees, Q. Douhard, *et al.* 2008. Absence of gonadotropin-releasing hormone 1 and Kiss1 activation in alpha-fetoprotein knockout mice: prenatal estrogens defeminize the potential to show preovulatory luteinizing hormone surges. *Endocrinology* **149:** 2333–2340.
54. Corbier, P. 1985. Sexual differentiation of positive feedback: effect of hour of castration at birth on estradiol-induced luteinizing hormone secretion in immature male rats. *Endocrinology* **116:** 142–147.
55. Kauffman, A.S., V.M. Navarro, J. Kim, *et al.* 2009. Sex differences in the regulation of Kiss1/NKB neurons in juvenile mice: implications for the timing of puberty. *Am. J. Physiol. Endocrinol. Metab.* **297:** E1212–E1221.
56. Olster, D.H. & J.D. Blaustein. 1991. Progesterone facilitates lordosis, but not LH release, in estradiol pulse-primed male rats. *Physiol. Behav.* **50:** 237–242.
57. Elsaesser, F. & N. Parvizi. 1979. Estrogen feedback in the pig: sexual differentiation and the effect of prenatal testosterone treatment. *Biol. Reprod.* **20:** 1187–1193.
58. Karsch, F.J. & D.L. Foster. 1975. Sexual differentiation of the mechanism controlling the preovulatory discharge of luteinizing hormone in sheep. *Endocrinology* **97:** 373–379.
59. Karsch, F.J., D.J. Dierschke & E. Knobil. 1973. Sexual differentiation of pituitary function: apparent difference between primates and rodents. *Science* **179:** 484–486.
60. Kineman, R.D., L.S. Leshin, J.W. Crim, *et al.* 1988. Localization of luteinizing hormone-releasing hormone in the forebrain of the pig. *Biol. Reprod.* **39:** 665–672.
61. Lehman, M.N., J.E. Robinson, F.J. Karsch, *et al.* 1986. Immunocytochemical localization of luteinizing hormone-releasing hormone (LHRH) pathways in the sheep brain during anestrus and the mid-luteal phase of the estrous cycle. *J. Comp. Neurol.* **244:** 19–35.
62. Silverman, A.J., R. Silverman, M.N. Lehman, *et al.* 1987. Localization of a peptide sequence contained in the precursor to gonadotropin releasing hormone (GnRH). *Brain Res.* **402:** 346–350.
63. Standish, L.J., L.A. Adams, L. Vician, *et al.* 1987. Neuroanatomical localization of cells containing gonadotropin-releasing hormone messenger ribonucleic acid in the primate brain by in situ hybridization histochemistry. *Mol. Endocrinol.* **1:** 371–376.
64. King, J.C., E.L. Anthony, D.M. Fitzgerald, *et al.* 1985. Luteinizing hormone-releasing hormone neurons in human preoptic/hypothalamus: differential intraneuronal localization of immunoreactive forms. *J. Clin. Endocrinol. Metab.* **60:** 88–97.
65. Goodman, R.L. 1978. The site of the positive feedback action of estradiol in the rat. *Endocrinology* **102:** 151–159.
66. Knobil, E. 1980. The neuroendocrine control of the menstrual cycle. *Recent Prog. Horm. Res.* **36:** 53–88.
67. Spies, H.G. & R.L. Norman. 1975. Interaction of estradiol and LHRH on LH release in rhesus females: evidence for a neural site of action. *Endocrinology* **97:** 685–692.
68. Caraty, A., C. Fabre-Nys, B. Delaleu, *et al.* 1998. Evidence that the mediobasal hypothalamus is the primary site of action of estradiol in inducing the preovulatory gonadotropin releasing hormone surge in the ewe. *Endocrinology* **139:** 1752–1760.
69. Smith, J.T., Q. Li, A. Pereira, *et al.* 2009. Kisspeptin neurons in the ovine arcuate nucleus and preoptic area are involved in the preovulatory luteinizing hormone surge. *Endocrinology* **150:** 5530–5538.
70. Tomikawa, J., T. Homma, S. Tajima, *et al.* 2010. Molecular characterization and estrogen regulation of hypothalamic KISS1 gene in the pig. *Biol. Reprod.* **82:** 313–319.
71. Estrada, K.M., C.M. Clay, S. Pompolo, *et al.* 2006. Elevated KiSS-1 expression in the arcuate nucleus prior

to the cyclic preovulatory gonadotrophin-releasing hormone/lutenising hormone surge in the ewe suggests a stimulatory role for kisspeptin in oestrogen-positive feedback. *J. Neuroendocrinol.* **18:** 806–809.

72. Ramaswamy, S., K.A. Guerriero, R.B. Gibbs, *et al.* 2008. Structural interactions between kisspeptin and GnRH neurons in the mediobasal hypothalamus of the male rhesus monkey (Macaca mulatta) as revealed by double immunofluorescence and confocal microscopy. *Endocrinology* **149:** 4387–4395.
73. Bocklandt, S. & E. Vilain. 2007. Sex differences in brain and behavior: hormones versus genes. *Adv. Genet.* **59:** 245–266.
74. Gorski, R.A. 1979. The neuroendocrinology of reproduction: an overview. *Biol. Reprod.* **20:** 111–127.

Ann. N.Y. Acad. Sci. ISSN 0077-8923

ANNALS OF THE NEW YORK ACADEMY OF SCIENCES
Issue: *Phylogenetic Aspects of Neuropeptides*

Neuromedin U is necessary for normal gastrointestinal motility and is regulated by serotonin

Yoshiki Nakashima,[1,2] Takanori Ida,[1] Takahiro Sato,[1] Yuki Nakamura,[1] Tomoko Takahashi,[1] Kenji Mori,[3] Mikiya Miyazato,[3] Kenji Kangawa,[3] Jingo Kusukawa,[2] and Masayasu Kojima[1]

[1]Molecular Genetics, Institute of Life Science, Kurume University, Kurume, Fukuoka, Japan. [2]Dental and Oral Medical Center, Kurume University School of Medicine, Kurume, Fukuoka, Japan. [3]Department of Biochemistry, National Cardiovascular Center Research Institute, Suita, Osaka, Japan

Address for correspondence: Masayasu Kojima, M.D., Ph.D., Molecular Genetics, Institute of Life Science, Kurume University, Hyakunenkohen 1-1, Kurume, Fukuoka 839-0864, Japan. mkojima@lsi.kurume-u.ac.jp

Neuromedin U (NMU) was originally isolated from porcine spinal cord and shown to be distributed in numerous tissues, including the gastrointestinal tract. However, little is known about the role of NMU in the regulation of gastrointestinal functions. We established a radioimmunoassay system that is exceptionally specific for mouse NMU and found high NMU content in the gastrointestinal tract, particularly in the Auerbach's and Meissner's plexi, suggesting a possible role of NMU in gastrointestinal motility. NMU promoted small intestinal transit, and NMU deficiency resulted in lowered intestinal motility rate and diminished the effect of serotonin-induced defecation and diarrhea. These results indicate that NMU promotes intestinal transit and maintains intestinal homeostasis.

Keywords: neuromedin U; serotonin; acetylcholine; intestinal transit

Introduction

The peptide neuromedin U (NMU), named for the strong contractile activity that it evokes in the smooth muscle of the uterus, was first isolated from porcine spinal cord, but is found in many other tissues, including the gastrointestinal tract, pituitary, and hypothalamus.[1–3] Two subtypes of the NMU receptor (NMUR) have been identified: NMU1R (FM-3) and NMU2R (FM-4[4–6]). While NMU1R is abundantly expressed in peripheral tissues, such as the small and large intestines, skin, spinal cord, and pituitary gland, the expression of NMU2R is restricted to specific regions in the hypothalamic paraventricular nucleus of the central nervous system.[4] A variety of NMU functions have been revealed: NMU modulates pain sensation in the spinal cord,[7,8] induces inflammation in the skin,[9,10] induces loss of bone density by stimulating the sympathetic nervous system,[11] suppresses food intake, and stimulates energy expenditure.[4,5,12]

Distribution analysis of NMU in rat and porcine tissues indicates that NMU is expressed abundantly in gastrointestinal tissues, particularly in the jejunum, duodenum, and ileum.[13] NMU is exclusively localized to nerve fibers, mainly in the myenteric and submucosal plexi of the rat.[3] It is generally accepted that the enteric nervous system plays a major role in the complex regulation of intestinal motility, an integrated process that includes myoelectrical function, contractility, tone, compliance, and transit.[14] In addition, it has been reported that NMU induces gastrointestinal contractile activity through the NMU1R and regulates ion transport in the gastrointestinal mucosa.[15] The specific role of NMU in the control of gastrointestinal function, and the relationships between NMU and regulators of gastrointestinal function such as serotonin and acetylcholine, have not been elucidated. To determine whether NMU is essential for normal gastrointestinal motility, we created NMU-deficient mice and examined their gastrointestinal function.

doi: 10.1111/j.1749-6632.2010.05504.x

Materials and methods

Animals

Generation of $NMU^{-/-}$ mice was previously described in detail.[12] $NMU^{-/-}$ mice and C57BL/6J male mice (9–16 weeks old; Japan SLC, Shizuoka, Japan) were maintained under controlled temperature (23 ± 2° C) and light conditions (lights on from 0700-1900). Animals were fed standard rodent chow pellets with *ad libitum* water intake. All procedures were conducted in compliance with protocols approved by the Ethical Committee for the Research of Life Science of Kurume University.

Preparation of anti-mouse NMU serum

[Cys17]-mouse NMU [1–16] peptide was synthesized by the Fmoc solid-phase method on a peptide synthesizer (433A; Applied Biosystems, Foster City, CA). [Cys17]-mouse NMU [1–16] (2 mg) was conjugated to 2 mg of maleimide-activated mariculture keyhole limpet hemocyanin (mcKLH, Pierce, Rockfold, IL) in conjugation buffer. Conjugate was emulsified with an equal volume of Freund's complete adjuvant (Sigma-Aldrich, St. Louis, MO). Corresponding batches of antiserum were obtained from immunization of New Zealand white rabbits by subcutaneous injection.

Radioiodination of mouse NMU

Mouse NMU [1–23] was synthesized and used as a tracer ligand for anti-mouse NMU. The ligand was radioiodinated by the lactoperoxidase method. After radioiodination, the monoiodinated ligand was purified by RP-HPLC on a Symmetry C18 column (3.9 × 150 mm, Waters, Milford, MA). The radioiodinated tracer was stored at 4° C in the presence of 0.1% bovine serum albumin (BSA). The tracer was stable for 2 months under these conditions.

Preparation of tissue samples and RIA for mouse NMU

To prepare tissue samples, tissue was quickly removed after the mice were sacrificed. Each tissue was diced and boiled for 5 min in a 10-fold volume of water to inactivate intrinsic proteases. The solutions were adjusted to a final concentration of 1 M AcOH after cooling. Tissues were then homogenized with a polytron mixer and centrifuged. The supernatants were obtained as tissue samples. All assay procedures were performed in duplicate at 4° C. The RIA incubation mixture was composed of 100 μL standard NMU or unknown samples, and 100 μL of anti-mouse NMU antiserum diluted with RIA buffer (50 mmol/L sodium phosphate buffer (pH 7.4), 0.5% BSA, 0.5% Triton-X 100, 80 mmol/L NaCl, 25 mmol/L EDTA-2Na, and 0.05% NaN_3) containing 0.5% normal rabbit serum. The anti-mouse NMU antiserum was used at a final dilution of 1/80,000. After 12 h incubation, 100 μL of ^{125}I-labeled tracer (18,000 cpm) was added. After an additional 24 h incubation, 500 μL of 1% γ-globulin and 16% polyethylene glycol were added and incubated for 15 min. Free and bound tracers were separated by centrifugation at 3,000 rpm for 20 min. After aspiration of supernatant, radioactivity in the pellet was assessed with a gamma counter (ARC-1000M, Aloka, Tokyo, Japan).

Immunohistochemistry

The mouse gastrointestinal tract was immersed in 4% paraformaldehyde/PBS solution, then immersed in a 30% sucrose solution. Tissues were then embedded in OCT compound (Tissue-Tek Miles, Elkhart, IN). Sections were cut at a thickness of 10 μm using a cryostat (CM3050S; Leica Microscopy and Scientific Instruments Group, Heerburgg, Switzerland) and mounted on Matsunami adhesive MAS-coated slides (Matsunami, Osaka, Japan). Immunohistochemical staining for NMU was performed by the avidin-biotinylated-peroxidase complex (ABC) system using a VECTASTAIN ABC-PO kit (Vector Laboratories Inc., Burlingame, CA) as previously described.[16] Briefly, sections were pretreated with 3% hydrogen peroxide in methanol for 5 min to inactivate endogenous peroxidase activity. After rinsing with PBS, sections were treated for 30 min with 1.5% normal goat serum, then incubated with polyclonal rabbit anti-mouse NMU serum (diluted to 1:1000) for 16 h at 4° C. Sections were rinsed in PBS and incubated with biotinylated anti-rabbit IgG for 40 min. After rinsing in PBS, sections were incubated with avidin-biotinylated regents for 1 h. Sections were visualized with DAB solution (DAKO, Kyoto, Japan). The sections were counterstained with Mayer's haematoxylin for 3 min.

Measurement of gastrointestinal movement

Marker solution (300 μL) was administered to mice perorally through a stainless tube after 16 h fasting.

The marker solution contained 2.5% carboxymethyl cellulose sodium salt (Wako, Osaka, Japan) and Evans blue (Nacalai Tesque, Kyoto, Japan). We measured intestinal transport rate at 10 min after administration of the marker solution, because after 10 min the marker solution entered into large intestine and it was difficult to measure correctly the distance travelled by the marker. Ten minutes after administration of the marker solution, mice were sacrificed by cervical dislocation and the gastrointestinal tract was removed immediately. We measured the length of the small intestine (from the pylorus to the ileum orifice) and the distance travelled by the marker over 10 min. The index of transit (%) was calculated as the ratio of the distance passed by the marker to the total length of the intestine. The anti-motility agents ramosetron hydrochloride (5-HT_3 receptor antagonist), 100 μg/kg BW (Astellas, Tokyo, Japan), RS 23597-190 hydrochloride (5-HT_4 receptor antagonist), 10 mg/kg BW (TOCRIS, Bristol, UK) and atropine sulfate (muscarinic acetylcholine receptor antagonist), 1 mg/kg BW (Mitsubishi Tanabe Pharma, Osaka, Japan), were dissolved in physiological saline and administered to experimental animals (8 mice per group). The prokinetic drugs carbamylcholine chloride (muscarinic acetylcholine receptor agonist) 100 μg/kg BW (Nacalai Tesque, Kyoto, Japan) and NMU were dissolved in physiological saline and administered to experimental animals (10 $NMU^{-/-}$ and 10 C57BL/6J mice per group). To investigate the effect of NMU on serotonin-induced defecation, we recorded the first evacuation time and counted the excreted fecal number for 12 min after subcutaneous injection of 10 mg/kg BW serotonin (Sigma, St. Louis, MO). In all experiments assessing gastrointestinal motility, mice were kept on wire mesh to prevent the intake of fecal matter and floor sawdust.

Real-time PCR analysis of NMU mRNA contents

To examine the effect of serotonin on NMU mRNA expression, mice were subcutaneously injected with 10 mg/kg BW of serotonin. Total RNA was extracted from the mouse colon using TRIzol reagent (Invitrogen, Carlsbad, CA) according to the manufacturer's instructions. Single-stranded cDNA was synthesized using a Superscript preamplification system (Invitrogen, Carlsbad, CA). Real-time quantitative PCR was performed on an ABI PRISM 7000 Sequence Detection System (PE Applied Biosystems, Foster City, CA) using a SYBR Green PCR Kit (QIAGEN, Tokyo, Japan). Primers used were as follows: mouse NMU sense, 5′-gtcctctgttgtgcatccgtt-3′, mouse NMU antisense, 5′-gcgtggcctgaataaaaagta3′; mouse RPS18 sense, 5′-ttctggccaacggtctagacaac-3′; mouse RPS18 antisense, ccagtggtcttggtgtgctga. Each sample was analyzed in duplicate in addition to standards and no-template controls. Reactions contained 500 ng cDNA, 0.15 μM primers, and 25 μL 2 × QuantiTect SYBR Green PCR Master Mix (QIAGEN, Tokyo, Japan) in a final volume of 50 μL. After an initial 15 min at 95° C to activate the HotStarTaq DNA polymerase, PCR fragments were amplified by 40 cycles of denaturation for 30 sec at 94° C, annealing for 30 sec at 60° C, and extension for 60 sec at 72° C. The relative mRNA levels were standardized to a housekeeping gene, glyceraldehyde-3-phosphate dehydrogenase.

Statistical analysis

Results are presented as means ± S.E.M. for each group. Experimental means were compared by one-way ANOVA followed by Tukey's multiple range test. The criterion for statistical significance was $P < 0.05$ for all tests. We used the statistical software program GraphPad PRISM (GraphPad software, CA) for all analyses.

Results

Characterization of mouse NMU antiserum and the distribution of immunoreactive (ir)-NMU in mouse tissues

To investigate the distribution of NMU in mouse tissues, we obtained rabbit antiserum specific for mouse NMU by immunization with synthetic mouse NMU [1–23], the portion that showed no homologous sequence with NMU in other species. This antiserum did not cross-react with rat NMU (Fig. 1A) or NMS, a neuropeptide homologous to NMU (data not shown). Half-maximal binding of radioiodinated mouse NMU to the anti-mouse NMU [1–23] antiserum was inhibited by mouse NMU at 9.8 fmol/tube (Fig. 1A). These results indicate that this antiserum specifically recognizes the mouse NMU.

We investigated the distribution and concentration of NMU in mouse tissues using the mouse NMU radioimmunoassay system. The regional

A

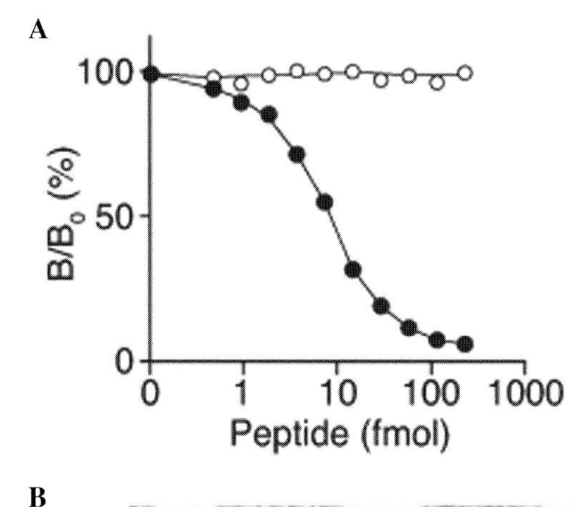

B

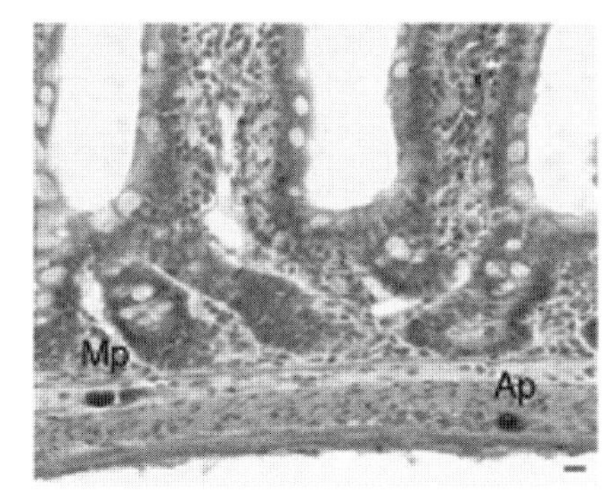

C

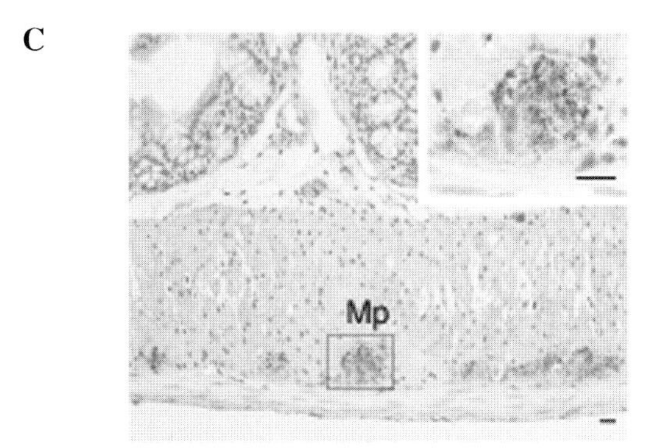

Figure 1. Specificity of mouse NMU antibody and immunohistochemistry of mouse NMU in intestines. (A) Standard NMU RIA curves and cross-reactivity of antiserum. Inhibition of ^{125}I-labeled mouse NMU binding to anti-mouse NMU [1–16] serum by serial dilution of mouse NMU (closed circle) and rat NMU (open circle). (B) and (C) Immunohistochemical detection of NMU in jejunum (B) and colon (C). Mp: Meissner's plexus, Ap: Auerbach's plexus. Part (C, *top right*) is magnification of the rectangular Meissner's plexus region outlined at *bottom*. Bars = 20 μm.

Table 1. Regional distribution of ir-NMU in mice

Region	ir-mouse NMU (fmol/mg wet tissue)
Hypothalamus	1.2 ± 0.1
Medulla oblongata	3.9 ± 0.4
Spinal cord	4.3 ± 0.4
Pituitary gland	1.9 ± 0.1
Tongue	0.9 ± 0.1
Stomach	5.1 ± 0.6
Duodenum	64.3 ± 8.5
Jejunum	79.3 ± 4.6
Ileum	42.0 ± 5.4
Cecum	58.0 ± 7.8
Colon	83.8 ± 16.1
Rectum	56.3 ± 3.9
Ovary	0.7 ± 0.2
Skin	0.7 ± 0.1

Data are means ± S.E.M. ($n = 3$).

distribution of NMU in various mouse tissues is summarized in Table 1. Significant amounts of mouse NMU were distributed in the hypothalamus, medulla oblongata, spinal cord, pituitary gland, and gastrointestinal tract. In particular, high levels of NMU were present in the gastrointestinal tract, including the duodenum, jejunum, ileum, caecum, colon, and rectum (Table 1). In addition, NMU in the gastrointestinal tissues does not localize in endocrine cells but localizes in the Auerbach's and Meissner's plexi (Fig. 1B and 1C). These results indicate that NMU may regulate intestinal transit and bowel fluid secretion by modulating intestinal nervous activity.

NMU stimulates small intestinal transit in wild-type mice

Fasted wild-type (WT) mice were intraperitoneally administered with NMU at doses of 0, 0.05, 0.1, 0.2, and 1 nmol. Ten minutes after NMU administration, the mice were administered with intragastric marker solution. Ten minutes later, the mice were sacrificed for measurement of small intestinal transit (SIT). We found that the intestinal transport rate was increased by NMU administration at concentrations of 0.1 and 0.2 nmol. However, the highest dose of NMU (1.0 nmol) did not change the transport rate (Fig. 2A).

A

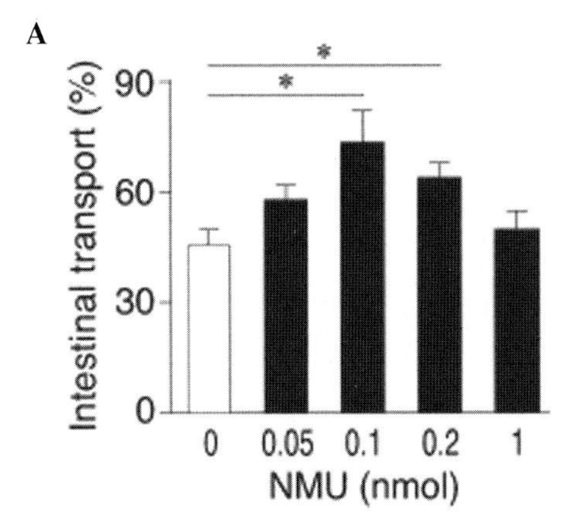

B

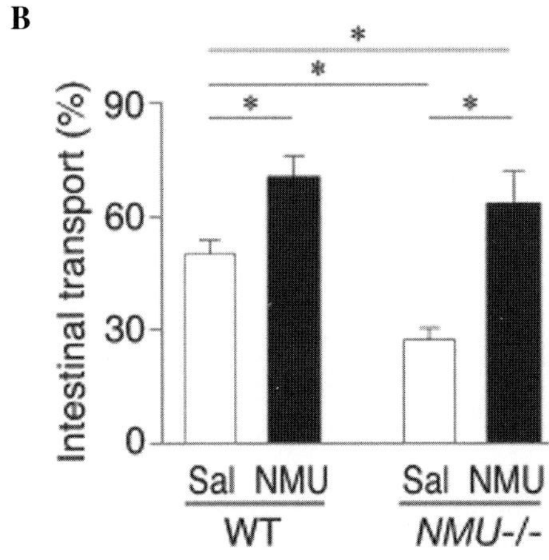

Figure 2. Effects of NMU on intestinal transit. (A) Intestinal transport rate (%) after NMU injection (0 to 1.0 nmol) in WT mice. This index was calculated as the ratio of length filled with marker to the total intestinal length (from pylorus to the ileum orifice). (B) Intestinal transport rate (%) in WT and $NMU^{-/-}$ mice. Mice in both groups were injected with saline and NMU (0.2 nmol). Sal: saline. Error bars indicate S.E.M. $^{*}P < 0.05$.

Small intestinal transit in NMU deficient mice

We observed no macroscopic abnormality of the gastrointestinal tract in NMU deficient ($NMU^{-/-}$) mice. Groups of fasted WT and $NMU^{-/-}$ mice were intraperitoneally administered with NMU at the dose of 0.2 nmol. Ten minutes after NMU administration, intragastric marker solution was given to both groups. The mice were sacrificed 10 min after marker solution injection for measurement of SIT. The intestinal transport rate of saline-injected $NMU^{-/-}$ mice was low compared with that of saline injected WT mice (intestinal transport rate; WT: 44.9 ± 2.3%, $NMU^{-/-}$: 24.4 ± 2.2%) (Fig. 2B). Furthermore, NMU (0.2 nmol) administration clearly induced bowel peristalsis in both WT and $NMU^{-/-}$ mice (intestinal transport rate after NMU administration; WT: 63.1 ± 4.1%, $NMU^{-/-}$: 56.0 ± 5.9%) (Fig. 2B). $NMU^{-/-}$ mice showed an impairment of intestinal transport ability and increased tendency to peristalsis. Thus, peripheral NMU treatment improves the functional loss of intestinal transit in $NMU^{-/-}$ mice.

Effects of NMU on serotonin-induced defecation and diarrhea

Many peptides are expressed in the gastrointestinal tract for maintenance of functions such as gastrointestinal movement and exocrine secretion. Serotonin is an important gastrointestinal monoamine neurotransmitter that regulates intestinal transit. Because NMU has a powerful contractile effect on gastrointestinal smooth muscle, we investigated the effects of NMU on serotonin-induced defecation and diarrhea using $NMU^{-/-}$ mice. After subcutaneous injection of serotonin, the evacuation time of both normal and soft feces were delayed in $NMU^{-/-}$ mice relative to WT mice (Fig. 3A). In addition, the feces number of $NMU^{-/-}$ mice was approximately two thirds that of WT mice (WT: 7.9 ± 2.3, $NMU^{-/-}$: 5.1 ± 2.0) (Fig. 3B). These results indicate that NMU is necessary for the induction of normal gastrointestinal movement and evacuation by serotonin.

Serotonin increases NMU mRNA expression in colon

We next investigated NMU expression in the colon at 30 min after serotonin injection in WT mice, and found that NMU mRNA expression was increased approximately two times higher than after saline injection (saline injection: 100.0 ± 37.4%, serotonin injection: 228.5 ± 42.1%) (Fig. 3C).

Effect of a 5-hydroxytryptamine (5-HT)$_3$/5-HT$_4$ receptor antagonist on small intestinal transit

As shown above, NMU stimulates small intestinal transit. We next examined the effects of a 5-HT$_3$/5-HT$_4$ antagonist on intestinal transport rate. Fasted WT mice were intraperitoneally administered with a mixture of 5-HT$_3$ receptor antagonist and 5-HT$_4$ receptor antagonist. Sixty minutes after the injection of the 5-HT$_3$/5-HT$_4$ antagonist mixture,

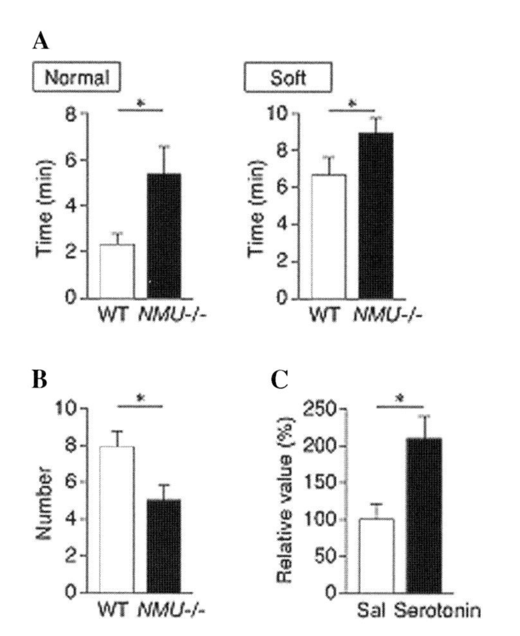

Figure 3. Effects of NMU on serotonin-induced defecation and diarrhea. Evacuation times of normal and soft feces (A) and feces number after serotonin injection (B) in WT and $NMU^{-/-}$ mice. (C) NMU mRNA expression in the colon of WT mice after serotonin injection. Sal: saline. The error bars indicate S.E.M. $^*P < 0.05$.

NMU was administered to the mice. After NMU administration, the mice were given intragastric marker solution and were sacrificed 10 min later for measurement of SIT as before. We found that SIT was decreased in the mice pretreated with a mixture of 5-HT_3/5-HT_4 receptor antagonists, in comparison with saline pretreated control mice (intestinal transport rate; saline pretreatment: 38.7 ± 1.8%, pretreatment with mixture of 5-HT_3/5-HT_4 receptor antagonists: 31.5 ± 2.0%) (Fig. 4A). However, pretreatment with the 5-HT_3/5-HT_4 receptor antagonist mixture had no effects on SIT induced by NMU injection (intestinal transport rate; saline pretreatment: 60.5 ± 2.4%, pretreatment with mixture of 5-HT_3/5-HT_4 receptor antagonists : 62.8 ± 2.4%) (Fig. 4A).

Effect of a muscarinic acetylcholine receptor antagonist on small intestinal transit

We next examined the effect of atropine sulfate, a muscarinic acetylcholine receptor antagonist. Twenty minutes after atropine sulfate administration, NMU was administered. The mice were then given intragastric marker solution and sacrificed

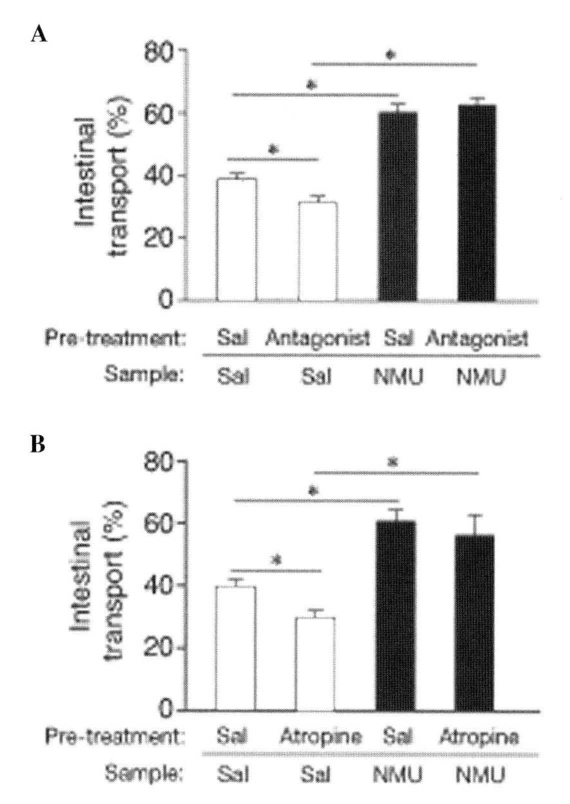

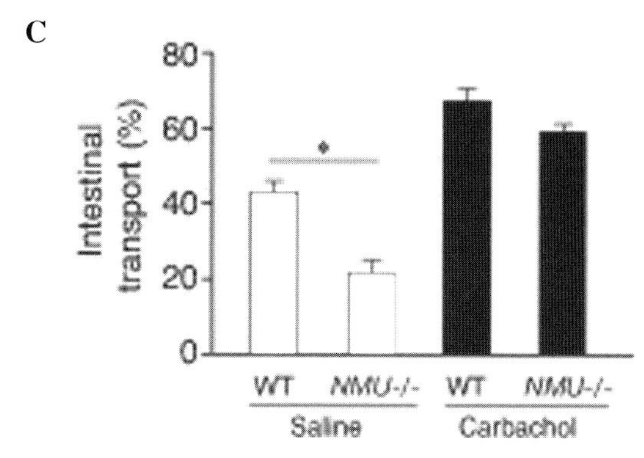

Figure 4. Effects of serotonin antagonist and muscarinic acetylcholine receptor antagonist and agonist on intestinal transit. (A) Effect of pretreatment with serotonin antagonist (mixture of 100 μg/kg Ramosetron hydrochloride and 10.0 mg/kg RS 23597-190) on intestinal transport rate. (B) Effect of pretreatment with muscarinic antagonist atropine sulfate (1 mg/kg BW) on intestinal transport rate. (C) Effect of pretreatment with muscarinic agonist carbamylcholine chloride (0.1 mg/kg) on intestinal transport rate (%) in WT and $NMU^{-/-}$ mice. Sal: saline. The error bars indicate S.E.M. $^*P < 0.05$. The abbreviations used are: NMU, neuromedin U; SIT, small intestinal transit; WT, wild-type.

10 min later for measurement of SIT. We found that SIT was decreased in the mice pretreated with atropine sulfate, in comparison with saline pretreated controls (intestinal transport rate; saline pretreatment: 39.8 ± 2.3%, pretreatment with atropine sulfate: 29.8 ± 2.4%) (Fig. 4B). Moreover, pretreatment with atropine sulfate did not affect NMU-induced SIT (intestinal transport rate; saline pretreatment: 60.8 ± 3.7%, pretreatment with atropine sulfate: 56.2 ± 6.6%) (Fig. 4B).

Effect of carbamylcholine chloride on small intestinal transit

We next examined the effect of carbamylcholine chloride, a muscarinic acetylcholine receptor agonist. Ten minutes after carbamylcholine chloride administration mice were given intragastric marker solution and sacrificed 10 min later for measurement of SIT. We obtained the same results as in Figure 2B: that SIT was retarded in $NMU^{-/-}$ mice in comparison to WT mice (intestinal transport rate; WT: 42.8 ± 3.0%, $NMU^{-/-}$: 21.5 ± 6.3%) (Fig. 4C). Both $NMU^{-/-}$ and WT mice with carbamylcholine chloride treatment showed increased SIT when compared to vehicle control mice; however, carbamylcholine chloride administration induced the same level of bowel peristalsis in both WT and $NMU^{-/-}$ mice (intestinal transport rate after carbamylcholine chloride administration; WT: 67.5 ± 3.3%, $NMU^{-/-}$: 59.2 ± 2.0%) (Fig. 4C). Thus, carbamylcholine chloride treatment improves the loss of small intestinal transit function in $NMU^{-/-}$ mice.

Discussion

In this study, we first examined NMU distribution in mouse gastrointestinal tissues using an antibody specific to mouse NMU. We confirmed that mouse NMU immunoreactivity was localized exclusively to nerve fibers in the myenteric and submucous plexi, distribution patterns that are similar to those reported in earlier studies in the rat and pig.[3,13] These results suggest that NMU exerts its contractile activity through the enteric nervous system.

To examine the physiological functions of NMU, we created NMU-deficient mice and examined their endocrine, metabolic, and immune phenotypes. We previously reported that NMU deficiency resulted in suppression of energy expenditure and led to obesity.[12] Moreover, we observed early onset of puberty and inflammatory abnormality in $NMU^{-/-}$ mice.[10,17] Because NMU is most abundant in gastrointestinal tissues, we investigated phenotypes concerning gastrointestinal function and found that $NMU^{-/-}$ mice showed impaired intestinal motility, measured as a lower intestinal transport rate. Taken together with the result that NMU administration increases intestinal transport rate, these results suggest that NMU regulates normal gastrointestinal movement.

Exogenous 5-HT is known to evoke intestinal irritations such as diarrhea and motility disorders.[18] We next examined the relationship between NMU and serotonin in the regulation of gastrointestinal movement. In $NMU^{-/-}$ mice the initiation time to display defecation and diarrhea after serotonin injection was prolonged and the number of feces was decreased. Moreover, serotonin administration increased the level of NMU mRNA expression in gastrointestinal tissues. These results indicate that the gastrointestinal actions of serotonin are mediated in part through NMU.

Both serotonin and acetylcholine are important regulators of gastrointestinal motility and intestinal electrolyte transport.[19] We next examined whether NMU regulated gastrointestinal movement independently of acetylcholine action. Pretreatment with the muscarinic acetylcholine receptor antagonist, atropine, decreased the intestinal transport rate. However, pretreatment with atropine had no effects on NMU-induced stimulation of the intestinal transport rate. Moreover, although the intestinal transport rate in $NMU^{-/-}$ mice was decreased in comparison to that of WT mice, administration of the muscarinic acetylcholine receptor agonist carbachol had similar stimulatory effects on intestinal transport rate in both WT and $NMU^{-/-}$ mice. These results suggest that NMU and acetylcholine independently regulate gastrointestinal motility.

In summary, using $NMU^{-/-}$ mice we have demonstrated that NMU is an important regulator of normal gastrointestinal motility. We have also shown that serotonin augments the expression of NMU in the colon, and that the gastrointestinal actions of serotonin are mediated in part through NMU. These results suggest NMU as a potential target for the treatment of gastrointestinal motility disorders.

Acknowledgments

We thank K. Shirouzu, M. Naito, K. Tsuchiyama, Y. Yamashita, and S. Koga for their technical assistance. This study was supported by Grants-in-Aid for Scientific Research (B) from the Ministry of Education, Culture, Sports, Science and Technology of Japan, Research on Measures for Intractable Diseases from the Health and Labour Sciences Research Grants, and by the Program for Promotion of Basic and Applied Research for Innovations in Bio-oriented Industry (BRAIN).

Conflicts of interest

The authors declare no conflicts of interest.

References

1. Minamino, N., K. Kangawa & H. Matsuo. 1985. Neuromedin U-8 and U-25: novel uterus stimulating and hypertensive peptides identified in porcine spinal cord. *Biochem. Biophys. Res. Commun.* **130:** 1078–1085.
2. Domin, J. *et al.* 1986. Characterization of neuromedin U like immunoreactivity in rat, porcine, guinea-pig and human tissue extracts using a specific radioimmunoassay. *Biochem. Biophys. Res. Commun.* **140:** 1127–1134.
3. Ballesta, J. *et al.* 1988. Occurrence and developmental pattern of neuromedin U-immunoreactive nerves in the gastrointestinal tract and brain of the rat. *Neuroscience* **25:** 797–816.
4. Howard, A.D. *et al.* 2000. Identification of receptors for neuromedin U and its role in feeding. *Nature* **406:** 70–74.
5. Kojima, M. *et al.* 2000. Purification and identification of neuromedin U as an endogenous ligand for an orphan receptor GPR66 (FM3). *Biochem. Biophys. Res. Commun.* **276:** 435–438.
6. Fujii, R. *et al.* 2000. Identification of neuromedin U as the cognate ligand of the orphan G protein-coupled receptor FM-3. *J. Biol. Chem.* **275:** 21068–21074.
7. Yu, X.H. *et al.* 2003. Pro-nociceptive effects of neuromedin U in rat. *Neuroscience* **120:** 467–474.
8. Nakahara, K. *et al.* 2004. Neuromedin U is involved in nociceptive reflexes and adaptation to environmental stimuli in mice. *Biochem. Biophys. Res. Commun.* **323:** 615–620.
9. Moriyama, M. *et al.* 2006. The neuropeptide neuromedin U activates eosinophils and is involved in allergen-induced eosinophilia. *Am. J. Physiol. Lung. Cell. Mol. Physiol.* **290:** L971–L977.
10. Moriyama, M. *et al.* 2005. The neuropeptide neuromedin U promotes inflammation by direct activation of mast cells. *J. Exp. Med.* **202:** 217–224.
11. Sato, S. *et al.* 2007. Central control of bone remodeling by neuromedin U. *Nat. Med.* **13:** 1234–1240.
12. Hanada, R. *et al.* 2004. Neuromedin U has a novel anorexigenic effect independent of the leptin signaling pathway. *Nat. Med.* **10:** 1067–1073.
13. Timmermans, J.P. *et al.* 1989. Neuromedin U-immunoreactivity in the nervous system of the small intestine of the pig and its coexistence with substance P and CGRP. *Cell. Tissue Res.* **258:** 331–337.
14. Hansen, M.B. 2003. Neurohumoral control of gastrointestinal motility. *Physiol. Res.* **52:** 1–30.
15. Dass, N.B. *et al.* 2007. Neuromedin U can exert colon-specific, enteric nerve-mediated prokinetic activity, via a pathway involving NMU1 receptor activation. *Br. J. Pharmacol.* **150:** 502–508.
16. Sato, T. *et al.* 2005. Molecular forms of hypothalamic ghrelin and its regulation by fasting and 2-deoxy-d-glucose administration. *Endocrinology* **146:** 2510–2516.
17. Fukue, Y. *et al.* 2006. Regulation of gonadotropin secretion and puberty onset by neuromedin U. *FEBS Lett.* **580:** 3485–3488.
18. Sikander, A., S.V. Rana & K.K. Prasad. 2009. Role of serotonin in gastrointestinal motility and irritable bowel syndrome. *Clin. Chim. Acta* **403:** 47–55.
19. Mandl, P. & J.P. Kiss. 2007. Role of presynaptic nicotinic acetylcholine receptors in the regulation of gastrointestinal motility. *Brain Res. Bull.* **72:** 194–200.

Ann. N.Y. Acad. Sci. ISSN 0077-8923

ANNALS OF THE NEW YORK ACADEMY OF SCIENCES
Issue: *Phylogenetic Aspects of Neuropeptides*

Functional interaction of regulator of G protein signaling-2 with melanin-concentrating hormone receptor 1

Mayumi Miyamoto-Matsubara,[1] Shinjae Chung,[2] and Yumiko Saito[1]

[1]Graduate School of Integrated Arts and Sciences, Hiroshima University, Hiroshima, Japan. [2]Department of Pharmacology, University of California, California, USA

Address for correspondence: Yumiko Saito, Graduate School of Integrated Arts and Sciences, Hiroshima University, 1-7-1 Kagamiyama, Higashi-Hiroshima 739-8521, Japan. yumist@hiroshima-u.ac.jp

Melanin-concentrating hormone receptor 1 (MCHR1) is a G protein–coupled receptor (GPCR) highly expressed in the central nervous system. MCHR1 mediates many physiological functions including energy homeostasis and emotional processing. By acting as GTPase-activating proteins, regulators of G protein–signaling (RGS) proteins are negative modulators of GPCRs. We previously elucidated that RGS8 of the B/R4 RGS subfamily potently inhibits the action of both Gαq- and Gαi/o-dependent MCHR1 signaling. In the present study of living cells, we provide evidence that another B/R4 protein, RGS2, is an efficient regulator of MCHR1-mediated calcium signaling exclusively via the Gαq-dependent pathway. This effect was not observed for RGS4 and RGS5 proteins. Cotransfection of RGS2 with RGS8 additively increased the potency for inhibition of MCHR1 signaling. Truncation experiments revealed that an internal sequence within the N-terminal region of RGS2 (amino acids 28–80) was involved in the RGS2 modulation of MCHR1 activity. Our data suggest that RGS2 and RGS8 differentially associate with MCHR1 and may represent two distinct modes of signaling mechanisms *in vivo*.

Keywords: feeding; anxiety; G protein; calcium

Introduction

Regulators of G protein–signaling (RGS) proteins are members of the RGS family of proteins able to attenuate and modulate the signaling of heterotrimeric G protein–coupled receptors (GPCRs) by acting as GTPase-activating proteins toward Gα subunits.[1] To date, more than 30 different RGS proteins have been identified, and the family of RGS proteins has been classified into six subfamilies based on sequence similarities.[1,2] The B/R4 subfamily containing RGS1, 2, 4, 5, 8, 13, 16, and 18 is relatively small. Its members consist of a conserved N-terminal cationic amphipathic α-helix and the RGS domain. Most members of the B/R4 subfamily can selectively interact with Gαi/o or Gαq class proteins as purified proteins in solution in reconstituted systems, and the biochemical specificities of the RGS and Gα contact faces and the structures of the B/R4 RGS proteins have been well characterized.[1]

Accumulating evidence supports the notion that RGS proteins can distinguish GPCRs coupled to the same class of Gα protein in a living cellular system.[3] For example, electrophysiological analyses in Xenopus oocytes revealed that RGS8 decreased the response upon activation of Gαq-coupled M1 muscarinic acetylcholine receptor (M1 mAChR) or substance P receptor, but did not remarkably attenuate the signaling from Gαq-coupled M3 mAChR.[4] RGS1, 2, 4, and 16 displayed different relative potencies of their inhibitory activities toward calcium signaling from Gαq/11-linked muscarinic, bombesin, or cholecystokinin receptors in pancreatic acinar cells.[5] These studies suggest that the specificities of RGS protein actions depend on their selective interactions with GPCR complexes to selectively modulate the signaling.

As a mechanism for the receptor specificities, direct interactions of RGS proteins with GPCRs are reported to be mediated through the third intracellular (i3) loop[6–9] or C-terminal domain[9,10] of the

doi: 10.1111/j.1749-6632.2010.05507.x

 © 2010 New York Academy of Sciences.

receptors. Moreover, the N-termini of RGS2, RGS4 and RGS8, which do not contain the RGS domain, were reported to play roles in the formation of RGS-GPCR-G protein–ternary complexes.[6–8,10,11] However, understanding of the precise molecular interactions between GPCRs and RGS proteins requires more information based on evaluation of the different receptor type-specific attenuations mediated by RGS proteins.

Melanin-concentrating hormone (MCH) is an attractive neuropeptide involved in different physiological functions including energy homeostasis and mood determination.[12] Genetic and pharmacological approaches toward MCH and MCH receptor 1 (MCHR1) inactivation result in a lean phenotype with increased energy metabolism as well as anxiolytic and antidepressant phenotypes.[13] When expressed in heterologous cell lines, MCHR1 is able to activate multiple signaling pathways, such as calcium mobilization, cyclic AMP inhibition, and ERK activation through Gαq- and Gαi/o-coupled pathways.[14–16] Recently, we showed that RGS8 significantly suppresses MCHR1-mediated calcium mobilization in human embryonic kidney cells (HEK293) via both Gαq- and Gαi/o-dependent pathways.[17] The N-terminus of RGS8 mediates functional interactions with Arg^{253} and Arg^{256} of the i3 loop of MCHR1. Both RGS8 and MCHR1 are highly expressed in the shell of the nucleus accumbens,[18,19] while MCHR1 is also highly expressed in other areas, such as the piriform cortex, cerebral cortex, olfactory bulb, hippocampal formation, and amygdala.[19] Therefore, we investigated whether another RGS protein that is highly expressed in the brain is involved in MCHR1 function as a means of targeting its specificity and G protein selectivity.

In the present study, we obtained experimental evidence that RGS2, which has a broad localization in the brain, negatively modulates MCHR1 signaling in an exclusively Gαq-dependent manner. Our data suggest that RGS2 and RGS8 differentially limit MCHR1-induced signaling in terms of the G protein selectivity and may represent two distinct modes of signaling mechanisms.

Materials and methods

DNA construction

The generation of a cDNA encoding a Flag epitope tag before the first methionine of rat MCHR1 was described previously.[20] Wild-type MCHR1 and Flag-tagged MCHR1 (Flag-MCHR1) have similar EC50 values for MCH in calcium mobilization assays in HEK293T cells.[18] Rat M1 mAChR was a kind gift from Dr. T. Haga (Gakushuin University, Tokyo, Japan). RGS2, RGS4, and RGS5 were obtained from Missouri S&T cDNA Resource Center (Rolla, MO). DNA fragments for deletion mutants of RGS2, namely RGS2[28–211] (residues 28–211), RGS2[54–211] (residues 54–211), and RGS2[80–211] (residues 80–211), were amplified by polymerase chain reaction (PCR), respectively. The fragments were isolated and subcloned in-frame into the pcDNA3.1 vector. The open reading frames of all the PCR-generated constructs were verified by nucleotide sequence analysis.

Cell culture and transfection

HEK293T cells in 6-well plates were transiently transfected with Flag-MCHR1 with or without a vector or RGS2 (total amount, 1.0 μg/well). The DNA was mixed with the LipofectAMINE PLUS transfection reagent (Invitrogen, Carlsbad, CA), and the mixture was diluted with OptiMEM (Invitrogen) and added to the HEK293T cells. The transfected cells were cultured in DMEM containing 10% fetal bovine serum. For calcium influx assays, the cells were plated on 96-well plates at 24 h after transfection and cultured for a further 18 h with or without 260 ng/mL of pertussis toxin (PTX).

Western blotting analyses for RGS proteins and Flag-MCHR1 expressed in HEK293T cells

To generate whole-cell extracts, transiently transfected HEK293T cells were lysed with ice-cold SDS sample buffer (50 mmol/L Tris-HCl pH 6.8, 2% SDS, 50 mmol/L β-mercaptoethanol, 10% glycerol), and then homogenized by sonication at 4° C. Aliquots of the total proteins were separated by SDS-PAGE and electrotransferred to PVDF membranes. After blocking with 5% skim milk, RGS2 on the membranes was detected using an anti-RGS2 antibody (sc-9103; Santa Cruz Biotechnology, Santa Cruz, CA; dilution 1:200) overnight at 4° C, followed by horseradish peroxidase-conjugated donkey anti-goat IgG (sc-2020; Santa Cruz Biotechnology, Santa Cruz, CA; dilution 1:4000). The reactive bands were visualized with ECL substrates (GE Healthcare, UK). After immunoblotting with the anti-RGS2 antibody, the membranes were stripped and reprobed with an anti-actin antibody (Chemicon International Inc., Temecula, CA).

Measurement of intracellular Ca^{2+}

Transiently transfected HEK293T cells seeded on 96-well plates (Becton Dickinson, Franklin Lakes, NJ) were loaded with a nonwash calcium dye (Calcium Assay Kit; Molecular Devices, Sunnyvale, CA) in Hank's balanced salt solution containing 20 mmol/L HEPES (pH 7.5) for 1 h at 37° C. For each concentration of MCH or carbachol examined, the level of intracellular Ca^{2+} ($[Ca^{2+}]i$) was detected using a Flexstation II imaging plate reader (Molecular Devices).[20] Data were expressed as the fluorescence (arbitrary units) versus time. The EC50 values for MCH or carbachol were obtained from sigmoidal fits using a nonlinear curve-fitting program (Prism version 3.0; GraphPad Software, San Diego, CA).

Results

RGS2 attenuates MCHR1-mediated calcium signaling

To identify potential B/R4 RGS proteins that modulate calcium mobilization via MCHR1, we transiently expressed Flag-MCHR1 in HEK293T cells with or without expression of RGS proteins. Among the B/R4 RGS proteins, RGS2, RGS4, and RGS5 were selected because they are expressed at high levels in several distinct regions of the brain.[18,21] As shown in Fig. 1, coexpression of RGS2 with MCHR1 significantly attenuated calcium signaling by MCH. This effect was not caused by RGS4 or RGS5. Transfection of graded amounts of the RGS2 plasmid led to graded expression of RGS2 protein in the cells, and RGS2 was able to dose-dependently inhibit the extent of MCH-mediated calcium signaling (Fig. 2A, B). Table 1 shows the EC50 values for MCH in cells expressing MCHR1 and RGS2. Cells transfected with 0.8 μg of RGS2 and 0.2 μg of Flag-MCHR1 showed a 5.6-fold higher EC50 value and a 24% decrease in the maximal response. Cells transfected with 0.2 μg of RGS2 still showed significant inhibition of MCHR1-mediated calcium signaling with a 2.7-fold higher EC50 value.

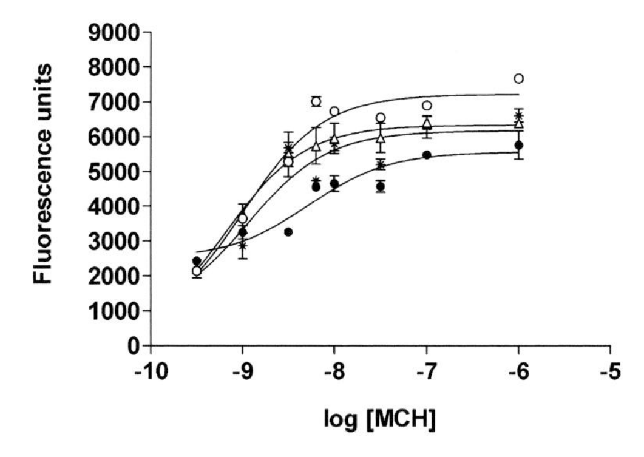

Figure 1. RGS2 inhibits MCHR1-mediated calcium signaling. Dose–response relationships of MCH-stimulated calcium mobilization in HEK293T cells expressing Flag-MCHR1 in combination with RGS proteins. HEK293T cells on 6-well plates were transfected with Flag-MCHR1 (0.2 μg) and RGS2, RGS4, RGS5, or vector plasmid pcDNA3.1 (0.8 μg). Vector plasmid was added to keep the total amount of DNA constant (1 μg). At 20 h after transfection, the cells were reseeded on 96-well plates. After loading with a nonwash calcium dye, MCH was added for 80 s and alterations in the calcium mobilization were monitored using a Flexstation II imaging plate reader. Data represent the means ± SEM of triplicate determinations.

RGS2 differentially modifies MCHR1-mediated calcium signaling from RGS8

MCHR1-induced calcium mobilization is known to be mediated through both Gαq- and Gαi/o-dependent pathways.[16,17] In fact, after PTX pretreatment, which uncouples Gαo/i from receptors, the EC50 value was 8.3 ± 0.9 nmol/L (mean ± SEM of three independent experiments) (Fig. 3A). In untreated cells, the EC50 value was 2.6 ± 0.6 nmol/L, and the +PTX (Gαq-dependent response)/–PTX (Gαo/i and Gαq-dependent response) ratio was 3.2, indicating that a large component of the calcium mobilization was dependent on Gαq activity rather than Gαi/o activity. RGS2 is known to exhibit specificity for Gαq in biochemical binding assays and single turnover GTPase acceleration assays, while an effect on Gαi was observed after reconstitution of the protein in phospholipid

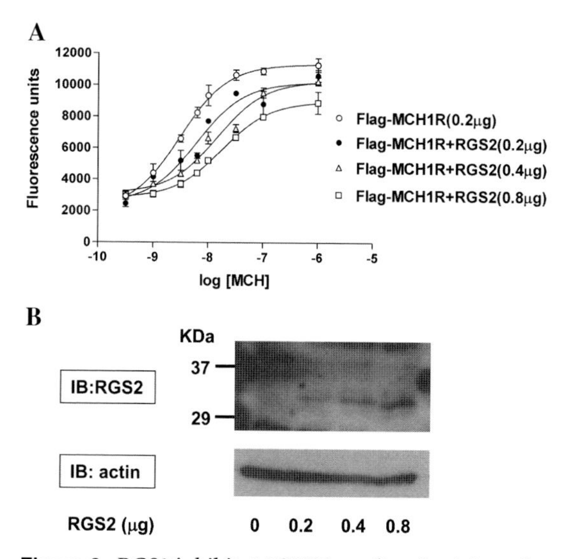

Figure 2. RGS2 inhibits MCHR1-mediated calcium signaling in a dose-dependent manner. (A) Dose–response relationships of MCH-stimulated calcium mobilization in HEK293T cells expressing Flag-MCHR1 in combination with RGS2. HEK293T cells on 6-well plates were transfected with Flag-MCHR1 (0.2 μg) and the indicated amounts of RGS2 or vector plasmid pcDNA3.1. Vector plasmid was added to keep the total amount of DNA constant (1 μg). Data represent the means ± SEM of triplicate determinations. (B) Western blot analysis of RGS2 expressed in the cells used for the functional assays shown in (A). Total protein extracts from cell lysates were resolved by SDS-PAGE using a 15% gel and transferred to a PVDF membrane. After immunoblotting with an anti-RGS2 antibody (IB:RGS2), the membrane was stripped and reprobed with an anti-actin antibody (IB:actin).

vesicles containing M2 mAchR.[22] In a cellular context, RGS2 negatively modulates receptor-mediated signaling that is dependent on Gαq-dependent pathways, such as α1A-adrenergic receptor, M1 mAchR, and cholecystokinin receptor-2 (CCK2R) signaling.[6,7,10,23] and the Gαi-dependent pathway for M2 mAchR.[22] Therefore, we examined whether RGS2 inhibited MCHR1-mediated signaling via the Gαq- or Gαq-dependent pathway. When cells transfected with 0.4 μg of RGS2 and 0.2 μg of Flag-MCHR1 were pretreated with PTX, the EC50 value was 65.8 ± 5.5 nmol/L. Since the EC50 value for PTX-treated cells transfected with Flag-MCHR1 alone was 8.3 nmol/L, RGS2 inhibited MCHR1-mediated calcium signaling with a 7.8-fold higher EC50 value under conditions that blocked Gαo/i-mediated signaling. On the other hand, cells co-transfected with RGS2 and Flag-MCHR1 without PTX treatment had a 3.9-fold higher EC50 value (Fig. 3A, Table 1). If RGS2 attenuated both Gαq- and Gαo/i-mediated calcium signaling similar to the case for RGS8, the extent of inhibition should be decreased by PTX treatment. However, RGS2 elicited a stronger efficacy of MCHR1-induced calcium signaling under a Gαq-dominant cellular environment. These observations suggest that RGS2 negatively modulates MCHR1 signaling by exclusively inhibiting the actions of Gαq protein.

We previously showed that RGS8 negatively modulates MCHR1-mediated calcium signaling.[17] To observe whether RGS2 affected the efficacy of RGS8 toward MCHR1, we cotransfected MCHR1 with RGS2 or RGS8 into HEK293T cells and examined the MCHR1-induced calcium mobilization. Cells transfected with MCHR1/RGS2 and MCHR1/RGS8 showed 3.8- and 7.0-fold higher EC50 values, respectively (Fig. 3B), indicating that RGS8 has a more potent effect than RGS2 for attenuating MCHR1 signaling. Since triple transfection of MCHR1, RGS2, and RGS8 into cells led to an 11.2-fold higher EC50 value, cotransfection of RGS2 with RGS8 additively increases the potency to inhibit MCHR1

Table 1. Dose–response relationships of MCH-stimulated calcium mobilization in HEK293T cells expressing Flag-MCH1R in combination with RGS2

Combination of MCHR1 with RGS2	EC50 value of ligand (nmol/L) (fold)	Max response (%)
Flag-MCHR1 (0.2 μg)	3.7 ± 1.8 (1)	100
Flag-MCHR1 + RGS2 (0.2 μg)	10.0 ± 2.0 (2.7)	94.8 ± 13.0
Flag-MCHR1 + RGS2 (0.4 μg)	14.4 ± 3.2 (3.9)	89.1 ± 3.4
Flag-MCHR1 + RGS2 (0.8 μg)	18.5 ± 3.0 (5.6)	76.4 ± 4.5

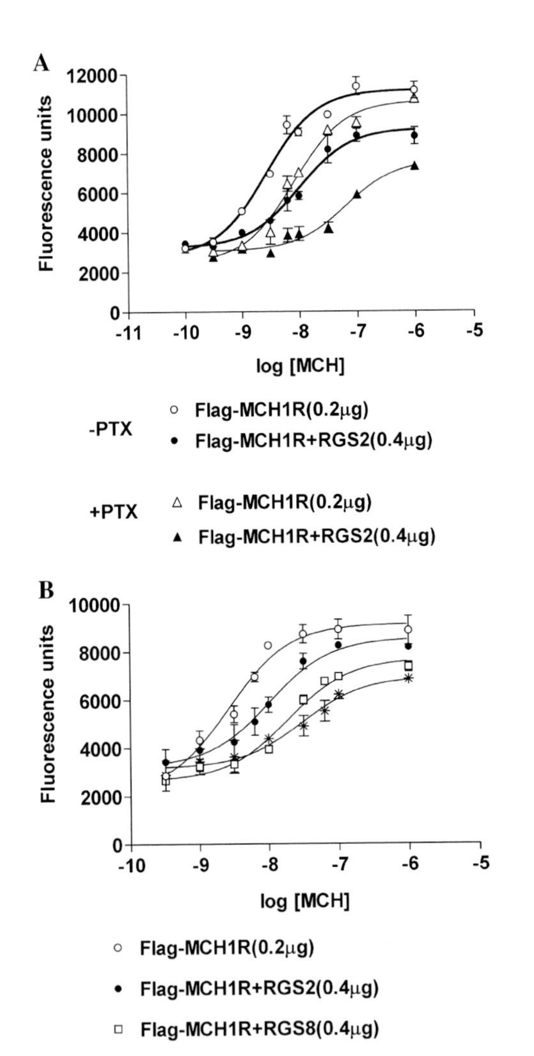

Figure 3. RGS2 and RGS8 differentially associate with MCHR1. (A) Dose–response relationships of MCH-stimulated calcium mobilization in HEK293T cells expressing Flag-MCHR1 and RGS2 with or without PTX pretreatment (260 ng/mL, 18 h). HEK293T cells on 6-well plates were transfected with Flag-MCHR1 (0.2 μg) and RGS2 (0.4 μg) or vector plasmid pcDNA3.1. Vector plasmid was added to keep the total amount of DNA constant (1 μg). Data represent the means ± SEM of triplicate determinations. (B) Dose–response relationships of MCH-stimulated calcium mobilization in HEK293T cells expressing Flag-MCHR1, RGS2, and RGS8. HEK293T cells on 6-well plates were transfected with Flag-MCHR1 (0.2 μg), RGS2 (0.4 μg), RGS8 (0.4 μg), or vector plasmid pcDNA3.1. Vector plasmid was added to keep the total amount of DNA constant (1 μg). Data represent the means ± SEM of triplicate determinations.

signaling. These observations suggest that interactions of MCHR1 with RGS2 and RGS8 are not competitive, and that MCHR1 contains two non-overlapping structural sites for interactions with RGS2 and RGS8.

Domain of RGS2 is necessary for RGS2 regulation of MCHR1-mediated calcium mobilization

RGS2 is composed of an N-terminal region (residues 1–80) comprising an amphipathic helix, a central RGS box (residues 80–200) responsible for accelerating GTP hydrolysis by Gα and a short C-terminal domain (residues 201–211).[23] To establish which regions of RGS2 other than the RGS box are important for the inhibition of MCHR1 signaling, we created a series of deletions along the length of the first 80 amino acids of the protein (Table 2). As shown in Table 2, intact RGS2 and RGS2[28–211] equally inhibited the MCH-induced signaling with 5.8- and 5.4-fold higher EC50 values, respectively. Deletion of the first 53 amino acids (RGS2[54–211]) partially reduced the inhibitory effect, and further deletion of the first 79 amino acids (RGS2[80–211]) caused an even greater reduction in terms of the EC50 value. Therefore, the internal region between amino acids 28 and 80 is necessary for potent inhibition and significantly contributes to the RGS2 interaction with MCHR1. A previous study using deletion and chimeric RGS2 constructs showed that the first 78 amino acids of the N-terminal region of RGS2 contained the sites of interaction with the i3 loop of α1A-adrenergic receptor or M1 mAchR.[6,7] To evaluate the sequence of RGS2 responsible for the interaction with M1 mAchR in our HEK293T system, we performed calcium mobilization assays stimulated by carbachol. Coexpression of RGS2 caused inhibition of M1 mAchR-mediated signaling with a 5.4-fold higher EC50 value without significantly changing the maximal response (Table 2). Intact RGS2 and RGS2[28–211] produced equivalent inhibitory effects on the carbachol-induced calcium signaling, whereas RGS2[54–211] completely abolished the RGS2 inhibitory effect. These findings suggest that the requirement for the internal sequence within the N-terminus of RGS2 for modulating the receptor-mediated calcium signaling partially overlaps but is not identical between MCHR1 and M1 mAchR.

Table 2. The N-terminus region of RGS2 differentially regulates receptor signaling via MCHR1 and M1 mAChR

Combination of receptor with truncated RGS2	EC50 value of ligand (nmol/L) (fold)	Max response (%)
Flag-MCHR1	4.5 ± 1.8 (1)	100
Flag-MCHR1 + RGS2	26.1 ± 8.7 (5.8)	83.5 ± 5.7
Flag-MCHR1 + RGS2[28–211]	24.3 ± 8.6 (5.4)	78.1 ± 2.3
Flag-MCHR1 + RGS2[54–211]	11.7 ± 2.3 (2.6)	86.7 ± 10.4
Flag-MCHR1 + RGS2[80–211]	6.3 ± 1.7 (1.4)	76.8 ± 7.3
M1 mAChR	28.4 ± 7.0 (1)	100
M1 mAChR + RGS2	153.4 ± 8.9 (5.4)	110.3 ± 4.9
M1 mAChR + RGS2[28–211]	159.0 ± 12.3 (5.6)	97.4 ± 2.9
M1 mAChR + RGS2[54–211]	30.2 ± 6.0 (1.1)	98.3 ± 11.5
M1 mAChR + RGS2[80–211]	26.1 ± 8.2 (0.92)	91.2 ± 18.1

Discussion

Our earlier study showed that RGS8 belonging to the B/R4 subfamily is a potent negative modulator of MCHR1-induced signaling. In the present study, we examined the possibility of significant functional roles for other B/R4 RGS proteins via MCHR1 in HEK293T cells. Our results demonstrated that RGS2 is another efficient regulator of MCHR1-mediated calcium signaling. RGS2 is known to act as a negative modulator for α1A-adrenergic receptor, M1 mAchR and CCK2R via Gαq-dependent pathways,[6,7,10,23] and for M2 mAchR via a Gαi-dependent pathway.[22] Furthermore, RGS2 as well as RGS1 and RGS3L were reported to attenuate angiotensin II- and sphingosine 1-phosphate-induced ERK phosphorylation, possibly via Gαq- and/or Gαi/o-dependent pathways.[24] Importantly, the present study has provided the first evidence that two members of the B/R4 RGS family, RGS2 and RGS8, can limit MCHR1-induced calcium signaling via the inhibition of different G proteins. That is, MCHR1 signaling involving Gαq and Gαi/o proteins is impaired by RGS8; however, MCHR1 signaling involving Gαq protein is preferentially inhibited by RGS2.

The RGS N-terminal domain has been shown to be critical for conferring receptor selectivity. We previously showed that the N-terminal nine amino acids of RGS8 are necessary for its optimal capacity to downregulate MCHR1 signaling.[17] In the present study, we have demonstrated that an internal sequence within the N-terminal region of RGS2 (amino acids 28–80) is deeply involved in MCHR1-mediated signaling. The N-terminal region of RGS2 has an amphipathic α-helix (amino acids 33–53) and a coil-turn loop structure (amino acids 21–31 and 55–71),[23] and the amphipathic α-helix was demonstrated to participate in the anchoring of RGS2 to the plasma membrane.[25,26] Our data have shown that RGS2[54–211] lacking the amphipathic α-helix still partially retains the capacity to attenuate MCHR1-mediated calcium signaling. This observation suggests that both the lipid bilayer-targeting region and the expected loop in RGS2 are simultaneously required for potent inhibition of MCHR1 signaling. Our cellular assays also indicated that an internal region between amino acids 28 and 54 of RGS2 is critical for the association with M1 mAchR. Interestingly, a different internal region of RGS2 (between amino acids 54 and 79) has been reported to be responsible for the ability to inhibit CCK2R signaling.[10,23] These data raise the possibility that RGS2 possesses different intrinsic regions that can display specific interactions with different GPCRs.

A recent study indicated that RGS2 utilizes direct mechanisms to form stable functional pairs with preferred GPCRs to selectively modulate the signaling; it is reported that direct interactions of RGS2 are mediated through the i3 loop of α1A-adrenergic receptor or M1 mAchR[6,7] or the C-terminal domain in CCK2R.[10] RGS2 has also been shown to indirectly interact with several GPCRs through the scaffolding protein spinophilin.[3] At present, it remains to be clarified which region within MCHR1 is responsible for the functional interaction with RGS2 (or the RGS2 complex) and how RGS2 would show its inhibitory effects. Since an HEK293T cell lysate containing HA-tagged RGS2 did not bind to a GST-fusion protein with the i3 loop of MCHR1 in pull-down assays,[17] a selective interaction between MCHR1 and RGS2 is unlikely to occur through the i3 loop. We further performed functional assays of combinations of RGS2 with serial C-terminal deletion mutants of MCHR1 in living cells. However, the inhibitory effects of RGS2 in cells expressing three mutants (ΔS325, ΔQ333, and ΔT342)[20] were identical to that in cells expressing intact MCHR1 (data not shown). Analyses of further combinations between a variety of MCHR1 mutants and RGS2 will elucidate a possible explanation regarding the association of RGS2 with MCHR1.

Although we showed the regulatory role of RGS2 in HEK293T cells expressing exogenous MCHR1, preferential role of RGS2 in Gαq-mediated signaling path under "natural" cellular environment is unclear. Several cell lines have been found to express functional MCHR1, with their signaling to the Gαq pathway being observed to have a lower affinity than to the Gαi/o pathway.[27,28] Considerable effort will be required to identify new cell lines activating MCH-induced Gαq-coupled signaling, and to evaluate the actual regulatory role of RGS2. However, it may be possible that multiple G protein–mediated pathways are involved the functions of MCH in various brain regions. It will, therefore, be necessary to demonstrate the physiological consequences by assessing the co-expression of RGS2 and MCHR1 within the same cells in the brain.

In conclusion, the present study raises the possibility that RGS2 is a novel component of MCHR1-mediated calcium signaling in addition to RGS8. The MCH–MCHR1 system is a viable target for obesity-related therapy and for mood disorders, while RGS2 is known to be an essential signaling molecule in disease states including hypertension and anxiety.[29–31] RGS2 is highly expressed in the neocortex, amygdale, and hippocampal formation[21,22] where RGS8 is expressed at low to moderate levels.[18] By contrast, RGS8 is especially highly expressed in the nucleus accumbens shell[18] where RGS2 is expressed at a moderate level.[21] Therefore, one can speculate that the inhibitory effects elicited by these two different RGS proteins could represent two distinct modes of MCHR1 modulation in the nervous system. Further study incorporating neurophysiological analyses will be a new research avenue and provide an important link between RGS proteins and MCHR1 activity.

Conflicts of interest

The authors declare no conflicts of interest.

References

1. Hollinger, S. & J.R. Hepler. 2002. Cellular regulation of RGS proteins: modulators and integrators of G protein signaling. *Pharmacol. Rev.* **54:** 527–559.
2. Willars, G.B. 2006. Mammalian RGS proteins: multifunctional regulators of cellular signaling. *Sem. Cell Dev. Biol.* **17:** 363–376.
3. Neitzel, K.L. & J.R. Hepler. 2006. Cellular mechanisms that determine selective RGS protein regulation of G protein-coupled receptor signaling. *Sem. Cell Dev. Biol.* **17:** 383–389.
4. Saitoh, O. *et al.* 2002. Alternative splicing of RGS8 gene determines inhibitory function of receptor type-specific Gq signaling. *Proc. Natl. Acad. Sci. USA* **99:** 10138–10143.
5. Xu, X. *et al.* 1999. RGS proteins determine signaling specificity of Gq-coupled receptors. *J. Biol. Chem.* **274:** 3549–3556.
6. Bernstein, L.S. *et al.* 2004. RGS2 binds directly and selectively to the M1 muscarinic acetylcholine receptor third intracellular loop to modulate Gq/11α signaling. *J. Biol. Chem.* **279:** 21248–21256.
7. Hague, C. *et al.* 2005. Selective inhbitin of α1A-adrenergic receptor signaling by RGS2 association with the receptor third intracellular loop. *J. Biol. Chem.* **280:** 27289–27295.
8. Itoh, M. *et al.* 2006. Alternative splicing of RGS8 gene changes the binding property to the M1 muscarinic receptor to confer receptor type-specific Gq regulation. *J. Neurochem.* **99:** 1505–1516.
9. Georgoussi, L. *et al.* 2006. Selective interactions between G protein subunits and RGS4 with the C-terminal domains of the μ-and δ-opioid receptors regulate opioids receptor signaling. *Cell Signal.* **18:** 771–782.
10. Langer, I. *et al.* 2009. Evidence for a direct and functional interactin between the regulators of G protein signaling-2 and phosphorylated C terminus of cholecystokinin-2 receptor. *Mol. Pharma.* **75:** 502–513.
11. Zeng, W. *et al.* 1998. The N-terminal domain of RGS4 confers receptor-selective inhibition of G protein signaling. *J. Biol. Chem.* **273:** 34687–34690.
12. Pissios, P., R.L Bradley & E. Maratos-Flier. 2006. Expanding the scales: the multiple roles of MCH in regulating

energy balance and other biological functions. *Endocri. Rev.* **27:** 606–620.
13. Antal-Zimanyi, I. & X. Khawaja. The role of melanin-concentrating hormone in energy homeostasis and mood disorders. *J. Mol. Neurosci.* In press.
14. Chambers, J. *et al.* 1999 Melanin-concentrating hormone is the cognate ligand for the orphan G-protein-coupled receptor SLC-1. *Nature* **400:** 261–265.
15. Saito, Y. *et al.* 1999. Molecular characterization of the melanin-concentrating-hormone receptor. *Nature* **400:** 265–269.
16. Hawes, B.E. *et al.* 2000. The melanin-concentrating hormone couples to multiple G proteins to activate diverse intracellular signaling pathways. *Endocrinology* **141:** 4524–4532.
17. Miyamoto-Matsubara, M. *et al.* 2008. Regulation of melanin-concentrating hormone receptor 1 signaling by RGS8 with the receptor third intracellular loop. *Cell. Signal.* **20:** 2084–2094.
18. Gold, S.J. *et al.* 1997. Regulators of G-protein signaling (RGS) proteins: Region-specific expression of nine subtypes in rat brain. *J. Neurosci.* **15:** 8024–8037.
19. Saito, Y. *et al.* 2001. Expression of the melanin-concentrating hormone (MCH) receptor mRNA in the rat brain. *J. Comp. Neurol.* **435:** 26–40.
20. Tetsuka, M. *et al.* 2004. The basic residues in the membrane-proximal C-terminal tail of the rat melanin-concentrating hormone receptor 1 are required for receptor function. *Endocrinology* **145:** 3712–3723.
21. Grafstein-Dunn, E. *et al.* 2001. Regional distribution of regulators of G-protein signaling (RGS) 1,2,13,14,16 and GAIP messenger ribonucleic acids by in site hybridization in rat brain. *Mol. Brain. Res.* **88:** 113–123.
22. Ingi, T. *et al.* 1998. Dynamic regulation of RGS2 suggests a novel mechanism in G-protein signaling and neuronal plasticity. *J. Neurosci.* **15:** 7178–7188.
23. Tikhonova, I.G. *et al.* 2006. Modeled structure of the whle regulator G-protein signaling-2. *Biochem. Biophys. Res. Commun.* **341:** 715–720.
24. Cho, H. *et al.* 2003. The aorta and heart differentially express RGS (regulators of G-protein signaling) proteins that selectively regulate sphigoshine 1-phosphate, angiotensin II and endothelin-1 signaling. *Biochem. J.* **317:** 973–980.
25. Heximer, S.P. *et al.* 2001. Mechanisms governing subcellular localization and function of human RGS2. *J. Biol. Chem.* **276:** 14195–14203.
26. Gu, S. *et al.* 2007. Unique hydrophobic extension of the RGS2 amphipathic hekix domain imparts increased plasma membrane binding and function relative to other RGS R4/B subfamily members. *J. Biol. Chem.* **282:** 33064–33075.
27. Eberle, A.N. *et al.* 2004. Expression and characterization of melanin-concentrating hormone receptor on mammalian cell lines. *Peptides* **25:** 1585–1595.
28. Cotta-Grand, N. *et al.* 2009. Melanin-concentrating hormone induces neurite outgrowth in human SH-SY5Y cells through p53 and MAPkinase signaling pathways. *Peptides* **30:** 2014–2024.
29. Heximer, S.P. *et al.* 2003. Hypertension and prolonged vasoconstrictor signaling in RGS2-deficient mice. *J. Clin. Invest.* **111:** 445–452.
30. Oliveira-Dos-Santos, A.J. *et al.* 2000. Regulation of T cell activation, anxiety, and male aggression by RGS2. *Proc. Natl. Acad. Sci. USA* **97:** 12272–12277.
31. Yalcin, B. *et al.* 2004. Genetic dissection of a behavioral quantitative trait locus shows that Rgs2 modulates anxiety in mice. *Nat. Genet.* **36:** 1197–1202.

Ann. N.Y. Acad. Sci. ISSN 0077-8923

ANNALS OF THE NEW YORK ACADEMY OF SCIENCES
Issue: *Phylogenetic Aspects of Neuropeptides*

Translational research of ghrelin

Hiroaki Ueno, Tomomi Shiiya, and Masamitsu Nakazato

Neurology, Respirology, Endocrinology and Metabolism, Department of Internal Medicine, Faculty of Medicine, University of Miyazaki, Miyazaki, Japan

Address for correspondence: Masamitsu Nakazato, Neurology, Respirology, Endocrinology and Metabolism, Department of Internal Medicine, Faculty of Medicine, University of Miyazaki, Miyazaki 889-1692, Japan. nakazato@med.miyazaki-u.ac.jp

Gastrointestinal peptides play important roles regulating feeding and energy homeostasis. Most gastrointestinal peptides including glucagon like peptide-1, peptide YY, amylin, and oxytomodulin are anorectic, and only ghrelin is an orexigenic peptide. Ghrelin increases appetite, modulates energy balance, suppresses inflammation, and enhances growth hormone secretion. Given its diversity of functions, ghrelin is expected be an effective therapy for lean patients with cachexia caused by chronic heart failure, chronic respiratory disease, anorexia nervosa, functional dyspepsia, and cancer. Clinical trials have demonstrated that ghrelin effectively increases lean body mass and activity in cachectic patients. Ghrelin interrupts the vicious cycle of the cachectic paradigm through its orexigenic, anabolic, and anti-inflammatory effects, and ghrelin administration may improve the quality of life of cachectic patients. We discuss the significant roles of ghrelin in the pathophysiology of cachectic diseases and the possible clinical applications of ghrelin.

Keywords: ghrelin; GHS-R; cachexia; growth hormone; obesity

Introduction

Feeding and energy homeostasis are controlled by the hypothalamus, limbic system, and peripheral organs, including the stomach, intestine, liver, adipose tissue, pancreas, and muscle. When food enters the gastrointestinal (GI) tract, stretch and macronutrients initiate the secretion of peptide hormones that transmit feeding signals via the afferent vagus nerve to the hypothalamus. Alternatively, some information is directly transmitted to the central nervous system through the blood–brain barrier. Nearly all GI peptides are anorectic and decrease feeding (e.g., cholecystokinin, glucagon like peptide-1, peptide YY, amylin, oxyntomodulin, and insulin), but only ghrelin is orexigenic. Recently, many questions regarding the role of GI peptides in regulating feeding and other functions have been clarified, and studies of the clinical utility of GI peptides as therapeutic agents are advancing. Our efforts have focused on ghrelin, and we previously identified roles for ghrelin in regulating cachexia, obesity, gastrointestinal diseases, and chronic obstructive pulmonary disease (COPD). In this review, we summarize the current status of translational research using the physiological functions of ghrelin, and we further discuss possible future directions for ghrelin research.

Biosynthesis of ghrelin

Ghrelin is a 28-amino acid peptide first isolated from the stomachs of humans and rats in 1999.[1] Acylation of its third residue, Ser-3, by the addition of middle-chain fatty acid, *n*-octanoic acid, is essential for its biological activity, including the binding and activation of the ghrelin receptor (formally known as the growth-hormone-secretagogue receptor (GHS-R)).[1] This acyl modification is performed by the polytopic membrane-bound enzyme ghrelin O-acyltransferase (GOAT).[2] Ghrelin mRNA is predominantly expressed in the stomach, but lower amounts are seen in the bowel, pituitary gland, kidney, lung, placenta, testis, pancreas, leukocytes, and hypothalamus. In addition, small amounts of ghrelin mRNA have been detected by real-time PCR in the adrenal gland, adipocytes, gall bladder, skeletal muscle, myocardium, skin, spleen, liver, ovary, and prostate.[3] Plasma ghrelin levels are higher in subjects with low body mass index (BMI) compared

doi: 10.1111/j.1749-6632.2010.05509.x

with normal- or high-BMI subjects,[4] and circulating ghrelin levels are elevated in underweight patients with COPD or those with anorexia nervosa.[5,6] Moreover, in patients with malignancy-associated cachexia, plasma ghrelin levels are increased by approximately 25% compared to normal subjects.[7] Finally, the high plasma ghrelin levels seen in the presence of an empty stomach rapidly decrease after feeding. These results suggest that ghrelin production is stimulated to maintain energy homeostasis and provide a defense against starvation.

Physiological functions of ghrelin

Ghrelin stimulates growth hormone (GH) secretion, food intake, and weight gain, but it has much broader physiologic functions, including roles in circulation, digestion, inflammation, and cell proliferation (Fig. 1).

Ghrelin stimulates GH secretion *in vitro* and *in vivo* in many species including humans.[1,8] Centrally or peripherally administered ghrelin strongly stimulates food intake in animals,[9] and intravenous administration of ghrelin to healthy humans increases energy intake from a buffet lunch by 28 ± 3.9%. Visual analogue scores for appetite are also increased under these conditions.[10] The orexigenic activity of ghrelin is independent of its stimulatory effects on GH secretion.[9] Continuous intracerebroventricular administration of ghrelin induces food intake and increased fat mass by selectively using carbohydrates, leading to weight gain.[11]

In addition to its ability to protect against cardiac dysfunction, ghrelin has other effects on the cardiovascular system. Intravenous infusion of ghrelin significantly decreases mean arterial pressure and increases cardiac output without affecting heart rate in healthy volunteers[12] and patients with chronic heart failure (CHF).[13] In addition, peripheral arterial infusion of ghrelin in normal subjects increases blood flow in a dose-dependent manner.

Ghrelin stimulates gastric acid secretion and gastric motility. These effects are lost following both vagotomy and administration of atropine, but a histamine H2-receptor antagonist has no effect on ghrelin-induced gastric functions. Thus, the effects of ghrelin on gastric function are transmitted via the vagus nerve.[14] Ghrelin also accelerates transit of a liquid meal through the small intestine in rats,[15] and it also ameliorates the delayed GI transit seen in alloxan-induced diabetic mice or streptozotocin-induced diabetic guinea pigs, suggesting that ghrelin

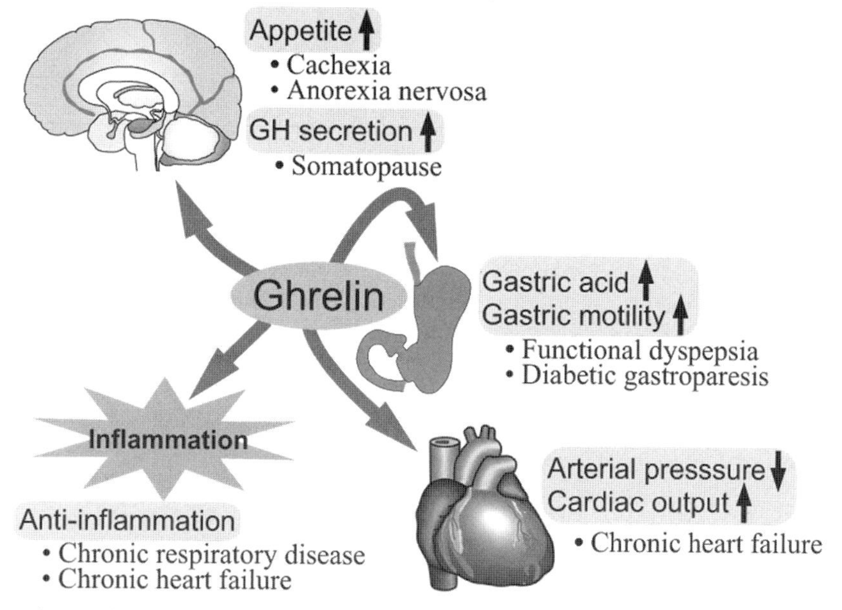

Figure 1. Physiological functions and potential for clinical applications of ghrelin. Ghrelin and ghrelin receptor agonists may have roles as novel therapeutics to affect multiple organ systems. Ghrelin-targeted therapy has been shown to be clinically beneficial in several diseases in humans.

may be a potential therapy for diabetic patients with gastroparesis.[16]

Ghrelin has anti-inflammatory activity and protects against endotoxin-induced shock by inhibiting the expression of proinflammatory cytokines by activated monocytes and endothelial cells.[17,18] Ghrelin attenuates sepsis-induced acute lung injury and reduces mortality by increasing pulmonary blood flow, downregulating proinflammatory cytokines, and inhibiting NF-kappaB activation in rats.[19] In addition to increasing appetite in cachectic patients with chronic respiratory infections, ghrelin's anti-inflammatory functions may protect against airway inflammation in patients with chronic pulmonary disease.

Translational research of ghrelin

Chronic heart failure. Plasma ghrelin levels are significantly higher in CHF patients with cachexia than in those without cachexia.[20] Chronic administration of ghrelin for 3 weeks improved left ventricular function and concomitantly increased plasma GH and insulin like growth factor-I (IGF-I) levels; it also attenuated left ventricular remodeling and cardiac cachexia in rats with CHF.[21] Moreover, treatment of CHF patients with ghrelin led to increased exercise capacity, food intake, and lean body mass.[20] Ghrelin also significantly decreased plasma norepinephrine in patients with CHF. In cachectic patients with CHF, ghrelin administration increased muscle strength and body weight but did not affect plasma levels of IL-6 or TNF-α,[22] despite the increased levels of proinflammatory cytokines seen in CHF patients with cardiac cachexia.[20] Taken together, these data suggest that treatment with ghrelin can ameliorate muscle wasting associated with CHF, but its anti-inflammatory properties in cardiac disease remain unclear.

Because both GH and IGF-1 are essential for myocardial growth and homeostasis,[23] beneficial effects of ghrelin in patients with cardiac cachexia may be partially explained by the ghrelin-induced GH and IGF-1 secretion. GH supplementation has beneficial effects on myocardial structure and function in CHF patients,[24] and both GH and IGF-1 lead to improved cardiac function through enhanced compensatory hypertrophy and decreased left ventricular wall stress in rats with CHF.[25] Hexarelin, a GHS-R agonist, prevents cardiac damage after ischemia-reperfusion in GH-independent mechanisms.[26] Ghrelin directly inhibits apoptosis of cardiomyocytes and endothelial cells through activation of extracellular signal-regulated kinase-1/2 and Akt serine kinases.[27] Thus, the benefits of ghrelin in cachectic patients with CHF are mediated by both GH-dependent and GH-independent mechanisms.

Gastrointestinal dysfunction. Disordered gastric emptying and dysregulation of gastroduodenal motility give rise to functional dyspepsia and gastroparesis,[28] and several studies have examined the safety and efficacy of ghrelin administration to patients affected by these disorders. Because ghrelin enhances gastric motility and secretion, it was hypothesized that treatment with ghrelin could lead to physiologic improvements in patients with gastroparesis. Indeed, a double-blind, placebo-controlled, crossover study showed that ghrelin increases gastric emptying in patients with diabetic gastroparesis,[29] and, in patients with idiopathic gastroparesis, ghrelin treatment enhances gastric emptying, and alleviates meal-related symptoms.[30] In addition, repeated intravenous infusions of ghrelin (3 μg/kg) twice a day for 2 weeks to patients with functional dyspepsia-related anorexia led to increased daily food intake compared to levels before and after ghrelin treatment. In this study, sensations of hunger were significantly elevated following ghrelin administration.[31]

Total gastrectomy can lead to significant weight loss and decreased appetite, with associated lower quality of life (QOL) after surgery. Plasma ghrelin levels fell to 10–20% of their preoperative levels immediately after total gastrectomy, and they remained significantly depressed for the entirety of the monitoring period.[32] Thus, ghrelin supplementation after gastrectomy may be an effective therapy to correct these defects, but further clinical studies are needed.

Chronic respiratory disease. Patients with late-stage chronic respiratory disease often have dramatically increased work-of-breathing requiring greater energy expenditure, and elevated levels of inflammatory cytokines are often seen in these patients. Together, these phenomena induce a relatively cachectic state in affected patients. A 3-week treatment with recombinant human ghrelin (2 μg/kg, intravenously, twice a day) improved the exercise tolerance of patients with COPD, and ghrelin increased muscle strength and nutritional status of these patients.[33] Thus, even underweight patients with COPD may benefit from ghrelin treatment.

Chronic respiratory infections in patients can lead to a state of near constant inflammation characterized by a neutrophil-predominant airway infiltrate, and this can promote an end-stage cachectic state. The constant activation of neutrophils can cause substantial cytotoxicity to bronchial and alveolar epithelial cells, leading to impaired pulmonary function and requirements for excess energy expenditure with associated weight loss. We showed that treatment with ghrelin (2 μg/kg, intravenously, twice a day) for 3 weeks reduced both sputum volume as well as sputum neutrophil counts in patients with chronic respiratory infections.[34] In addition, treated patients demonstrated increased food intake, body weight, and walk tolerance. Decreased sputum concentrations of the inflammatory cytokines IL-6 and TNF-α were also seen.[34] Since neutrophil-induced cellular damage promotes further bacterial proliferation in the lungs, ghrelin may be able to interrupt this vicious cycle to promote healing in these patients. Further studies are needed to examine the possible therapeutic efficacy of ghrelin for refractory chronic respiratory disease.

Anorexia nervosa (AN). AN is a psychiatric illness and eating disorder characterized by disordered body image with an obsessive fear of gaining weight; it is often associated with extremely low body weight. The high plasma ghrelin levels seen in lean patients with AN reflects the negative energy balance of affected individuals.[4,35,36] No changes in appetite were reported following a single administration of ghrelin (300-min intravenous infusion, 5 pmol/kg) to patients with AN.[36] However, a bolus intravenous injection of 1 μg/kg ghrelin caused sensations of hunger in six of nine patients with AN, and this was similar to the effects seen in normal subjects (five of seven responders).[37] Recently, another study administered ghrelin (3 μg/kg twice a day) to five AN patients, and this led to decreased gastrointestinal symptoms and increased sensations of hunger and daily energy intake (12–36%) without serious adverse events.[38] Further clinical studies, including randomized controlled trials, are needed to further assess the use of ghrelin in the treatment of AN.

Cancer cachexia. Cancer, irrespective of the organ affected, often leads to a cachectic state. Cancer patients with cachexia survive for a shorter period of time and respond more poorly to anticancer therapies compared to those without cachexia.[39,40] Up to 50% of cancer patients show changes in eating at the time of diagnosis, and this likely contributes to weight loss.[39] Cachexia and anorexia are closely associated with decreased QOL in cancer patients. Ghrelin administration induced significant increases in body weight and food intake in rodent models of cancer cachexia.[41,42] Several randomized, double-blind placebo-controlled trials have examined the safety and efficacy of ghrelin or GHS-R agonist in patients with cancer-related cachexia. In these studies, food intake and sensation of hunger were significantly increased after ghrelin treatment.[43,44]

Ghrelin is a mitogen for hepatoma cells,[45] but it exhibits antiproliferative effects in lung carcinoma cells *in vitro*.[46] Thus, there are conflicting data about the possible role of ghrelin in oncogenesis, and studies are ongoing to further identify the function of ghrelin in carcinogenesis.[41,47–49] Large-scale long-term clinical trials are needed to determine whether ghrelin treatment promotes tumor growth.

Antiaging. GH secretion peaks during puberty and gradually decreases reaching 20–25% of peak levels by age of 60. The normal changes in GH hyposecretion state are referred to as somatopause, and this is thought to promote aging by reducing muscle mass, strength and bone mineral density, and increase fat mass.[50] GH supplement therapy improves muscle, bone, and fat mass, but its functional effects and long-term safety remain unclear.[51]

Administration of ghrelin to aged rodents stimulates GH secretion, food intake and weight gain, while maintaining low adiposity.[52,53] We confirmed the ability of ghrelin to increase soleus muscle mass in a disuse atrophy model in mice (unpublished data). A double-blind, crossover study using MK-677, a GHS-R agonist, recruited healthy men and women between 60 and 81 years, and the daily administration of MK-677 for 12 months significantly increased GH and IGF-I levels to those seen in healthy young adults. Treated subjects also had increased total lean body mass and intracellular water (a biomarker of fat-free mass).[54] Total body fat, abdominal fat, muscle strength, and QOL did not change. This was a small-scale trial ($n = 65$), and further clinical studies are needed.

Chronic kidney disease. Malnutrition is a common and early feature of chronic kidney disease, and it is strongly associated with increased mortality in dialysis patients.[55,56] In a rat model of chronic renal disease, treatment with ghrelin increased food intake and lean body mass, and decreased circulating inflammatory cytokines.[57] A randomized double-blind crossover study showed that subcutaneous ghrelin administration (12 μg/kg) to malnourished dialysis patients for 1 week increased daily energy intake and decreased blood pressure.[58]

Diabetic neuropathy. Ghrelin has neuroprotective properties and prevents neuron apoptosis.[59] We recently reported that ghrelin, but not des-acyl ghrelin, reverses experimental diabetic peripheral neuropathy in mice,[60] suggesting that ghrelin may serve as a new therapeutic approach for diabetic neuropathy.

Conclusions and perspectives

The rates of obesity have been increasing worldwide, and obesity has become a major social and medical issue. Diet and exercise are important for the treatment of obesity, but these are difficult to implement and maintain for the duration of time needed to effect real weight change. To develop more effective pharmacologic treatments for obesity, the mechanisms regulating feeding and satiety must be fully understood, and GI peptides are becoming increasingly recognized for their contribution to energy homeostasis. Anorexigenic GI peptides, ghrelin receptor antagonists, and ghrelin neutralizing antibodies have been studied for their potential application as antiobesity drugs.[61,62] In addition, combination therapy with different GI peptides and leptin may also be effective for the treatment of obesity.[63] However, various adverse events, in particular nausea and vomiting, were often reported in clinical trials of GI peptides, and further research to determine the optimal administration route, dosage, and combination of drugs is needed.

Ghrelin has a number of different documented and putative physiologic functions, and it may be an effective therapy for multiple conditions (Fig. 1). Clinical trials examining the safety and efficacy of chronic ghrelin administration to patients after total gastrectomy, AN, diabetic gastroparesis, and disuse muscle atrophy are ongoing or in the planning stages. Although "acylated" ghrelin is the form currently in clinical use, the functions of des-acyl ghrelin should be examined in the future because important physiologic roles for des-acyl ghrelin have been reported.[27,64–66]

Acknowledgments

This work was supported in part by JSPS KAKENHI (No. 20659131, M.N.).

Conflicts of interest

The authors declare no conflicts of interest.

References

1. Kojima, M., H. Hosoda, Y. Date, *et al.* 1999. Ghrelin is a growth-hormone-releasing acylated peptide from stomach. *Nature* **402:** 656–660.
2. Yang, J., M.S. Brown, G. Liang, *et al.* 2008. Identification of the acyltransferase that octanoylates ghrelin, an appetite-stimulating peptide hormone. *Cell* **132:** 387–396.
3. Gnanapavan, S., B. Kola, S.A. Bustin, *et al.* 2002. The tissue distribution of the mRNA of ghrelin and subtypes of its receptor, GHS-R, in humans. *J. Clin. Endocrinol. Metab.* **87:** 2988.
4. Shiiya, T., M. Nakazato, M. Mizuta, *et al.* 2002. Plasma ghrelin levels in lean and obese humans and the effect of glucose on ghrelin secretion. *J. Clin. Endocrinol. Metab.* **87:** 240–244.
5. Itoh, T., N. Nagaya, M. Yoshikawa, *et al.* 2004. Elevated plasma ghrelin level in underweight patients with chronic obstructive pulmonary disease. *Am. J. Respir. Crit. Care Med.* **170:** 879–882.
6. Otto, B., U. Cuntz, E. Fruehauf, *et al.* 2001. Weight gain decreases elevated plasma ghrelin concentrations of patients with anorexia nervosa. *Eur. J. Endocrinol.* **145:** 669–673.
7. Shimizu, Y., N. Nagaya, T. Isobe, *et al.* 2003. Increased plasma ghrelin level in lung cancer cachexia. *Clin. Cancer Res.* **9:** 774–778.
8. Takaya, K., H. Ariyasu, N. Kanamoto, *et al.* 2000. Ghrelin strongly stimulates growth hormone release in humans. *J. Clin. Endocrinol. Metab.* **85:** 4908–4911.

9. Nakazato, M., N. Murakami, Y. Date, *et al.* 2001. A role for ghrelin in the central regulation of feeding. *Nature* **409:** 194–198.
10. Wren, A.M., L.J. Seal, M.A. Cohen, *et al.* 2001. Ghrelin enhances appetite and increases food intake in humans. *J. Clin. Endocrinol. Metab.* **86:** 5992–5995.
11. Tschop, M., D.L. Smiley & M.L. Heiman. 2000. Ghrelin induces adiposity in rodents. *Nature* **407:** 908–913.
12. Nagaya, N., M. Kojima, M. Uematsu, *et al.* 2001. Hemodynamic and hormonal effects of human ghrelin in healthy volunteers. *Am. J. Physiol. Regul. Integr. Comp. Physiol.* **280:** R1483–R1487.
13. Nagaya, N., K. Miyatake, M. Uematsu, *et al.* 2001. Hemodynamic, renal, and hormonal effects of ghrelin infusion in patients with chronic heart failure. *J. Clin. Endocrinol. Metab.* **86:** 5854–5859.
14. Date, Y., M. Nakazato, N. Murakami, *et al.* 2001. Ghrelin acts in the central nervous system to stimulate gastric acid secretion. *Biochem. Biophys. Res. Commun.* **280:** 904–907.
15. Trudel, L., C. Tomasetto, M.C. Rio, *et al.* 2002. Ghrelin/motilin-related peptide is a potent prokinetic to reverse gastric postoperative ileus in rat. *Am. J. Physiol. Gastrointest. Liver. Physiol.* **282:** G948–952.
16. Qiu, W.C., Z.G. Wang, R. Lv, *et al.* 2008. Ghrelin improves delayed gastrointestinal transit in alloxan-induced diabetic mice. *World J. Gastroenterol.* **14:** 2572–2577.
17. Li, W.G., D. Gavrila, X. Liu, *et al.* 2004. Ghrelin inhibits proinflammatory responses and nuclear factor-kappaB activation in human endothelial cells. *Circulation* **109:** 2221–2226.
18. Dixit, V.D., E.M. Schaffer, R.S. Pyle, *et al.* 2004. Ghrelin inhibits leptin- and activation-induced proinflammatory cytokine expression by human monocytes and T cells. *J. Clin. Invest.* **114:** 57–66.
19. Wu, R., W. Dong, M. Zhou, *et al.* 2007. Ghrelin attenuates sepsis-induced acute lung injury and mortality in rats. *Am. J. Respir. Crit. Care Med.* **176:** 805–813.
20. Nagaya, N., M. Uematsu, M. Kojima, *et al.* 2001. Elevated circulating level of ghrelin in cachexia associated with chronic heart failure: relationships between ghrelin and anabolic/catabolic factors. *Circulation* **104:** 2034–2038.
21. Nagaya, N., M. Uematsu, M. Kojima, *et al.* 2001. Chronic administration of ghrelin improves left ventricular dysfunction and attenuates development of cardiac cachexia in rats with heart failure. *Circulation* **104:** 1430–1435.
22. Nagaya, N., J. Moriya, Y. Yasumura, *et al.* 2004. Effects of ghrelin administration on left ventricular function, exercise capacity, and muscle wasting in patients with chronic heart failure. *Circulation* **110:** 3674–3679.
23. Dec, G.W. & V. Fuster. 1994. Idiopathic dilated cardiomyopathy. *N. Engl. J. Med.* **331:** 1564–1575.
24. Cittadini, A., J.D. Grossman, R. Napoli, *et al.* 1997. Growth hormone attenuates early left ventricular remodeling and improves cardiac function in rats with large myocardial infarction. *J. Am. Coll. Cardiol.* **29:** 1109–1116.
25. Le Corvoisier, P., L. Hittinger, P. Chanson, *et al.* 2007. Cardiac effects of growth hormone treatment in chronic heart failure: a meta-analysis. *J. Clin. Endocrinol. Metab.* **92:** 180–185.
26. Locatelli, V., G. Rossoni, F. Schweiger, *et al.* 1999. Growth hormone-independent cardioprotective effects of hexarelin in the rat. *Endocrinology* **140:** 4024–4031.
27. Baldanzi, G., N. Filigheddu, S. Cutrupi, *et al.* 2002. Ghrelin and des-acyl ghrelin inhibit cell death in cardiomyocytes and endothelial cells through ERK1/2 and PI 3-kinase/AKT. *J. Cell Biol.* **159:** 1029–1037.
28. Tack, J., R. Bisschops & G. Sarnelli. 2004. Pathophysiology and treatment of functional dyspepsia. *Gastroenterology* **127:** 1239–1255.
29. Murray, C.D., N.M. Martin, M. Patterson, *et al.* 2005. Ghrelin enhances gastric emptying in diabetic gastroparesis: a double blind, placebo controlled, crossover study. *Gut* **54:** 1693–1698.
30. Tack, J., I. Depoortere, R. Bisschops, *et al.* 2005. Influence of ghrelin on gastric emptying and meal-related symptoms in idiopathic gastroparesis. *Aliment. Pharmacol. Ther.* **22:** 847–853.
31. Akamizu, T., H. Iwakura, H. Ariyasu, *et al.* 2008. Repeated administration of ghrelin to patients with functional dyspepsia: its effects on food intake and appetite. *Eur. J. Endocrinol.* **158:** 491–498.
32. Takachi, K., Y. Doki, O. Ishikawa, *et al.* 2006. Postoperative ghrelin levels and delayed recovery from body weight loss after distal or total gastrectomy. *J. Surg. Res.* **130:** 1–7.
33. Nagaya, N., T. Itoh, S. Murakami, *et al.* 2005. Treatment of cachexia with ghrelin in patients with COPD. *Chest* **128:** 1187–1193.
34. Kodama, T., J. Ashitani, N. Matsumoto, *et al.* 2008. Ghrelin treatment suppresses neutrophil-dominant inflammation in airways of patients with chronic respiratory infection. *Pulm. Pharmacol. Ther.* **21:** 774–779.
35. Hotta, M., R. Ohwada, H. Katakami, *et al.* 2004. Plasma levels of intact and degraded ghrelin and their responses to glucose infusion in anorexia nervosa. *J. Clin. Endocrinol. Metab.* **89:** 5707–5712.

36. Miljic, D., S. Pekic, M. Djurovic, *et al.* 2006. Ghrelin has partial or no effect on appetite, growth hormone, prolactin, and cortisol release in patients with anorexia nervosa. *J. Clin. Endocrinol. Metab.* **91:** 1491–1495.
37. Broglio, F., L. Gianotti, S. Destefanis, *et al.* 2004. The endocrine response to acute ghrelin administration is blunted in patients with anorexia nervosa, a ghrelin hypersecretory state. *Clin. Endocrinol. (Oxf)* **60:** 592–599.
38. Hotta, M., R. Ohwada, T. Akamizu, *et al.* 2009. Ghrelin increases hunger and food intake in patients with restricting-type Anorexia Nervosa: a pilot study. *Endocr. J.* **56:** 1119–1128.
39. Dewys, W.D., C. Begg, P.T. Lavin, *et al.* 1980. Prognostic effect of weight loss prior to chemotherapy in cancer patients. Eastern Cooperative Oncology Group. *Am. J. Med.* **69:** 491–497.
40. Andreyev, H.J., A.R. Norman, J. Oates, *et al.* 1998. Why do patients with weight loss have a worse outcome when undergoing chemotherapy for gastrointestinal malignancies? *Eur. J. Cancer* **34:** 503–509.
41. Hanada, T., K. Toshinai, N. Kajimura, *et al.* 2003. Anti-cachectic effect of ghrelin in nude mice bearing human melanoma cells. *Biochem. Biophys. Res. Commun.* **301:** 275–279.
42. DeBoer, M.D., X.X. Zhu, P. Levasseur, *et al.* 2007. Ghrelin treatment causes increased food intake and retention of lean body mass in a rat model of cancer cachexia. *Endocrinology* **148:** 3004–3012.
43. Strasser, F., T.A. Lutz, M.T. Maeder, *et al.* 2008. Safety, tolerability and pharmacokinetics of intravenous ghrelin for cancer-related anorexia/cachexia: a randomised, placebo-controlled, double-blind, double-crossover study. *Br. J. Cancer* **98:** 300–308.
44. Neary, N.M., C.J. Small, A.M. Wren, *et al.* 2004. Ghrelin increases energy intake in cancer patients with impaired appetite: acute, randomized, placebo-controlled trial. *J. Clin. Endocrinol. Metab.* **89:** 2832–2836.
45. Murata, M., Y. Okimura, K. Iida, *et al.* 2002. Ghrelin modulates the downstream molecules of insulin signaling in hepatoma cells. *J. Biol. Chem.* **277:** 5667–5674.
46. Ghe, C., P. Cassoni, F. Catapano, *et al.* 2002. The antiproliferative effect of synthetic peptidyl GH secretagogues in human CALU-1 lung carcinoma cells. *Endocrinology* **143:** 484–491.
47. Yeh, A.H., P.L. Jeffery, R.P. Duncan, *et al.* 2005. Ghrelin and a novel preproghrelin isoform are highly expressed in prostate cancer and ghrelin activates mitogen-activated protein kinase in prostate cancer. *Clin. Cancer Res.* **11:** 8295 8303.
48. Volante, M., E. Allia, E. Fulcheri, *et al.* 2003. Ghrelin in fetal thyroid and follicular tumors and cell lines: expression and effects on tumor growth. *Am. J. Pathol.* **162:** 645–654.
49. Duxbury, M.S., T. Waseem, H. Ito, *et al.* 2003. Ghrelin promotes pancreatic adenocarcinoma cellular proliferation and invasiveness. *Biochem. Biophys. Res. Commun.* **309:** 464–468.
50. Fanciulli, G., A. Delitala & G. Delitala. 2009. Growth hormone, menopause and ageing: no definite evidence for 'rejuvenation' with growth hormone. *Hum. Reprod Update* **15:** 341–358.
51. Giordano, R., L. Bonelli, E. Marinazzo, *et al.* 2008. Growth hormone treatment in human ageing: benefits and risks. *Hormones (Athens)* **7:** 133–139.
52. Ariyasu, H., H. Iwakura, G. Yamada, *et al.* 2008. Efficacy of ghrelin as a therapeutic approach for age-related physiological changes. *Endocrinology* **149:** 3722–3728.
53. Toshinai, K., M.S. Mondal, T. Shimbara, *et al.* 2007. Ghrelin stimulates growth hormone secretion and food intake in aged rats. *Mech. Ageing Dev.* **128:** 182–186.
54. Nass, R., S.S. Pezzoli, M.C. Oliveri, *et al.* 2008. Effects of an oral ghrelin mimetic on body composition and clinical outcomes in healthy older adults: a randomized trial. *Ann. Intern. Med.* **149:** 601–611.
55. Marcen, R., J.L. Teruel, M.A. de la Cal, *et al.* 1997. The impact of malnutrition in morbidity and mortality in stable haemodialysis patients. Spanish Cooperative Study of Nutrition in Hemodialysis. *Nephrol. Dial. Transplant.* **12:** 2324–2331.
56. Garg, A.X., P.G. Blake, W.F. Clark, *et al.* 2001. Association between renal insufficiency and malnutrition in older adults: results from the NHANES III. *Kidney Int.* **60:** 1867–1874.
57. Deboer, M.D., X. Zhu, P.R. Levasseur, *et al.* 2008. Ghrelin treatment of chronic kidney disease: improvements in lean body mass and cytokine profile. *Endocrinology* **149:** 827–835.
58. Ashby, D.R., H.E. Ford, K.J. Wynne, *et al.* 2009. Sustained appetite improvement in malnourished dialysis patients by daily ghrelin treatment. *Kidney Int.* **76:** 199–206.
59. Chung, H., E. Kim, D.H. Lee, *et al.* 2007. Ghrelin inhibits apoptosis in hypothalamic neuronal cells during oxygen-glucose deprivation. *Endocrinology* **148:** 148–159.
60. Kyoraku, I., K. Shiomi, K. Kangawa, *et al.* 2009. Ghrelin reverses experimental diabetic neuropathy in mice. *Biochem. Biophys. Res. Commun.* **389:** 405–408.
61. Esler, W.P., J. Rudolph, T.H. Claus, *et al.* 2007. Small-molecule ghrelin receptor antagonists improve glucose

tolerance, suppress appetite, and promote weight loss. *Endocrinology* **148:** 5175–5185.

62. Helmling, S., C. Maasch, D. Eulberg, *et al.* 2004. Inhibition of ghrelin action in vitro and in vivo by an RNA-Spiegelmer. *Proc. Natl. Acad. Sci. USA* **101:** 13174–13179.
63. Roth, J.D., T. Coffey, C.M. Jodka, *et al.* 2007. Combination therapy with amylin and peptide YY[3–36] in obese rodents: anorexigenic synergy and weight loss additivity. *Endocrinology* **148:** 6054–6061.
64. Toshinai, K., H. Yamaguchi, Y. Sun, *et al.* 2006. Des-acyl ghrelin induces food intake by a mechanism independent of the growth hormone secretagogue receptor. *Endocrinology* **147:** 2306–2314.
65. Muccioli, G., N. Pons, C. Ghe, *et al.* 2004. Ghrelin and des-acyl ghrelin both inhibit isoproterenol-induced lipolysis in rat adipocytes via a non-type 1a growth hormone secretagogue receptor. *Eur. J. Pharmacol.* **498:** 27–35.
66. Thompson, N.M., D.A. Gill, R. Davies, *et al.* 2004. Ghrelin and des-octanoyl ghrelin promote adipogenesis directly in vivo by a mechanism independent of the type 1a growth hormone secretagogue receptor. *Endocrinology* **145:** 234–242.

Ann. N.Y. Acad. Sci. ISSN 0077-8923

ANNALS OF THE NEW YORK ACADEMY OF SCIENCES
Issue: *Phylogenetic Aspects of Neuropeptides*

Pituitary adenylate cyclase activating polypeptide in the retina: focus on the retinoprotective effects

T. Atlasz,[1] K. Szabadfi,[2] P. Kiss,[3] B. Racz,[4] F. Gallyas,[4] A. Tamas,[3] V. Gaal,[5] Zs. Marton,[6] R. Gabriel,[2] and D. Reglodi[3]

[1]Department of Sportbiology, University of Pecs, Pecs, Hungary. [2]Department of Experimental Zoology and Neurobiology, University of Pecs, Pecs, Hungary. [3]Department of Anatomy, University of Pecs, Pecs, Hungary. [4]Department of Biochemistry and Medical Chemistry, University of Pecs, Pecs, Hungary. [5]Department of Ophthalmology, University of Pecs, Pecs, Hungary. [6]Department of Environmental and Laser Spectroscopy, University of Pecs, Pecs, Hungary

Address for correspondence: Dora Reglodi, M.D., Ph.D., Department of Anatomy, University of Pecs, Medical School, 7624 Pecs, Szigeti u 12, Hungary. dora.reglodi@aok.pte.hu

Pituitary adenylate cyclase activating polypeptide (PACAP) is a neurotrophic and neuroprotective peptide that has been shown to exert protective effects against different neuronal injuries, such as traumatic brain and spinal cord injury, models of neurodegenerative diseases, and cerebral ischemia. PACAP and its receptors are present in the retina. In this study, we summarize the current knowledge on retinal PACAP with focus on the retinoprotective effects. Results of histological, immunohistochemical, and molecular biological analysis are reviewed. *In vitro*, PACAP shows protection against glutamate, thapsigargin, anisomycin, and anoxia. *In vivo*, the protective effects of intravitreal PACAP treatment have been shown in the following models of retinal degeneration in rats: excitotoxic injury induced by glutamate and kainate, ischemic injury, degeneration caused by UV-A light, optic nerve transection, and streptozotocin-induced diabetic retinopathy. Studying the molecular mechanism has revealed that PACAP acts by activating antiapoptotic and inhibiting proapoptotic signaling pathways in the retina *in vivo*. These studies strongly suggest that PACAP is an excellent candidate retinoprotective agent that could be a potential therapeutic substance in various retinal diseases.

Keywords: retinal injury; retinoprotection; neuropeptide; PACAP

Pituitary adenylate cyclase activating polypeptide

Pituitary adenylate cyclase activating polypeptide (PACAP) was isolated from hypothalamic extracts on the basis of its action on pituitary adenylate cyclase.[1,2] PACAP belongs to the vasoactive intestinal peptide (VIP)/secretin/glucagon peptide superfamily, and has a remarkably well conserved structure throughout evolution.[3] The receptors for PACAP are G protein–coupled receptors and can be basically divided into two main groups: PAC1 receptor, which binds PACAP with higher affinity than VIP, and VPAC receptors (VPAC1 and VPAC2), which bind PACAP and VIP with similar affinities.[3] PAC1 receptor is coupled to adenylate cyclase (AC) and phospholipase C (PLC). Through AC activation, it elevates cyclic 3′,5′-adenosine monophosphate (cAMP), and activates protein kinase A (PKA), which can activate the mitogen-activated protein kinase (MAPK) pathway. PAC1 receptor binding can also activate MAPK pathway independently of AC activation. PLC activation stimulates Ca^{2+} mobilization and protein kinase C (PKC) activaton. These, and other pathways regulated by PAC1 receptors are different in distinct cell types depending on the expressed splice variant, the PACAP concentration and other factors present. VPAC receptors couple to Gs proteins resulting in activation of AC, but other signaling pathways downstream of cAMP or independent of cAMP are associated with VPAC receptor activation depending on the tissues in which they are expressed.[3–7]

The biological actions of PACAP are very diverse. Among others, the neuropeptide influences

doi: 10.1111/j.1749-6632.2010.05512.x
 Ann. N.Y. Acad. Sci. 1200 (2010) 128–139 © 2010 New York Academy of Sciences.

reproductive functions, circadian rhythm, thermoregulation, feeding, stress pathways, memory, inflammatory reactions, and development.[3] One of the most extensively studied actions of PACAP is its potent neuroprotective effect. PACAP has been shown to protect neurons *in vitro* against various toxic agents, such as glutamate, 6-hydroxydopamine, HIV envelope protein, oxidative stress, and ceramide. Neuroprotective efficacy of PACAP *in vivo* has also been proven in numerous animal models of neurological diseases, such as cerebral ischemia, Parkinson's disease, and traumatic brain injury. The neuroprotective effects are thought to be mediated by PAC1 receptors, involving various downstream mechanisms of the PKA and PKC pathways. The neuroprotective effects of PACAP have been reviewed by several authors.[3–7]

The retina, an extension of the central nervous system constituted of three neuronal layers, provides an excellent model to assess neuronal degenerations. PACAP has been shown to protect the neurons of the retina in several *in vitro* and *in vivo* models of retinal degeneration. The present review summarizes the current knowledge on retinal PACAP with focus on the retinoprotective effects.

PACAP in the retina

Numerous studies have described the presence of PACAP and its receptors in the whole retina and in the different layers. Light microscopical and ultrastructural observations have revealed PACAP immunopositive nerve cell bodies in amacrine and horizontal cells and in the ganglion cell layer. Fibers and terminals in the nerve fiber layer and inner plexiform layer of the rat retina were also labeled.[8,9] At ultrastructural level, PACAP-like immunoreactivity is visible in plasma membranes, rough endoplasmic reticulum, and cytoplasmic matrix of the PACAP-positive neurons.[9] PACAP immunoreactivity has also been demonstrated in the mouse retina, where its expression pattern does not seem to be regulated by visual experience that is by covering developing eyes for various durations with plastic shells to induce myopia.[10] Presence of PACAP has also been shown in other species, such as teleost, turtle, and chicken retina.[11–13]

PACAP immunoreactivity displays circadian alteration in the chicken retina.[12] In mammals, the retinohypothalamic tract originates from a subset of retinal ganglion cells, and it mainly synapses in the suprachiasmatic nucleus of the hypothalamus, which is the biological master clock of mammalian species, as well as in thalamic structures associated with visual input. A subset of these ganglion cells is intrinsically photosensitive due to the expression of melanopsin. Melanopsin is exclusively expressed in PACAP-containing cells.[14] PACAP has been shown to be costored in this subset of retinal ganglion cells. The PACAP expression of these ganglion cells is partially under the control of dopamine, while it is not affected by photoreceptor degeneration.[15,16] PACAP has also been described in the efferent limboretinal pathways.[17]

PACAP immunoreactivity appears at early stages of retinal development.[18] PACAP can be detected in the inner nuclear layer of embryonic chicken retinas from embryonic day 8 (E8).[19] In the zebrafish, the first PACAP-immunoreactive elements appear in the superficial layer of the retina at 24 h postfertilization, and in the ganglion cell layer 72 h postfertilization.[20] At day 13 of development of the zebrafish, PACAP-immunoreactive elements were absent from the retina.[20,21] In the rat, PACAP mRNA appears in the ganglion cell layer at E20.[22]

PACAP receptors in the retina

The selective PACAP receptors are responsible for approximately 80% of PACAP binding in the retina.[23] Radioligand binding studies have revealed the existence of PACAP receptors also in the human fetal retina, which has been confirmed by the detection of PACAP receptor mRNA in RT-PCR experiments.[24] Retinoblastoma cells also contain PACAP receptors.[25] PAC1 receptor mRNA and protein expression have been described in all layers of neonatal rat retina.[26] Similarly, PAC1 receptor and its mRNA can be detected in chicken retinas already at E6.[19] The mRNA for all PACAP receptors has been reported in the retinal pigment epithelium.[27]

Detailed localization studies have revealed strong expression of PAC1 receptor mRNA in the ganglion cell layer (GCL), inner nuclear layer (INL) and nerve fiber layer, while weaker expression in the inner- and outer plexiform (IPL and OPL), the outer nuclear (ONL) layers and the outer segments of the photoreceptors.[28,29] *In situ* RT-PCR shows that both the short and the hop variants of PAC1 receptor mRNA are found in the ganglion and amacrine cells.[29] Other studies have confirmed the presence of VPAC receptors in the retina.[30] In culture, PAC1 receptor expression has been shown in the Muller cells, which are the major retinal glial cells.[31]

PACAP and its receptors occur in various other tissues of the eye. PACAP-immunoreactivity has been found in the lacrimal gland, choroid, iris, ciliary body, conjunctiva, sclera, cornea, and retroocular arteries.[32,33] In the ganglia giving rise to ocular innervation, PACAP is present in the trigeminal, sphenopalatine, and ciliary ganglia.[32] PACAP binding has been shown in the choroid and iris, where PACAP stimulates cAMP formation.[23]

Effects of PACAP in the retina

PACAP27 and –38 stimulate cAMP production by approximately fivefold in the retina.[23] This effect has been shown in various species, such as rat, rabbit, pig, and calf, where both PACAP forms produced a robust stimulation of AC activity, with PACAP38 being more potent than PACAP27, and both peptides being more effective than VIP.[34] Other studies have revealed similar potency of PACAP27 and –38 in cAMP formation, and found that PACAP forms stimulate also inositol monophosphate levels.[30] Similar actions have been found in developing rat, chicken, and human retinas.[19,24,26] A 100-fold increase in AC activity has been observed in human retinoblastoma cells.[25]

cAMP response element binding protein (CREB) is a transcriptional factor that may be activated by intracellular cAMP. PACAP induces phosphorylation of CREB already 5 min after treatment in neonatal rat retinas.[26] Other studies indicate that endogenous PACAP is important in maintaining the antiapoptotic mechanisms.[35,36] Functioning PACAP receptors in both the neural retina and pigment epithelial cells have been confirmed by electroretinogram examinations, showing that intraocularly infused PACAP27 and –38 induced changes in the electrical waves reflecting the activity of both the pigment layer and the neural retina.[37]

In retinal pigment epithelial cells, PACAP inhibits the interleukin-1(IL-1)-induced IL-6, -8, and monocyte chemotactic protein-1 expression.[27] Muller glial cells are one of the target cells for PACAP. They possess PAC1 receptors, and PACAP induces IL-6 release in cultured Muller cells at very low (pmol) concentrations.[31,38,39] Radial IL-6 expression can be observed throughout the retina 2 and 3 days after PACAP injection.[38] At higher (nanomolar) concentrations, PACAP stimulates the proliferation of Muller cells.[38] PACAP stimulates cAMP production in Muller cells, while it has no effect on inositol trisphosphate (IP_3) production.[31]

PACAP has been shown to play a role in retinal development: PACAP treatment promotes tyrosine hydroxylase activity in developing chicken retinas, implying that it acts as a determinant of dopaminergic neuronal development.[19] A recent study on PAC1 transgenic mice also indicates that an appropriate level of PACAP signaling is required for normal retinogenesis and visual function.[40] The transient expression of PACAP-like immunoreactive elements in the developing zebrafish retina suggests that the peptide could be implicated in neurotrophic activities and neurosensorial connections in the migration and differentiation processes.[20]

PACAP, in a complex interplay with glutamate, plays a very important role in light-entrainment and circadian rhythmical functions. The literature of these effects of PACAP is very extensive, showing that this PACAP-containing subpopulation of ganglion cells plays an important biological role.[14,41,42] PACAP is required, for example, for the normal integration of the phase-advancing light signal by the suprachiasmatic nucleus.[43] The pineal gland consists of modified photoreceptors, the pinealocytes, where the presence and actions of PACAP have also been shown.[44–48] For example, PACAP stimulates melatonin secretion without interfering with its circadian rhythm, it regulates p38 MAPK and 14-3-3 protein expression depending on the daily cycle, and regulates clock genes in the pineal gland.[44–48]

Retinal blood flow is influenced by appropriate contractility of retinal pericytes surrounding the capillaries. PACAP leads to relaxation of retinal pericytes via PKA- and PLC-mediated pathways.[49]

PACAP has several effects on other ocular tissues, the detailed discussion of which is beyond the scope of this paper. In brief, PACAP27 and –38 decreases uveal vascular resistance, resulting in an increased choroidal blood flow.[50] Both PACAP27 and –38 enhance cholinergic transmission in iris sphincter muscle, and PACAP27 induces relaxation of the dilator muscle.[51] The effect of PACAP on sphincter muscle response has also been shown in monkeys.[52] Intravitreal injection of PACAP induces hyperemia and mimicked the symptoms of ocular inflammation in the rabbit eye, indicating an involvement of PACAP in ocular inflammatory responses.[33] Systemic injection of PACAP has been shown to modify certain protein components of the rat tear film.[53]

Local PACAP treatment induces outgrowth of neuronal processes in injured corneal nerve trunks that is accompanied by accelerated recovery of corneal sensitivity.[54] This shows that the well-known effect of PACAP on neurite outgrowth and neuronal recovery following injuries is also present *in vivo*, in ocular tissues.

The *in vitro* retinoprotective effects of PACAP

The first report on the retinoprotective effects of PACAP was a study showing that PACAP protected cultured retinal neurons against glutamate toxicity.[55] This study was the logical continuation of a previous study of the same authors demonstrating similar protective effects of VIP, a peptide structurally related to PACAP.[56] Elevated glutamate concentrations lead to excitotoxic cell death in the nervous system, including apoptotic cell death in the retina.[35,36]

Shoge *et al.*[55] found that 10 nmol/L–1 μmol/L PACAP27 and –38 attenuated the 1 mmol/L glutamate-induced cell death in a dose-dependent manner. This was antagonized by cotreatment with the PACAP antagonist PACAP6–38, and the PKA inhibitor H-89. In addition, PACAP increased the cAMP levels and MAPK activation.

Anisomycin has been found to induce cell death within the neuroblastic layer of retinal explants from newborn rats.[26] A dose-dependent prevention of cell death was found by PACAP38 treatment: 1–10 nmol/L PACAP resulted in a complete inhibition of cell death indicated by reduced TUNEL positivity and pyknotic profiles in the neuroblastic layer.[26] A similar protective effect was observed with PACAP27 and the specific PAC1 receptor agonist, maxadilan. Photoreceptor cell death, induced by thapsigargin, was also prevented by PACAP.[26] The neuroprotective effects observed in this retinal explant model from newborn rats could be blocked by PACAP6–38 and the specific PAC1 receptor antagonist, maxd.4. These retinoprotective effects are most probably mediated by the PAC1 receptor activation, which involves the cAMP/cAMP-dependent protein kinase pathways, because the activation of IP_3 pathway could not be shown and inhibition of PLC did not inhibit the protective effects of PACAP.[26]

In a turtle eyecup preparation, the luminosity-type horizontal cells show robust light responses. It has been demonstrated that keeping the preparations in a solution containing PACAP attenuates the deterioration of light responses under anoxic conditions.[57] Intracellular responses recorded after 42 h are still stable and significantly larger that those of the control horizontal cells. These results show that PACAP has neuroprotective properties in hypoxia/anoxic conditions in the turtle retina.[57]

The *in vivo* retinoprotective effects of PACAP

Excitotoxic retinal injury

Glutamate, as the main excitatory transmitter in the central nervous system, is able to exert toxic effects when present at high concentrations. The excitotoxic injury of the retina is a major factor in several retinal diseases, such as glaucoma and ischemic retinopathy. Monosodium glutamate (MSG), when given systemically to newborn rats, passes the blood–retina barrier and leads to severe retinal degeneration.[58] The entire inner retina suffers a significant loss.[59] Such severe reduction in the inner retinal layers can be achieved by a minimum of three times MSG treatment on postnatal days 1, 5, and 9 (Table 1).[59,60] The internal to outer limiting membrane distance is approximately half of the normal retinas, measured 3 weeks after birth. With the almost entire disappearance of the IPL, the fusion of the INL and GCL can be observed (Figs. 1A and E). The number of cells/100 μm in the GCL is also approximately half that of the normal retinas, as measured in light microscopical sections from central retina areas of the same eccentricities, where GCL appears in one row (Table 1).

Although PACAP has been shown to cross the blood–brain barrier, systemic PACAP treatment leads only to a slight amelioration of the retinal morphology following MSG-induced degeneration.[61] However, local PACAP38 treatment, reached by intravitreal injections of 1–100 pmol PACAP, results in a significant attenuation of this degeneration.[59,60,62] While 1 pmol PACAP leads to a slight improvement, treatment with 100 pmol PACAP

Table 1. Morphometric quantification of the whole retina thickness (OLM-ILM) in μm and the cell number of GCL/100 μm in different conditions

	Adult Wistar rats							Wistar rat pups		
	Control	BCCAO	BCCAO + PACAP	UV (1 week – diffuse)	UV (1 week – diffuse) + PACAP	Diabetes	Diabetes + 3× PACAP	Control	3× MSG	3× MSG + 3× PACAP
Distance of OLM-ILM	120.87 ± 3.05	50.45 ± 7.46	118.33 ± 5.72	158.55 ± 1.23	101.8 ± 1.45	116.1 ± 6.06	116.3 ± 5.31	98.43 ± 1.59	52 ± 0.82	82.17 ± 3.72
Number of cells/100 μm GCL	9.50 ± 1.36	2.1 ± 0.83	6.33 ± 1.56	2.95 ± 0.29	4.95 ± 0.29	6.6 ± 1.43	6.8 ± 1.14	7.88 ± 1.42	4.56 ± 1.07	7.36 ± 1.44

Notes: Statistical analysis was done by ANOVA test and Tukey's B post-hoc analysis. Significant decreases were observed in 3× MSG- and BCCAO-induced retinas compared to control. The neuroprotective effects of PACAP treatment were analyzed by the thickness of different retinal layers and also the cell number of GCL/100 μm. The diffuse UV-A radiation-induced retinal degeneration was shown in the swelling of the whole retina and reduction in the cell number of GCL, which was ameliorated by the intravitreal injection of PACAP.

results in an almost intact appearance of the retina (Fig. 1F). Similar protection can be achieved by PACAP27 treatment.[63] The PACAP antagonists, PACAP6–38 and PACAP6–27 leads to a further aggravation of the MSG-induced degeneration, indicating that endogenous PACAP plays an important protective role in the natural defense of the retina against damage.[63] Analysis of the neurochemically identified neuronal cell types suffering MSG-induced degeneration reveals that in MSG-treated retinas, the cell bodies and processes display reduced immunoreactivity for vesicular glutamate transporter 1 (expressed in OPL and IPL), tyrosine hydroxylase (dopaminergic amacrine cells, INL), vesicular GABA transporter (OPL and IPL), parvalbumin (AII amacrine cells, INL, and GCL), calbindin (expressed mainly in horizontal cells, INL), calretinin (cholinergic amacrine cells, INL, and GCL) and PKCα (bipolar cells, INL) (Figs. 2A–C). There is a possibility that PACAP equally protects against MSG treatment in all these examined cell types suggesting that the protective effect of PACAP is not subtype-specific, but rather implies a common defense mechanism present in all neurons.[64]

Another glutamate receptor agonist, kainic acid, has also been shown to lead to excitotoxic cell death in the retina by intraocular treatment.[65] It has been found that *in vivo* treatment of adult rats with 5 nmol kainic acid led to a substantial loss of ganglion cells, which express AMPA/kainate receptors.[66] This can be attenuated by pretreatment with 10 pmol intravitreal PACAP injection.[65] These results, along with the above-mentioned data, indicate that PACAP is protective against excitotoxic retinal lesion both in newborn and in adult rats.

Ischemia-induced retinal degeneration

Bilateral common carotid artery occlusion (BCCAO) leads to a moderate reduction in the cerebral blood flow in rats, resulting in subtle biochemical and behavioral measures.[67] Depending on the rat strain, this model produces hypoperfusion-induced retinal degeneration. We have found that permanent BCCAO leads to a severe retinal damage, with all retinal layers bearing the marks of deterioration (Figs. 1A and D).[67,68] The most marked reduction in thickness can be observed in the plexiform layers, and as a consequence, the distance between outer limiting membrane and inner limiting membrane is significantly less than in control preparations (Table 1). The photoreceptor layer also suffers degeneration: the outer segments become shorter and the geometric arrangement is disturbed. Intraocular PACAP treatment (100 pmol) leads to a nearly intact appearance of the retina (Fig. 1G), which can be confirmed by morphometric analysis of the retinal layers.[68] Using cell-type-specific markers in ischemia-induced degeneration has revealed that, similarly to the effects of PACAP in MSG-induced degeneration, the retinoprotective action does not seem to be phenotype-specific, but rather reflects a general cytoprotective mechanism.[69] The changes observed in the immunostaining for all examined antisera (vesicular glutamate transporter 1, tyrosine hydroxylase, vesicular GABA transporter, parvalbumin, calbindin, calretinin, PKCα, and glial fibrillary

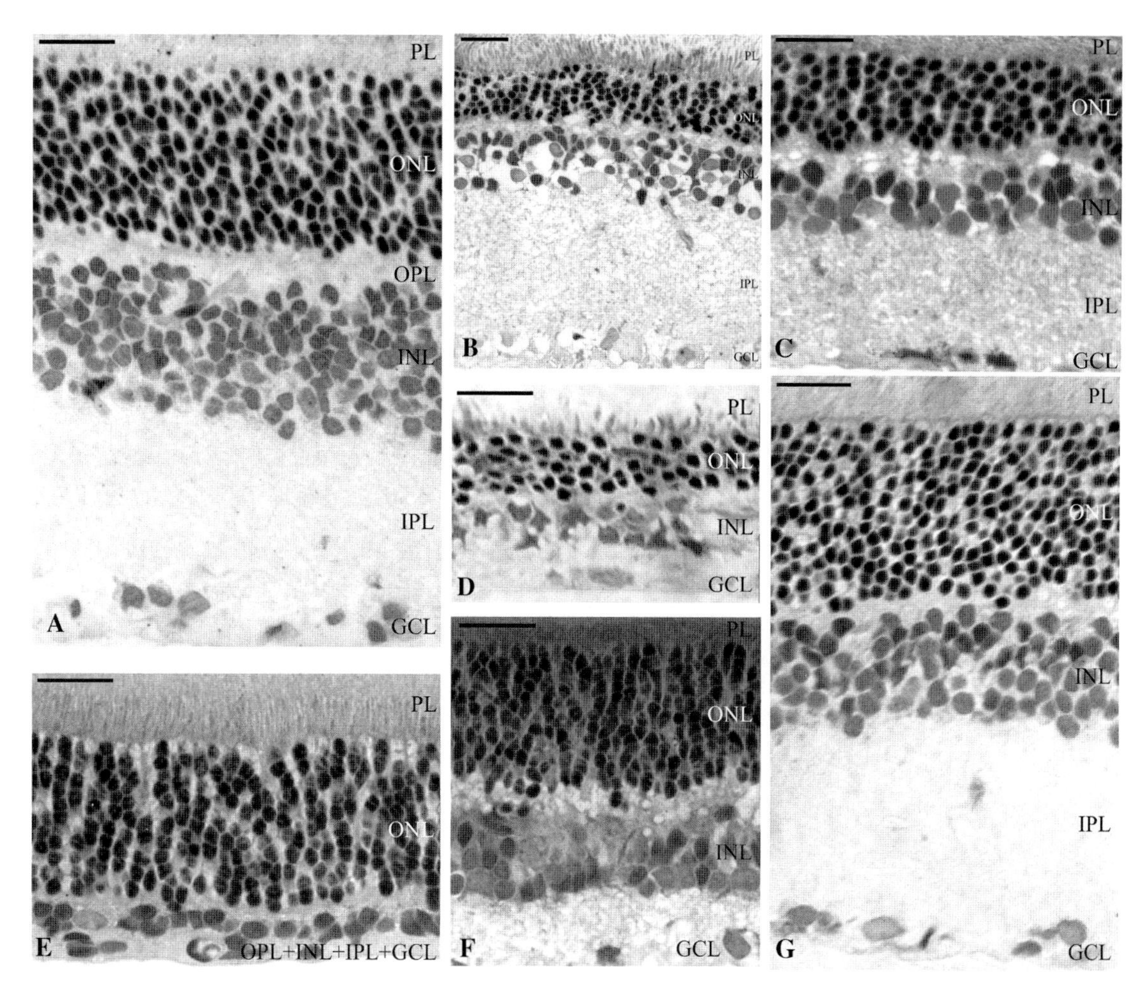

Figure 1. Histological sections from different retinal degenerations and treatments. Retinal tissue from control animal (A). Diffuse UV-A radiation (after 1 week) caused damaging effects: several neurodegenerative structures (empty cell body shapes, pyknotic cells) were observed in the INL; the IPL was swollen (B) compared to the intravitreal PACAP-treated retinas (C). BCCAO caused total retinal degeneration (D) ameliorated by intravitreal PACAP injection (G). 3× MSG induced severe retinal degeneration (E) attenuated by intravitreal 3× PACAP-treatment (F). Scale bar: 20 μm. PL, photoreceptor layer; ONL, outer nuclear layer; OPL, outer plexiform layer; INL, inner nuclear layer; IPL, inner plexiform layer; GCL, ganglion cell layer.

acidic protein) is counteracted by PACAP treatment (Figs. 2D–F).[69]

These actions of PACAP have been recently confirmed in another model of retinal ischemia. High intraocular pressure induces retinal ganglion cell death. A 60-min high pressure (110 mm Hg) results in a 45% cell death, which is significantly reduced by PACAP treatment.[70] These effects are mediated partly by cAMP- and MAPK-dependent pathways, because the inhibition of cAMP and MAPK leads to a concentration-dependent inhibition of the retinoprotective effects.[70]

These results are further supported by earlier observation with VIP. It had been found that in an ischemia/reperfusion injury of the rat retina, VIP counteracted the ischemia-induced elevations in malondialdehyde and catalase levels and VIP-treated rats had a smaller damage in the blood–retinal barrier.[71] Also, VIP-treated retinas had a more retained histological structure.[71]

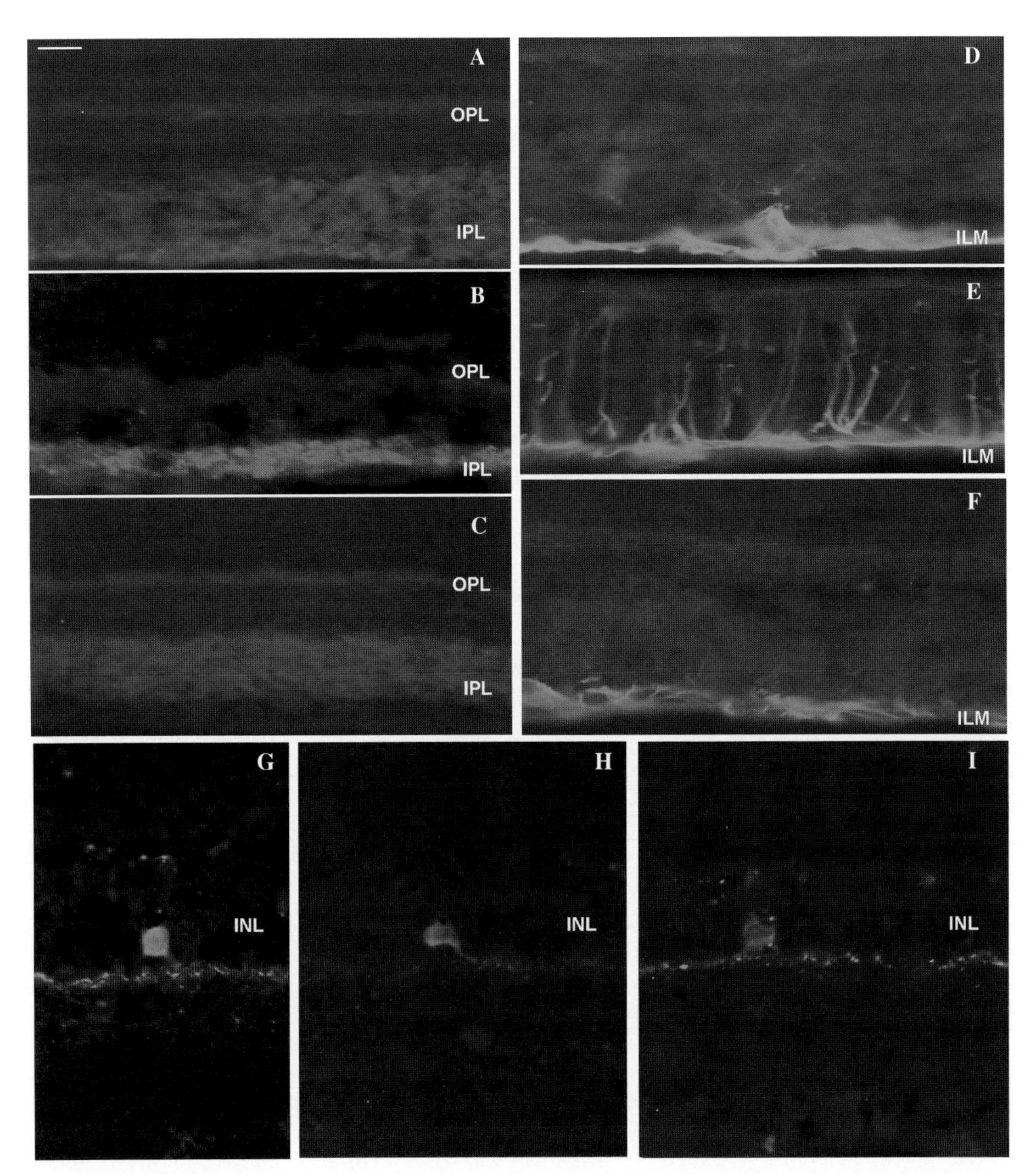

Figure 2. Representative fluorescent microphotographs of vesicular GABA transporter (VGAT), glial fibrillary acidic protein (GFAP), and tyrosine hydroxylase (TH) on cryosections of different conditions. Immunostaining for VGAT in control retina was apparent in the OPL and IPL (A). After MSG treatment, the immunoreaction was reduced (B), which was ameliorated by PACAP treatment (C). In control vertical retinal sections, the terminals of Muller glial cells showed weak GFAP immunostaining (D) compared to the observed strong immunolabeling in the entire of BCCAO retina (E). In PACAP-treated retinas GFAP levels were reduced in Muller glial cells and the retina was similar to that of the control ones (F). TH positive amacrine cells are normally observed in the innermost row of the INL (G). PACAP treatment counteracted (I) the degeneration caused by 1 month diabetes in the arborization and the synaptization of these TH-immunoreactive cells (H). Scale bar: 20 μm. PL, photoreceptor layer; ONL, outer nuclear layer; OPL, outer plexiform layer; INL, inner nuclear layer; IPL, inner plexiform layer; GCL, ganglion cell layer; ILM, inner limiting membrane.

Optic nerve transection

Optic nerve transection, performed by herniating the eye from the orbit and transecting the optic nerve without bleeding, has been reported to result in a significant, more than 50% loss of ganglion cells.[72] In a rat model of optic nerve transection, intravitreal 1–100 pmol PACAP treatment before the operation leads to a dose-dependent increase of ganglion cells, examined 14 days after transection.[72]

UV-A light-induced retinal degeneration

The protective effects of PACAP in the retina have been recently shown in two different kinds of UV-A radiation induced degeneration models: diffuse and focused UV-A-induced damage. Both types of UV-A irradiations are characterized by severe degeneration (pyknotic cells, empty cell body shapes) in the photoreceptor, the ONL and the INL. Interestingly, an increase in thickness could be detected in the IPL, which seemed to be swollen (Figs. 1A and B). Postirradiation intravitreal PACAP treatment significantly attenuates this light-induced degeneration (Fig. 1C; Table 1).[73]

Diabetic retinopathy

A recent study has shown that BDNF is able to counteract the deteriorating consequences of streptozotocin-induced diabetes in the adult rat retina.[74] In the early stages of diabetic retinopathy, the amacrine cells undergo characteristic degeneration. Severe damage of the dopaminergic amacrine cells can be observed in the INL, shown by the shape of their soma and their connections after streptozotocin treatment. Our observations, similarly to the results of Seki *et al.*,[74] show that intravitreal PACAP treatment is also able to protect the amacrine cell morphology (Figs. 2G–I).

Mechanisms of retino-protective effects of PACAP in excitotoxic retinal injury

Dozens of studies have examined the molecular mechanisms of PACAP-induced neuroprotection (Fig. 3).[3,6,75] Although cell-type specific mechanisms exist, the protective effects of PACAP seem to be mediated predominantly by PAC1 receptors, involving PKA and PKC pathways. A major contribution to this effect has been shown to come from the PKA/MAPK pathway and downstream, the inhibition of the apoptotis executor, caspase-3. In the retina, a part of this complex neuroprotective mechanism has been confirmed in an *in vivo* model, the MSG-induced degeneration. After neonatal MSG treatment, proapoptotic signaling pathways are upregulated and antiapoptotic pathways are downregulated. In brief, Western blot analysis has revealed the activation of the proapoptotic molecules c-Jun N-terminal kinase (JNK), apoptosis inducing factor (AIF), caspase-3, and the release of mitochondrial cytochrome c into the cytosol.[35,36,76] On the other hand, the protective PKA, CREB, and extracellular signal-regulated protein kinase (ERK) phosphorylation is strongly reduced. Intravitreal PACAP treatment can counteract all these changes, and this effect can be blocked by PACAP6–38. A recent study has revealed a formerly unknown mechanism in the neuroprotective effect of PACAP in this retinal model: the PKA/Bad/14-3-3 protein cascade (Fig. 3). 14-3-3 protein binds phosphorylated Bad, which, in turn is unable to bind, thus inactivate, the protective members of the Bcl family, Bcl-xL, and Bcl-2. MSG induces a strong reduction in the levels of 14-3-3 and the phosphorylation of Bad, which is significantly attenuated by PACAP treatment.[35] These studies together show that PACAP is able to counteract the glutamate-induced decrease in the antiapoptotic signaling and the increase in the proapoptotic signaling pathways, thus providing significant retinoprotection. These observations underlie the general cytoprotective effects of PACAP in the retina and explain the lack of cell subtype-specificity observed in other studies.

Further signaling pathways and mechanisms may add to the protective effects of PACAP in the retina. PACAP has been shown to stimulate IL-6 secretion in Muller glial cells.[38] IL-6, in turn, has neuroprotective effects in several nervous system pathologies, including brain ischemia and the retina, in spite of IL-6 generally playing a role as a proinflammatory cytokine.[4,38,77] In fact, PACAP has been shown to stimulate IL-6 in cerebral ischemic models, and it fails to exert neuroprotective effects in IL-6 knockout mice in a stroke model. To elucidate the exact protective mechanisms, by which PACAP is exerts such effective retinoprotective effects, await further investigation.

In summary, PACAP and its receptors are present in the retina and play important roles in retinal development, retinal light processing and blood supply. The well-known neuroprotective effects of

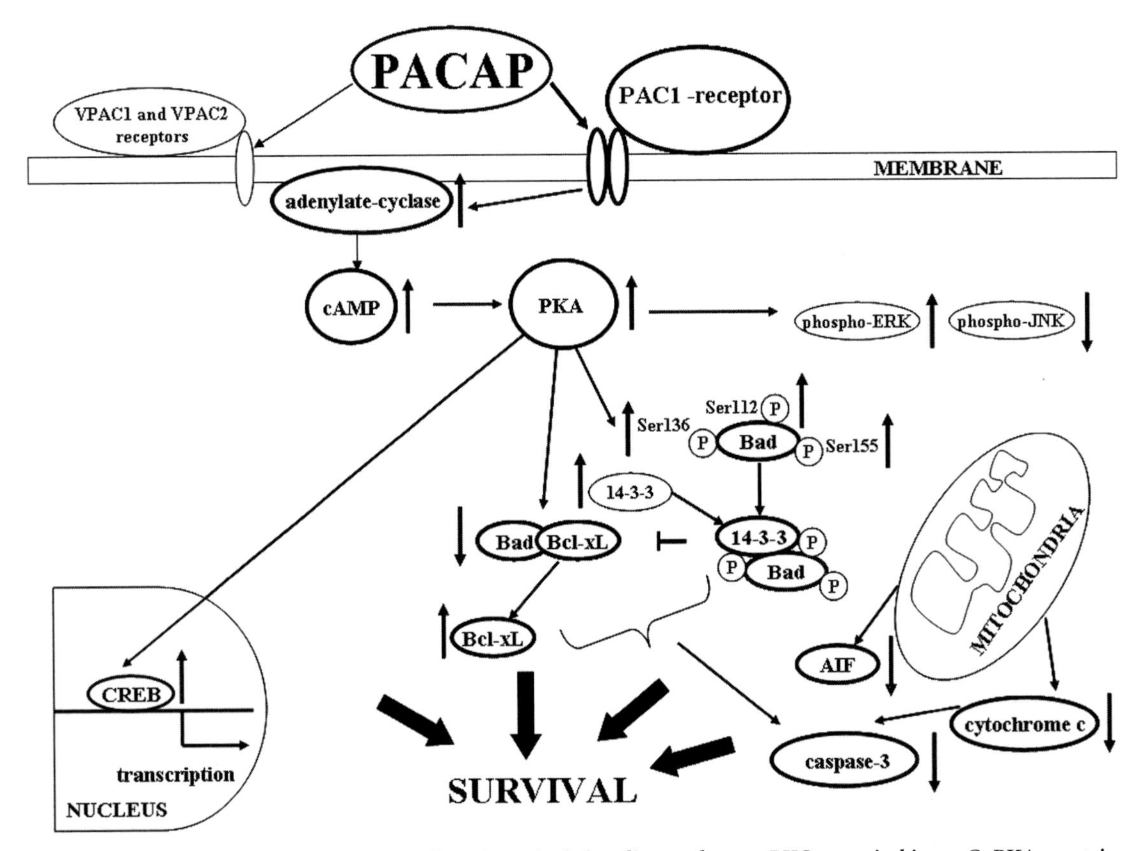

Figure 3. Schematic drawing of PACAP effects in retinal signaling pathways. PKC, protein kinase C; PKA, protein kinase A; cAMP, cyclic 3′,5′-adenosine monophosphate; PAC1, VPAC1, VPAC2, PACAP receptors; PLC, phosholipase C; JNK, caspase-3, AIF (apoptosis inducing factor), proapoptotic factors; ERK, Bad, Bcl-xL, antiapoptotic factors; CREB, cAMP response element binding protein.

PACAP are also present in the retina, as shown by numerous *in vitro* and *in vivo* retinal degeneration models. These studies strongly suggest that PACAP is an excellent candidate retinoprotective agent that could be a potential therapeutic substance in various retinal diseases.

Acknowledgments

This work was supported by Hungarian National Scientific Grants OTKA T061766, K72592, F67830, CNK 78480, ETT278–04/2009, Richter Gedeon Centenary Foundation, GVOP-3.2.1–2004-04–0172/3.0, Bolyai Scholarship, University of Pecs Medical School Research Grant 2009.

Conflicts of interest

The authors declare no conflicts of interest.

References

1. Miyata, A. *et al.* 1989. Isolation of a novel 38 residue-hypothalamic polypeptide which stimulates adenylate cyclase in pituitary cells. *Biochem. Biophys. Res. Commun.* **164:** 567–574.
2. Arimura, A. 2007. PACAP: the road to discovery. *Peptides* **28:** 1617–1619.
3. Vaudry, D. *et al.* 2009. Pituitary adenylate cyclase activating polypeptide and its receptors: 20 years after the discovery. *Pharmacol. Rev.* **61:** 283–357.
4. Ohtaki, H., T. Nakamachi, K. Dohi & S. Shioda. 2008. Role of PACAP in ischemic neural death. *J. Mol. Neurosci.* **36:** 16–25.
5. Shioda, S. *et al.* 2006. Pleiotropic functions of PACAP in the CNS: neuroprotection and neurodevelopment. *Ann. N.Y. Acad. Sci.* **1070:** 550–560.
6. Somogyvari-Vigh, A. & D. Reglodi. 2004. Pituitary adenylate cyclase activating polypeptide: a potential

neuroprotective peptide. Review. *Curr. Pharm. Des.* **10:** 2861–2889.
7. Waschek, J.A. 2002. Multiple actions of pituitary adenylyl cyclase activating peptide in nervous system development and regeneration. *Dev. Neurosci.* **24:** 14–23.
8. Izumi, S., T. Seki, S. Shioda *et al.* 2000. Ultrastructural localization of PACAP immunoreactivity in the rat retina. *Ann. N.Y. Acad. Sci.* **921:** 317–320.
9. Seki, T., S. Shioda, S. Izumi, *et al.* 2000. Electron microscopic observation of pituitary adenylate cyclase activating polypeptide (PACAP)-containing neurons in the rat retina. *Peptides* **21:** 109–113.
10. Mathis, U. & F. Schaeffel. 2007. Glucagon-related peptides in the mouse retina and the effects of deprivation of form vision. *Graefe's Arch. Clin. Exp. Ophthalmol.* **245:** 267–275.
11. Grone, B.P., S. Zhao, C.C. Chen & R.D. Fernald. 2007. Localization and diurnal expression of melanopsin, vertebrate ancient opsin, and pituitary adenylate cyclase activating peptide mRNA in a teleost retina. *J. Biol. Rhythms* **22:** 558–561.
12. Jozsa, R. *et al.* 2001. Distribution and daily variations of PACAP in the chicken brain. *Peptides* **22:** 1371–1377.
13. Reglodi, D., A. Somogyvari-Vigh, J. Vigh, *et al.* 2001. Pituitary adenylate cyclase activating polypeptide is highly abundant in the nervous system of the anoxia-tolerant turtle, Pseudemys scripta elegans. *Peptides* **22:** 873–878.
14. Hannibal, J. & J. Fahrenkrug. 2004. Target areas innervated by PACAP-immunoreactive retinal ganglion cells. *Cell Tissue Res.* **316:** 99–113.
15. Sakamoto, K., C. Liu, M. Kasamatsu *et al.* 2005. Dopamine regulates melanopsin mRNA expression in intrinsically photosensitive retinal ganglion cells. *Eur. J. Neurosci.* **22:** 3129–3136.
16. Wan, J. *et al.* 2006. Acute photoreceptor degeneration down-regulates melanopsin expression in adult rat retina. *Neurosci. Lett.* **400:** 48–52.
17. Vereczki, V. *et al.* 2006. Distribution of hypothalamic, hippocampal and other limbic peptidergic neuronal cell bodies giving rise to retinopetal fibers: anterograde and retrograde tracing and neuropeptide immunohistochemical studies. *Neuroscience* **140:** 1089–1100.
18. Bagnoli, P., M. Dal Monte & G. Casini. 2003. Expression of neuropeptides and their receptors in the developing retina of mammals. *Histol. Histopathol.* **18:** 1219–1242.
19. Borba, J.C. *et al.* 2005. Pituitary adenylate cyclase activating polypeptide (PACAP) can act as a determinant of the tyrosine hydroxylase phenotype of dopaminergic cells during retina development. *Dev. Brain Res.* **156:** 193–201.
20. Mathieu, M., L. Girosi, M. Vallarino & G. Tagliafierro. 2005. PACAP in developing sensory and peripheral organs of the zebrafish, Danio rerio. *Eur. J. Histochem.* **49:** 167–178.
21. Mathieu, M. *et al.* 2004. Pituitary adenylate cyclase activating polypeptide in the brain, spinal cord and sensory organs of the zebrafish, Danio rerio, during development. *Dev. Brain Res.* **151:** 169–185.
22. Skoglosa, Y., N. Takei & D. Lindholm. 1999. Distribution of pituitary adenylate cyclase activating polypeptide mRNA in the developing rat brain. *Mol. Brain Res.* **65:** 1–13.
23. Nilsson, F.E. *et al.* 1994. Characterization of ocular receptors for pituitary adenylate cyclase activating polypeptide (PACAP) and their coupling to adenylate cyclase. *Exp. Eye Res.* **58:** 459–467.
24. Olianas, M.C. *et al.* 1997. Expression of pituitary adenylate cyclase activating polypeptide (PACAP) receptors and PACAP in human fetal retina. *J. Neurochem.* **69:** 1213–1218.
25. Olianas, M.C., M.G. Ennas, G. Lampis & P. Onali 1996. Presence of pituitary adenylate cyclase activating polypeptide in Y-79 human retinoblastoma cells. *J. Neurochem.* **67:** 1293–1300.
26. Silveira, M.S. *et al.* 2002. Pituitary adenylate cyclase activating polypeptide prevents induced cell death in retinal tissue through activation of cyclic AMP-dependent protein kinase. *J. Biol. Chem.* **277:** 16075–16080.
27. Zhang, X.Y., S. Hayasaka, Z.L. Chi, *et al.* 2005. Effect of pituitary adenylate cyclase activating polypeptide (PACAP) on IL-6, and MCP-1 expression in human retinal pigment epithelial cell line. *Curr. Eye Res.* **30:** 1105–1111.
28. Seki, T., S. Izumi, S. Shioda, *et al.* 2000. Gene expression for PACAP receptor mRNA in the rat retina by in situ hybridization and in situ RT-PCR. *Ann. N.Y. Acad. Sci.* **921:** 366–369.
29. Seki, T., S. Shioda, D. Ogino, *et al.* 1997. Distribution and ultrastructural localization of a receptor for pituitary adenylate cyclase activating polypeptide and its mRNA in the rat retina. *Neurosci. Lett.* **238:** 127–130.
30. D'Agata, V. & S. Cavallaro. 1998. Functional and molecular expression of PACAP/VIP receptors in the rat retina. *Mol. Brain Res.* **54:** 161–164.
31. Kubrusly, R.C. *et al.* 2005. Expression of functional receptors and transmitter enzymes in cultured Muller cells. *Brain Res.* **1038:** 141–149.
32. Elsas, T., R. Uddmann & F. Sundler. 1996. Pituitary adenylate cyclase activating peptide-immunoreactive

nerve fibers in the cat eye. *Graefe's Arch. Clin. Exp. Ophthalmol.* **234:** 573–580.
33. Wang, Z.Y., P. Alm & R. Hakanson. 1995. Distribution and effects of pituitary adenylate cyclase activating peptide in the rabbit eye. *Neuroscience* **69:** 297–308.
34. Onali, P. & M.C. Olianas. 1994. PACAP is a potent and highly effective stimulator of adnylyl cyclase activity in the retinas of different mammalian species. *Brain Res.* **641:** 132–134.
35. Racz, B. *et al.* 2007. Effects of pituitary adenylate cyclase activating polypeptide (PACAP) on the PKA-Bad-14-3-3 signaling pathway in glutamate-induced retinal injury in neonatal rats. *Neurotox. Res.* **12:** 95–104.
36. Racz, B. *et al.* 2006. The neuroprotective effects of PACAP in monosodium glutamate-induced retinal lesion involves inhibition of proapoptotic signaling pathways. *Regul. Pept.* **137:** 20–26.
37. Jarkman, S., M. Kato & R. Bragadottir. 1998. Effects of pituitary adenylate cyclase activating polypeptide on the direct-current electroretinogram of the rabbit eye. *Ophthalmol. Res.* **30:** 199–206.
38. Nakatani, M. *et al.* 2006. Pituitary adenylate cyclase activating polypeptide (PACAP) stimulates production of interleukin-6 in rat Muller cells. *Peptides* **27:** 1871–1876.
39. Seki, T. *et al.* 2006. PACAP stimulates the release of interleukin-6 in cultured rat Muller cells. *Ann. N.Y. Acad. Sci.* **1070:** 535–539.
40. Lang, B. *et al.* 2010. GABAergic amacrine cells and visual function are reduced in PAC1 transgenic mice. *Neuropharmacology* **58:** 215–225.
41. Son, Y.J., J.W. Park & B.J. Lee. 2007. TTF-1 expression in PACAP-expressing retinal ganglion cells. *Mol. Cells* **23:** 215–219.
42. Koves, K. *et al.* 2000. PACAP and VIP in the photoneuroendocrine system. From the retina to the pituitary gland. *Ann. N.Y. Acad. Sci.* **921:** 321–326.
43. Beaule, C., J.W. Mitchell, P.T. Lindberg, *et al.* 2009. Temporally restricted role of retinal PACAP: integration of the phase-advancing light signal to the SCN. *J. Biol. Rhythms* **24:** 126–134.
44. Faluhelyi, N. *et al.* 2004. Development of the circadian melatonin rhythm and the effect of PACAP on melatonin production in the embryonic chicken pineal gland. An in vitro study. *Regul. Pept.* **123:** 23–28.
45. Csernus, V. *et al.* 2004. The effect of PACAP on rhythmic melatonin release of avian pineals. *Gen. Comp. Endocrinol.* **135:** 62–69.
46. Racz, B. *et al.* 2008. Effects of PACAP on the circadian changes of signaling pathways in chicken pinealocytes. *J. Mol. Neurosci.* **36:** 220–226.
47. Nagy, A.D. & V.J. Csernus. 2007. The role of PACAP in the control of circadian expression of clock genes in the chicken pineal gland. *Peptides* **28:** 1767–1774.
48. Rekasi, Z. & T. Czompoly. 2002. Accumulation of rat pineal serotonin N-acetyltransferase mRNA induced by pituitary adenylate cyclase activating polypeptide and vasoactive intestinal peptide in vitro. *J. Mol. Endocrinol.* **28:** 19–31.
49. Markhotina, N., G.J. Liu & D.K. Martin. 2007. Contractility of retinal pericytes grown on silicone elastomer substrates is through a protein kinase A-mediated intracellular pathway in response to vasoactive peptides. *IET Nanobiotechnol.* **1:** 44–51.
50. Nilsson, F.E. 1994. PACAP-27 and PACAP-38: vascular effects in the eye and some other tissues in the rabbit. *Eur. J. Pharmacol.* **253:** 17–25.
51. Yoshitomi, T. *et al.* 2002. Effect of pituitary adenylate cyclase activating peptide on isolated rabbit iris sphincter and dilator muscles. *Invest. Ophthalmol. Vis. Sci.* **43:** 780–783.
52. Yamaji, K., T. Yoshitomi & S. Usui. 2005. Action of biologically active peptide on monkey iris sphincter and dilator muscles. *Exp. Eye Res.* **80:** 815–820.
53. Gaal, V. *et al.* 2008. Investigation of the effects of PACAP on the composition of tear and endolymph proteins. *J. Mol. Neurosci.* **36:** 321–329.
54. Fukiage, C. *et al.* 2007. PACAP induces neurite outgrowth in cultured trigeminal ganglion cells and recovery of corneal sensitivity after flap surgery in rabbits. *Am. J. Ophthalmol.* **143:** 255–262.
55. Shoge, K. *et al.* 1999. Attenuation by PACAP of glutamate-induced neurotoxicity in cultured retinal neurons. *Brain Res.* **839:** 66–73.
56. Shoge, K. *et al.* 1998. Protective effects of vasoactive intestinal peptide against delayed glutamate neurotoxicity in cultured retina. *Brain Res.* **809:** 127–136.
57. Rabl, K. *et al.* 2002. PACAP inhibits anoxia-induced changes in physiological responses in horizontal cells in the turtle retina. *Regul. Pept.* **109:** 71–74.
58. Szabadfi, K. *et al.* 2009. Early postnatal enriched environment decreases retinal degeneration induced by monosodium glutamate treatment. *Brain Res.* **1259:** 107–112.
59. Babai, N. *et al.* 2006. Search for the optimal monosodium glutamate treatment schedule to study the neuroprotective effects of PACAP in the retina. *Ann. N.Y. Acad. Sci.* **1070:** 149–155.

60. Babai, N., T. Atlasz, A. Tamas, *et al.* 2005. Degree of damage compensation by various PACAP treatments in monosodium glutamate-induced retina degeneration. *Neurotox. Res.* **8:** 227–233.
61. Kiss, P. *et al.* 2006. Effects of systemic PACAP treatment in monosodium glutamate-induced behavioral changes and retinal degeneration. *Ann. N.Y. Acad. Sci.* **1070:** 365–370.
62. Tamas, A. *et al.* 2004. Effects of pituitary adenylate cyclase activating polypeptide in retinal degeneration induced by monosodium-glutamate. *Neurosci. Lett.* **372:** 110–113.
63. Atlasz, T. *et al.* 2009. Effects of pituitary adenylate cyclase activating polypeptide (PACAP1–38) and its fragments on retinal degeneration induced by neonatal MSG treatment. *Ann. N.Y. Acad. Sci.* **1163:** 348–352.
64. Atlasz, T. *et al.* 2008. PACAP-mediated neuroprotection of neurochemically identified cell types in MSG-induced retinal regeneration. *J. Mol. Neurosci.* **36:** 97–104.
65. Seki, T. *et al.* 2006. Neuroprotective effects of PACAP against kainic acid-induced neurotoxicity in rat retina. *Ann. N.Y. Acad. Sci.* **1070:** 531–534.
66. Zhang, D., Sucher, N.J. & Lipton, S.A. 1995. Co-expression of AMPA/kainate receptor-operated channels with high and low Ca^{2+} permeability in single rat retinal ganglion cells. *Neuroscience* **67:** 177–188.
67. Mester, L. *et al.* 2009. Protection against chronic hypoperfusion-induced retinal neurodegeneration by PARP inhibition via activation of PI3-kinase Akt pathway and suppression of JNK and p38 MAP kinases. *Neurotox. Res.* **16:** 68–76.
68. Atlasz, T. *et al.* 2007. Pituitary adenylate cyclase activating polypeptide is protective in bilateral carotid occlusion-induced retinal lesion in rats. *Gen. Comp. Endocrinol.* **153:** 108–114.
69. Atlasz, T. *et al.* 2010. Evaluation of the protective effects of PACAP with cell-specific markers in ischemia-induced retinal degeneration. *Brain Res. Bull.* **81:** 497–504.
70. Itoh, H. *et al.* 2009. Effects of pituitary adenylate cyclase activating polypeptide (PACAP) on the survival of ganglion cells in the rat retina after intraocular hypertension. Abstract NoP28, Phylogenetic Aspects of Neuropeptides/from Invertebrates to Humans, Satellite Symposium of the 9th International Symposium on VIP, PACAP and Related Peptides, Yakushima, Japan.
71. Tuncel, N. *et al.* 1996. Protection of rat retina from ischemia-reperfusion injury by vasoactive intestinal peptide (VIP): the effect of VIP on lipid peroxidation and antioxidant enzyme activity of retina and choroids. *Ann. N.Y. Acad. Sci.* **805:** 489–498.
72. Seki, T. *et al.* 2008. Suppression of ganglion cell death by PACAP following optic nerve transection in the rat. *J. Mol. Neurosci.* **36:** 57–60.
73. Atlasz, T. *et al.* 2009. PACAP protects rat retina from UV-A radiation-induced degeneration. Abstract NoP30, Phylogenetic Aspects of Neuropeptides/from Invertebrates to Humans, Satellite Symposium of the 9th International Symposium on VIP, PACAP and Related Peptides, Yakushima, Japan.
74. Seki, M. *et al.* 2004. Involvement of brain-derived neurotrophic factor in early retinal neuropathy of streptozotocin-induced diabetes in rats. *Diabetes* **53:** 2413–2419.
75. Falluel-Morel, A. *et al.* 2007. The neuropeptide pituitary adenylate cyclase-activating polypeptide exerts anti-apoptotic and differentiating effects during neurogenesis: focus on cerebellar granule neurones and embryonic stem cells. *J. Neuroendocrinol.* **19:** 321–327.
76. Racz, B. *et al.* 2006. Involvement of ERK and CREB signaling pathways in the protective effect of PACAP on monosodium glutamate-induced retinal lesion. *Ann. N.Y. Acad. Sci.* **1070:** 507–511.
77. Pucci, S., P. Mazzarelli, F. Missiroli, *et al.* 2008. Neuroprotection: VEGF, IL-6, and clusterin: the dark side of the moon. *Progr. Brain Res.* **173:** 555–573.

Ann. N.Y. Acad. Sci. ISSN 0077-8923

ANNALS OF THE NEW YORK ACADEMY OF SCIENCES
Issue: *Phylogenetic Aspects of Neuropeptides*

Ghrelin: more than endogenous growth hormone secretagogue

Masayasu Kojima[1] and Kenji Kangawa[2]

[1]Molecular Genetics, Institute of Life Science, Kurume University, Kurume, Fukuoka, Japan. [2]National Cardiovascular Center Research Institute, Suita, Osaka, Japan

Address for correspondence: Masayasu Kojima, Molecular Genetics, Institute of Life Science, Kurume University, Hyakunenkouen 1-1, Kurume, Fukuoka 839-0864, Japan. mkojima@lsi.kurume-u.ac.jp

Since its discovery 10 years ago, intensive research has been performed on ghrelin. The significance of ghrelin as a growth hormone–releasing hormone, appetite regulator, energy conservator, and sympathetic nerve suppressor has now been well established. In this review, we summarize recent topics on ghrelin, such as the processing protease of the ghrelin precursor, ghrelin *O*-acyl transaferase, ghrelin knockout and transgenic mice, and the molecular mechanism of ghrelin's orexigenesis.

Keywords: ghrelin; growth hormone releasing; appetite regulation

Introduction

It is just 10 years since the discovery of ghrelin.[1] During these years much research has been done to elucidate the physiological functions of ghrelin, not only a mere growth hormone–releasing hormone, but also an important appetite regulator, energy conservator, and sympathetic nerve suppressor.[2,3] At present, ghrelin is the only circulating orexigenic hormone secreted from the peripheral organ and acts on the hypothalamic arcuate nucleus, where is the regulatory region of appetite.

In our previous review in the *Trends in Endocrinology and Metabolism* at 2001,[4] we presented the prospection of ghrelin research for the next years, which were the regulatory mechanisms of ghrelin release, acyl-modification, and the search for physiological functions of ghrelin that are distinct from its GH-releasing activity. Some of these questions have now been answered and the results were published.

In this review, we summarize the recent topics on ghrelin and suggest the research direction on ghrelin in the next 10 years.

Processing steps from ghrelin precursor to mature ghrelin

Protease processing of ghrelin precursor

The amino acid sequences of mammalian ghrelin precursors are well conserved.[5] In mammalian ghrelin precursors, the 28-amino acid active ghrelin sequence immediately follows the signal peptide. The cleavage site for the signal peptide is the same in all mammalian ghrelins. Although propeptides are usually processed at dibasic amino acid sites by prohormone convertases, the C-terminus of the mammalian ghrelin peptide is processed at an uncommon Pro-Arg recognition site.

The processing proteases that participate for the processing of pro-ghrelin to 28-amino acid ghrelin peptide has been reported by Zhu *et al.*[6] They showed by using knockout (KO) mice of prohormone convertases, PC1/3, PC2 ,and PC5/6A, that the pro-ghrelin has not been processed only in PC1/3 KO mouse. They extracted the peptide fraction from the stomachs of prohormone convertase KO mice and examined the molecular weight of ghrelin by a western blotting. They found that in PC1/3 KO mouse the molecular weights of ghrelin

doi: 10.1111/j.1749-6632.2010.05516.x

 © 2010 New York Academy of Sciences.

was approximately 11 kDa and those in PC2 and PC5/6A KO mice were both approximately 3.4 kDa. These results suggest that pro-ghrelin is not processed in the stomach of PC1/3 KO mouse. Moreover, by immunohistochemical studies PC1/3 and ghrelin are colocalized in the same cells of the stomach.[7] Thus, PC1/3 is likely a processing protease responsible for the pro-ghrelin processing in the stomach.

Acyl-modification of ghrelin by ghrelin O-acyltransferase (GOAT)

Ghrelin is the first known and only case of a peptide hormone modified by a fatty acid and this acyl-modification of ghrelin is essential for the activity of ghrelin.[1] However, the enzyme that catalyses acyl transfer to ghrelin has not been identified until February 2008 when two independent groups reported the identification as Membrane-Bound *O*-Acyltransferase (MBOAT4) of ghrelin *O*-acyltransferase.[8,9] GOAT is a member of membrane-bound acyltransferases that comprises at least 16 acyltransferase enzymes. Among them, only GOAT shows ability to acyl-modify ghrelin. As like the existence of ghrelin in all vertebrate species, GOAT is found in mammalian, bird, and fish. Distribution of GOAT is similar to that of ghrelin: GOAT is predominantly distributed in gastrointestinal organs, in particular in the stomach.[9,10]

Substrate specificity of GOAT

GOAT specifically modifies the third amino acid serine and does not modify other serine residues (the second, sixth, and eighteenth amino acids in human ghrelin are serines).[9] When the serine residues at the second, sixth, and eighteenth amino acids in ghrelin are replaced into alanine, the third serine residue is correctly modified by *n*-octanoic acid. However, when the third serine was replaced into alanine, the third amino acid alanine was not modified by GOAT. The third amino acid of frog ghrelin is threonine and modified by *n*-octanoic acid such as Ser3 ghrelins.[11] When the third amino acid serine of rat ghrelin was replaced to threonine, GOAT modified the replaced threonine by *n*-octanoic acid. Thus, the position of serine or threonine in the third amino acid is important for *n*-octanoyl modification of ghrelin and GOAT modify not only serine but also threonine residue.

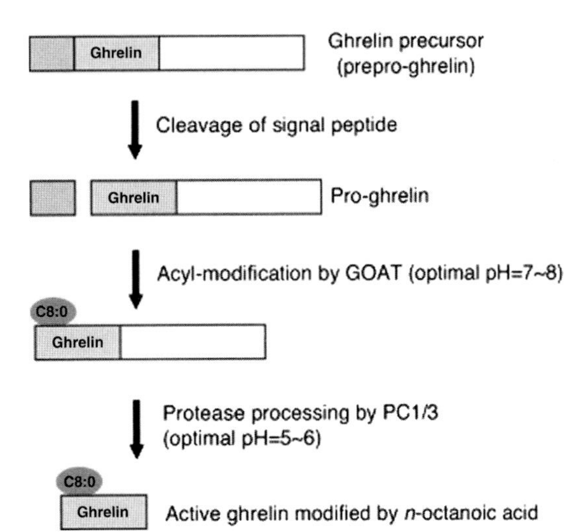

Figure 1. Diagram of a typical secretory cell and proposed processing steps of ghrelin. Considering the optimal pHs of GOAT (pH 7–8) and PC1/3 (pH 5–6), the acyl-modification of ghrelin should occur before protease processing. After removing the signal peptide, ghrelin precursor is modified with an acyl acid (mainly *n*-octanoic acid) by GOAT and cleaved by a processing protease PC1/3 to produce active ghrelin.

The optimal pH of GOAT is 7–8,[12] and the optimal pH of the processing protease PC1/3 is rather acidic (pH 6). Because the pH value becomes more acidic from the golgi to the secretary granules, the acyl-modification of ghrelin should occur before protease processing. In other word, ghrelin seems to be modified in the forms of prepro-ghrelin and pro-ghrelin. In fact, pro-ghrelin has been modified by *n*-octanoic acid.[6] Thus, the acyl-modification precedes the protease cleavage (Fig. 1).

Hypothalamic action of ghrelin

From peripheral tissues to central nervous system

Circulating ghrelin stimulates hypothalamic neurons and stimulates food intake. In general, the rate at which circulating peptides pass the blood–brain barrier is low and the rate at which circulating ghrelin passes the barrier has been shown to be very low. Thus, peripheral ghrelin should activate the appropriate hypothalamic regions in part via an indirect pathway.

The detection of ghrelin receptors on vagal afferent neurons in the rat nodose ganglion suggests that ghrelin signals from the stomach are transmitted to the brain via the vagus nerve (Fig. 2).[13–15] Moreover, the observation that ICV administration of ghrelin induces c-Fos in the dorsomotor nucleus of the vagus and stimulates gastric acid secretion indicates that ghrelin activates the vagus system.[16]

Vagotomy inhibits the ability of ghrelin to stimulate food intake and GH release.[13] A similar effect was also observed when capsaicin, a specific afferent neurotoxin, was applied to vagus nerve terminals to induce sensory denervation. However, the basal level of ghrelin concentration is not affected after vagotomy. On the other hand, fasting-induced elevation of plasma ghrelin is completely abolished by subdiaphragmatic vagotomy or atropine treatment.[17]

Moreover, peripheral ghrelin signaling, which travels to the nucleus tractus solitarius (NTS) via the vagus nerve, increases noradrenaline (NA) in the arcuate nucleus of the hypothalamus (Fig. 2).[18] Bilateral midbrain transections rostral to the NTS, or toxin-induced loss of neurons in the hindbrain that express dopamine β-hydroxylase (an NA synthetic enzyme), abolished ghrelin-induced feeding. Thus, the noradrenergic system is necessary in the central control of feeding behavior by peripherally administered ghrelin. These results indicate that the response of ghrelin to fasting is transmitted through vagal afferent transmission.

In the hypothalamic appetite regulatory cells

Recently, AMP-activated protein kinase (AMPK) has been shown to be involved in hypothalamic appetite regulation.[19,20] Injection of 5-amino-4-imidazole carboxamide riboside (AICAR), an activator of AMPK, significantly increases food intake. Administration of ghrelin *in vivo* increases AMPK activity in the hypothalamus.[21] By contrast, injection of leptin decreases hypothalamic AMPK activity and suppresses food intake.[19]

Lipid metabolism in the hypothalamus is thought to be important for appetite regulation.[22] After binding of ghrelin, the ghrelin receptor activates hypothalamic AMPK, and then it suppresses lipid synthesis.[23] By these results, malonyl-CoA level in the hypothalamus decreases and activity of carnitine palmitoyltransferase 1 (CPT1) is increased. Activated CPT1 then accelerates lipid transport into mitochondria to catabolize lipid.

UCP2 in the hypothalamus may be the key protein for appetite stimulation by ghrelin, because ghrelin regulates mitochondrial oxidation in the hypothalamic cells through UCP2.[24] Ninety percent of UCP2 expressing cells in the hypothalamus also expresses the ghrelin receptor. By using isolated hypothalamic synaptosomes, ghrelin increased oxygen consumption and total respiration.

It has been known that the orexigenic effect by ghrelin is exerted through NPY, and NPY mRNA level in the arcuate nucleus is increased by ghrelin. However, in UCP2-deficient mice ghrelin administration did not increase NPY level and the number of c-Fos protein positive cells in the arcuate nucleus NPY neurons were decreased.

Moreover, in the UCP2-deficient mice AMPK activity was not changed by ghrelin administration. AICAR stimulates mitochondrial respiration in the hypothalamus and increases appetite, whereas UCP2-deficient mice did not show any effects by AICAR. In contrast, AMPK inhibitor, compound C, suppressed appetite stimulation by ghrelin, however, in UCP2-deficient mice appetite suppression by compound C was not observed. These results suggest that AMPK activation by ghrelin is upstream of UCP2. In other words, ghrelin stimulates UCP2 through AMPK.

By lipid consumption in the hypothalamic mitochondria, free radicals production was increased. The produced free radicals are degradated by UCP2. In UCP2-deficient mice, ghrelin produced more free radical than in wild-type (WT) mice.

Taken together these results, lipid metabolism and mitochondrial free radical production in the hypothalamic neurons are key factors for the regulation of appetite (Fig. 2).

Life without ghrelin

Ghrelin and ghrelin receptor KO mice

It seems to be a curious thing that ablation of ghrelin causes no obvious phenotypes. Ghrelin KO mice showed normal size, growth rate, food intake, body composition, reproduction, and gross behavior, without any pathological changes (Table 1).[25–27] Not only ghrelin KO mouse, but also NPY or AgRP KO mice shows no abnormality in body weights and feeding behaviors.[28,29]

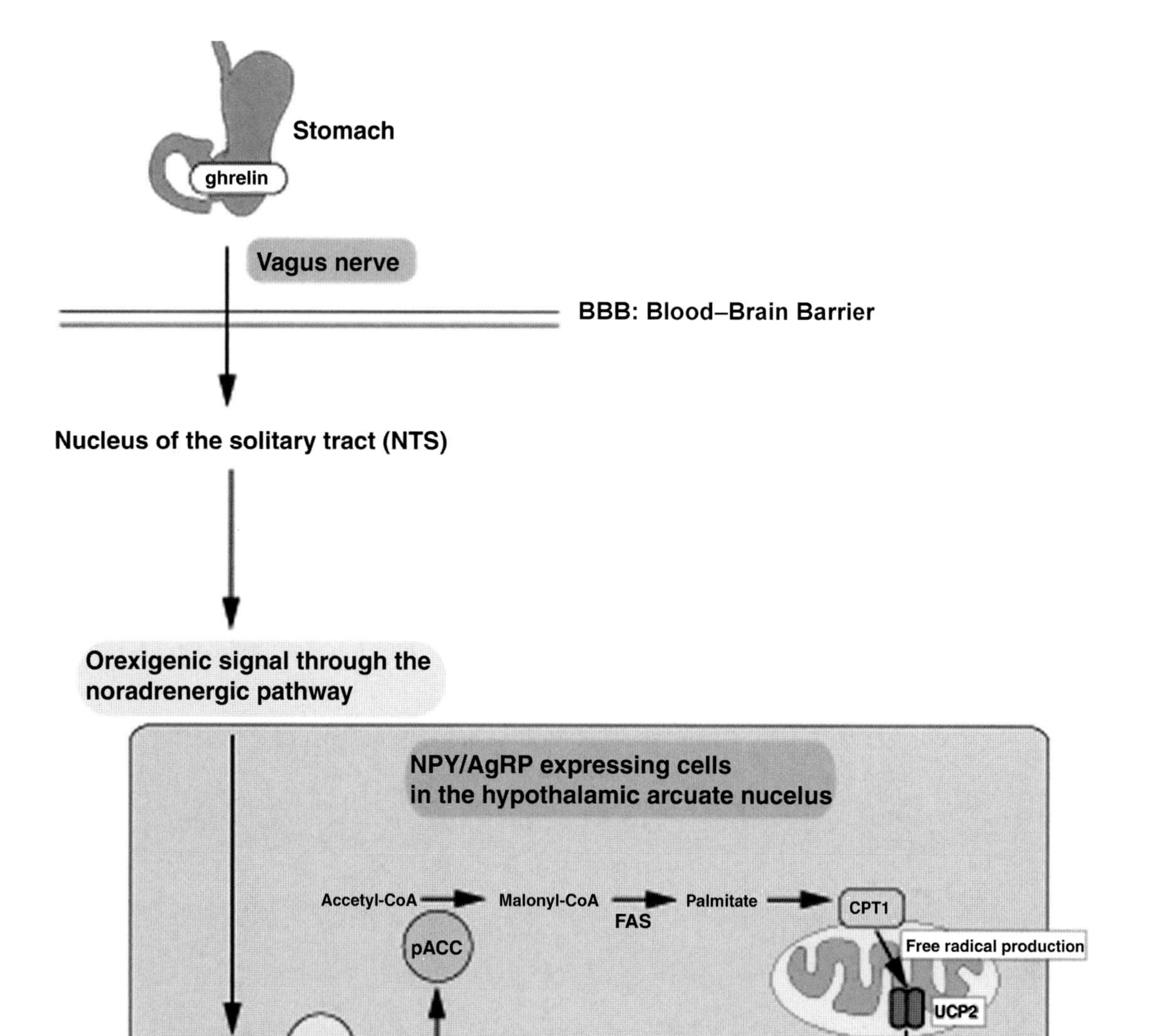

Figure 2. Orexigenic signal transduction of ghrelin from the stomach to the NPY/AgRP cells in the hypothalamic arcuate nucleus. Most of peptide hormones cannot directly pass through the blood–brain barrier. Ghrelin secreted from the stomach stimulates the vagus nerve and transmits a hunger signal to the nucleus of the solitary tract (NTS) in the medulla. The signal is transmitted to the hypothalamic appetite regulatory region through the noradrenergic pathway. In NPY/AgRP cells in the hypothalamic arcuate nucleus, ghrelin stimulates AMPK and regulates fatty acid metabolism, leading to CPT1 activation. Ghrelin's effects on mitochondrial respiration are dependent on the function of UCP2, and hence result in an increase of food intake.

Table 1. Summary of the phenotype of ghrelin and GHS-R knockout mice

Gene	Growth	Feeding behavior	Metabolic phenotype	Other phenotype	References
Ghrelin	Normal appearance Normal organ weights Normal tissue pathology Normal body composition Normal fertility	Normal No impairment of reflexive hyperphagia after fasting	Normal serum parameters Normal level of hypothalamic appetite regulatory peptides Decrease of respiratory quotient on a high-fat diet Low blood glucose level under caloric restriction Resistance to diet-induced obesity, when the diet was given just after weaning	Slightly elevated spontaneous wakefulness and rapid-eye-movement sleep (REMS) Reduced non-rapid-eye-movement sleep (NREMS) Loss of thymic epithelial cells Increase of adipogenic fibroblasts in the thymus Reduced number of naive T cells	25–27, 32
GHS-R (ghrelin receptor)	Normal appearance Modestly reduced body weights	Normal	Reduced serum insulin-like growth factor 1 levels Low blood glucose level under caloric restriction Resistance to diet-induced obesity, when the diet was given just after weaning	Loss of thymic epithelial cells Increase of adipogenic fibroblasts in the thymus Reduced number of naive T cells	31–32

Wortley *et al.*[27] reported that ghrelin KO mouse showed a reduction in respiratory quotient and a trend for lower body fat mass when the mouse was fed with a high-fat diet. Inconsistent with this results, Sun *et al.*[26] reported that no differences in body weight and respiratory quotient were observed between WT and ghrelin KO mice fed with a high-fat diet. However, they observed that under caloric restriction blood glucose levels were lower in ghrelin KO mice.[30] These discrepancy could be due to the differences in genetic background of the mice or the ages of mice used.

Mice lacking GHS-R do not show the typical increases in GH release and food intake upon ghrelin administration, indicating that GHS-R is indeed the primary biologically relevant ghrelin receptor.[31] Growth and development of GHS-R KO mice are normal, and their appetite and body composition are not different from those of their WT littermates (Table 1). Thus, ghrelin and its receptor are not critical for growth and appetite regulation.

Recently, an interesting phenotype in ghrelin and GHSR KO mice has been reported. Younm *et al.*[32] demonstrated the expression of ghrelin and its receptor in thymic stromal cells and found that their expression declines with aging. Genetic ablation of ghrelin and GHSR causes loss of thymic epithelial cells (TECs), increase of adipogenic fibroblasts in the thymus and reduction of naive T cells. These facts suggest that ghrelin may inhibit age-related EMT and thymic adipogenesis to maintain thymic function.

Anyway, ghrelin is not a critically required factor for the regulations of energy intake and energy expenditure, but may function in nutrient sensing and anti-aging.

Life with too much ghrelin

Ghrelin transgenic mice

Several lines of ghrelin transgenic mice have already been produced (Table 2). However, early attempts to overexpress ghrelin by using under the control of chicken β-actin, rat insulin II, or rat glucagon promoters produced high concentration of des-acyl ghrelin and failed to overexpress active *n*-octanoyl ghrelin.[33,34] The high des-acyl ghrelin mice with the chicken β-actin promoter exhibited lower body weights, shorter nose-to-anus lengths, and low serum IGF-I levels.[33] It has been reported that pancreatic ghrelin colocalizes with glucagon in islet α cells and *n*-octanoyl ghrelin is detected in pancreatic peptide extracts. However, the transgenic mice with the glucagon promoter show no increase of *n*-octanoyl ghrelin.[34]

Recently, by a bacterial artificial chromosome transgenetic method, new ghrelin transgenic mice were produced.[35] These mice expressed high concentrations of active *n*-octanoyl ghrelin in the stomach and brain, and showed hyperphagia, glucose intolerance, and reduced leptin sensitivity. Thus, high concentrations of *n*-octanoyl ghrelin are thought to stimulate orexigenic behavior.

Ghrelinoma

Two reports have been published on patients with ghrelin-producing tumors.[36,37] One case was a metastasizing gastric neuroendocrine tumor with high ghrelin concentration.[37] The patient developed type 2 diabetes mellitus (T2DM) with high blood sugar levels. Another case was a pancreatic ghrelinoma and the patient also suffered from T2DM.[36] The body mass indexes of these two patients were 32 and 29.9 kg/m^2, respectively. Thus, these patients with hyperghrelinemia were overweight or obese and suffered from T2DM, suggesting ghrelin's role in inducing hyperphagia and glucose intolerance.

Abnormal ghrelin secretion in Prader–Willi syndrome

Prader–Willi syndrome (PWS) is a complex genetic disorder characterized by mild mental retardation, hyperphagia, short stature, muscular hypotonia, and distinctive behavioral features.[38] Excessive appetite in PWS causes progressive severe obesity, which in turn leads to an increase of cardiovascular morbidity and mortality. The PWS genotype is characterized by a loss of one or more paternal genes in region q11–13 on chromosome 15. It has been suggested that this genetic alteration leads to dysfunction of several hypothalamic areas, including appetite regulatory regions.

High plasma ghrelin concentration is observed in PWS patients.[39,40] The mean plasma concentration of ghrelin was higher by three- to fourfold in PWS than in a reference population. Thus, ghrelin may be responsible, at least in part, for the hyperphagia seen in PWS. However, Feigerlova *et al.*[41] reports that high ghrelin concentration was observed even during the first year of life and precedes the onset of obesity. This result indicates that genetic abnormality affects on the production of ghrelin. It is unclear, however, what underlies the increased ghrelin levels in these patients. Imprinting of paternal genes in region q11–13 on chromosome 15 may induce the production of excessive amounts of transcription factors that increase ghrelin expression or, alternatively, a loss of a transcription inhibitory factor that normally suppresses ghrelin expression. In this context, it is noteworthy that the deletion of small nucleolar RNA Snord116 region, which is the comparable region of imprinted chromosome 15q11.2 in PWS, causes high circulating ghrelin concentration, a similar phenotype of PWS.[42] Elucidation of the precise mechanism by which ghrelin gene expression is regulated may reveal the genetic cause of hyperphagia in PWS.

Table 2. Summary of the phenotype of ghrelin transgenic mice

Promoter	Transgene expression	Acyl-ghrelin level	Des-acyl ghrelin level	Growth	Feeding behavior	Metabolic phenotype	References
Chicken β-actin	Stomach, small intestine, cerebrum, hypothalamus, pituitary, liver, kidney, lung, heart, skeletal muscle	→	↑	Reduced body weight Short nose-to-anus length	Normal daily food intake	Low GH level Low serum IFG-1 level Low response of GH release by ghrelin	33
Rat insulin II	Pancreatic β cells, plasma	↑ (Slightly)	↑	Normal growth No changes in pancreatic architecture and β-cell mass	Normal daily food intake	Suppression of glucose-stimulated insulin secretion Lower tendency of blood glucose level by insulin tolerance test Normal insulin level Lower tendency of plasma total cholesterol and triglyceride levels	34
Rat glucagon	Pancreatic α cells, plasma	↑ (Slightly in pancreas)	↑	Normal growth No changes in pancreatic architecture and β-cell mass	Normal daily food intake	Normal insulin level Normal plasma total cholesterol and triglyceride levels	34
Mouse ghrelin	Stomach, hypothalamus, plasma	↑	↑	No change in body weight No change in nose-to-anus length	Hyperphagia	Increase of energy expenditure Glucose intolerance. Decreased glucose-stimulated insulin release Reduced leptin sensitivity	35

Ghrelin and pancreatic cell development

Ghrelin is expressed in islet alpha (α) cells in the pancreas, where it has also been shown to colocalize with glucagon.[43] Recent findings indicate that ghrelin is also produced in a newly identified pancreatic islet cell, the ε cell.[44,45] The homeodomain protein Nkx2.2 is essential for the differentiation of all insulin-producing β cells and a subset of glucagon-producing α cells.[45,46] Nkx2.2 deficient mice show normal islet size, but a large number of cells fail to produce the four major islet hormones: insulin, glucagon, somatostatin, and pancreatic polypeptide (PP). An intriguing observation is that the pancreatic endocrine cells in *Nkx2.2*$^{-/-}$ mice are replaced by ε cells that produce ghrelin. Another transcription factor, Pax4, is necessary for the differentiation of β cells from the progenitor pancreatic cells and lack of Pax4 thereby decreases β-cell number and increases the number of ghrelin-producing ε cells. These results indicate that insulin- and ghrelin-producing cells are derived from a common progenitor cell and Nkx2.2 and Pax4 are involved in switching the cell fate to differentiate into either a β or an ε cell.

Future

Is ghrelin really necessary? Because ghrelin deficiency does not cause any obvious phenotypes in mice, it is a reasonable question whether ghrelin has indispensable roles *in vivo*. One possible explanation for the lack of observable altered phenotypes in ghrelin KO mice is that compensatory mechanisms may exist to make up for ghrelin's loss. Total ghrelin deficiency during fetal development may generate a robust compensatory system to avoid feeding abnormality or metabolic disorders. Time-dependent conditional ghrelin KO mice should help to elucidate the phenotypes of adult onset ghrelin deficiency.

A second unanswered question is the mechanism by which ghrelin secretes from the stomach. Fasting indeed stimulates ghrelin secretion, however, it is still not clear what humoral or nervous factors are involved in its regulation. Moreover, it is also unknown which tissue is responsible for detecting and relaying fasting signals to stimulate ghrelin release. It has been suggested that genetic alteration of 15q11–13 in PWS patients leads to dysfunction of several hypothalamic areas, including appetite regions, and the observed abnormal upregulation of ghrelin levels. Thus, the hypothalamus may play, in part, an important role in stimulating ghrelin secretion.

Finally, clinical applications of ghrelin to target anorexia and cachexia due to chronic diseases are currently under trial at Phase II. It may be in our very near future that ghrelin is used to effectively treat these diseases.

Conflicts of interest

The authors declare no conflicts of interest.

References

1. Kojima, M. *et al.* 1999. Ghrelin is a growth-hormone-releasing acylated peptide from stomach. *Nature* **402:** 656–660.
2. Litwack, G. (ed.). 2008. *Ghrelin*. Academic Press. Amsterdam.
3. Kojima, M. & K. Kangawa. 2005. Ghrelin: structure and function. *Physiol. Rev.* **85:** 495–522.
4. Kojima, M. *et al.* 2001. Ghrelin: discovery of the natural endogenous ligand for the growth hormone secretagogue receptor. *Trends Endocrinol. Metab.* **12:** 118–122.
5. Kojima, M., T. Ida & T. Sato. 2008. Structure of mammalian and nonmammalian ghrelins. *Vitam. Horm.* **77:** 31–46.
6. Zhu, X. *et al.* 2006. On the processing of proghrelin to ghrelin. *J. Biol. Chem.* **281:** 38867–38870.
7. Walia, P. *et al.* 2009. Ontogeny of ghrelin, obestatin, preproghrelin, and prohormone convertases in rat pancreas and stomach. *Pediatr. Res.* **65:** 39–44.
8. Gutierrez, J.A. *et al.* 2008. Ghrelin octanoylation mediated by an orphan lipid transferase. *Proc. Natl. Acad. Sci. USA* **105:** 6320–6325.
9. Yang, J. *et al.* 2008. Identification of the acyltransferase that octanoylates ghrelin, an appetite-stimulating peptide hormone. *Cell* **132:** 387–396.
10. Sakata, I. *et al.* 2009. Co-localization of ghrelin *O*-acyltransferase (GOAT) and ghrelin in gastric mucosal cells. *Am. J. Physiol. Endocrinol. Metab.* **297:** E134–E141.
11. Kaiya, H. *et al.* 2001. Bullfrog ghrelin is modified by *n*-octanoic acid at its third threonine residue. *J. Biol. Chem.* **276:** 40441–40448.
12. Ohgusu, H. *et al.* 2009. Ghrelin *O*-acyltransferase (GOAT) has a preference for n-hexanoyl-CoA over n-octanoyl-CoA as an acyl donor. *Biochem. Biophys. Res. Comm.* **386:** 153–158.
13. Date, Y. *et al.* 2002. The role of the gastric afferent vagal nerve in ghrelin-induced feeding and growth hormone secretion in rats. *Gastroenterology* **123:** 1120–1128.
14. Sakata, I. *et al.* 2003. Growth hormone secretagogue receptor expression in the cells of the stomach-projected afferent nerve in the rat nodose ganglion. *Neurosci. Lett.* **342:** 183–186.
15. Zhang, W. *et al.* 2004. Ghrelin stimulates neurogenesis in the dorsal motor nucleus of the vagus. *J. Physiol.* **559:** 729–737.
16. Date, Y. *et al.* 2001. Ghrelin acts in the central nervous system to stimulate gastric acid secretion. *Biochem. Biophys. Res. Commun.* **280:** 904–907.
17. Williams, D.L. *et al.* 2003. Vagotomy dissociates short- and long-term controls of circulating ghrelin. *Endocrinology* **144:** 5184–5187.
18. Date, Y. *et al.* 2006. Peripheral ghrelin transmits orexigenic signals through the noradrenergic pathway from the hindbrain to the hypothalamus. *Cell Metab.* **4:** 323–331.
19. Minokoshi, Y. *et al.* 2004. AMP-kinase regulates food intake by responding to hormonal and nutrient signals in the hypothalamus. *Nature* **428:** 569–574.

20. Xue, B. & B.B. Kahn. 2006. AMPK integrates nutrient and hormonal signals to regulate food intake and energy balance through effects in the hypothalamus and peripheral tissues. *J. Physiol.* **574:** 73–83.
21. Andersson, U. *et al.* 2004. AMP-activated protein kinase plays a role in the control of food intake. *J. Biol. Chem.* **279:** 12005–12008.
22. Lopez, M. *et al.* 2008. Hypothalamic fatty acid metabolism mediates the orexigenic action of ghrelin. *Cell Metab.* **7:** 389–399.
23. Kola, B. *et al.* 2005. Cannabinoids and ghrelin have both central and peripheral metabolic and cardiac effects via AMP-activated protein kinase. *J. Biol. Chem.* **280:** 25196–25201.
24. Andrews, Z.B. *et al.* 2008. UCP2 mediates ghrelin's action on NPY/AgRP neurons by lowering free radicals. *Nature* **454:** 846–851.
25. Sato, T. *et al.* 2008. Ghrelin deficiency does not influence feeding performance. *Regul. Pept.* **145:** 7–11.
26. Sun, Y., S. Ahmed & R.G. Smith. 2003. Deletion of ghrelin impairs neither growth nor appetite. *Mol. Cell Biol.* **23:** 7973–7981.
27. Wortley, K.E. *et al.* 2004. Genetic deletion of ghrelin does not decrease food intake but influences metabolic fuel preference. *Proc. Natl. Acad. Sci. USA* **101:** 8227–8232.
28. Erickson, J.C., K.E. Clegg & R.D. Palmiter. 1996. Sensitivity to leptin and susceptibility to seizures of mice lacking neuropeptide Y. *Nature* **381:** 415–421.
29. Qian, S. *et al.* 2002. Neither agouti-related protein nor neuropeptide Y is critically required for the regulation of energy homeostasis in mice. *Mol. Cell Biol.* **22:** 5027–5035.
30. Sun, Y. *et al.* 2008. Characterization of adult ghrelin and ghrelin receptor knockout mice under positive and negative energy balance. *Endocrinology* **149:** 843–850.
31. Sun, Y. *et al* 2004. Ghrelin stimulation of growth hormone release and appetite is mediated through the growth hormone secretagogue receptor. *Proc. Natl. Acad. Sci. USA* **101:** 4679–4684.
32. Youm, Y.H. *et al.* 2009. Deficient ghrelin receptor-mediated signaling compromises thymic stromal cell microenvironment by accelerating thymic adiposity. *J. Biol. Chem.* **284:** 7068–7077.
33. Ariyasu, H. *et al.* 2005. Transgenic mice overexpressing des-acyl ghrelin show small phenotype. *Endocrinology* **146:** 355–364.
34. Iwakura, H. *et al.* 2005. Analysis of rat insulin II promoter-ghrelin transgenic mice and rat glucagon promoter-ghrelin transgenic mice. *J. Biol. Chem.* **280:** 15247–15256.
35. Bewick, G.A. *et al.* 2009. Mice with hyperghrelinemia are hyperphagic and glucose intolerant and have reduced leptin sensitivity. *Diabetes* **58:** 840–846.
36. Corbetta, S. *et al.* 2003. Circulating ghrelin levels in patients with pancreatic and gastrointestinal neuroendocrine tumors: identification of one pancreatic ghrelinoma. *J. Clin. Endocrinol. Metab.* **88:** 3117–3120.
37. Tsolakis, A.V. *et al.* 2004. Malignant gastric ghrelinoma with hyperghrelinemia. *J. Clin. Endocrinol. Metab.* **89:** 3739–3744.
38. Nicholls, R.D. & J.L. Knepper. 2001. Genome organization, function, and imprinting in Prader-Willi and Angelman syndromes. *Annu. Rev. Genomics Hum. Genet.* **2:** 153–175.
39. Cummings, D.E. *et al.* 2002. Elevated plasma ghrelin levels in Prader Willi syndrome. *Nat. Med.* **8:** 643–644.
40. DelParigi, A. *et al.* 2002. High circulating ghrelin: a potential cause for hyperphagia and obesity in prader-willi syndrome. *J. Clin. Endocrinol. Metab.* **87:** 5461–5464.
41. Feigerlova, E. *et al.* 2008. Hyperghrelinemia precedes obesity in Prader-Willi syndrome. *J. Clin. Endocrinol. Metab.* **93:** 2800–2805.
42. Ding, F. *et al.* 2008. SnoRNA Snord116 (Pwcr1/MBII-85) deletion causes growth deficiency and hyperphagia in mice. *PLoS ONE.* **3:** e1709.
43. Date, Y. *et al.* 2002. Ghrelin is present in pancreatic alpha-cells of humans and rats and stimulates insulin secretion. *Diabetes* **51:** 124–129.
44. Heller, R.S. *et al.* 2005. Genetic determinants of pancreatic epsilon-cell development. *Dev. Biol.* **286:** 217–224.
45. Prado, C.L. *et al.* 2004. Ghrelin cells replace insulin-producing beta cells in two mouse models of pancreas development. *Proc. Natl. Acad. Sci. USA* **101:** 2924–2929.
46. Doyle, M.J., Z.L. Loomis & L. Sussel. 2007. Nkx2.2-repressor activity is sufficient to specify alpha-cells and a small number of beta-cells in the pancreatic islet. *Development* **134:** 515–523.

Ann. N.Y. Acad. Sci. ISSN 0077-8923

ANNALS OF THE NEW YORK ACADEMY OF SCIENCES
Issue: *Phylogenetic Aspects of Neuropeptides*

The orexin system: roles in sleep/wake regulation

Takeshi Sakurai, Michihiro Mieda, and Natsuko Tsujino

Department of Molecular Neuroscience and Integrative Physiology, Faculty of Medicine, Kanazawa University, Kanazawa, Ishikawa, Japan

Address for correspondence: Takeshi Sakurai, Department of Molecular Neuroscience and Integrative Physiology, Graduate School of Medical Science, Kanazawa University, Kanazawa, Ishikawa 920-8640, Japan. tsakurai@med.kanazawa-u.ac.jp

The neuropeptides orexin A and orexin B, produced in hypothalamic neurons, are critical regulators of sleep/wake states. Deficiency of orexin signaling results in narcoleptic phenotype in humans, dogs, and rodents. Recently, accumulating evidence has indicated that the orexin system regulates sleep and wakefulness through interactions with neuronal systems that are closely related with emotion, reward, and energy homeostasis. In this review, we will discuss the current understanding of the physiology of the orexin system especially focusing on its roles in the regulation of sleep/wakefulness states.

Key words: orexin; hypothalamus; narcolepsy; sleep; wakefulness

Introduction

Orexin was first identified as endogenous peptide ligands for two orphan G protein–coupled receptors (GPCRs) in 1998.[1] They were initially recognized as regulators of feeding behavior, first because of their exclusive production in the lateral hypothalamic area (LHA), a region known as the feeding center, and second owing to their pharmacological activity.[1–4] Subsequently, this peptide system was shown to be a critical regulator of sleep/wake states. Since then, this and other groups have been using a multidisciplinary approach to understand various aspects of the physiological roles of orexin peptides, and have uncovered crucial roles of the orexin system in the regulatory mechanisms in sleep/wake states, energy homeostasis, and the reward systems. Especially, the finding that an orexin deficiency causes narcolepsy in humans and other animals has had a huge impact on the studies of sleep and wakefulness and other areas.[5–9] Recent studies of orexin-producing neurons' efferent and afferent systems have suggested further roles for orexin in the coordination of emotion, energy homeostasis, reward, and arousal.[10–17] These findings suggest that orexin neurons are involved in sensing the body's external and internal environments, and regulate sleep/wake states accordingly, which is beneficial for survival. This review will discuss the mechanisms by which the orexin system regulates sleep/wake states, and how this mechanism relates to other systems that regulate emotion, reward, and energy homeostasis.

Orexin and orexin receptors

Orexins were identified by a strategy called reverse pharmacology. There are over 100 of GPCRs whose ligands are still unknown and are, therefore, referred to as orphan GPCRs. Many of these orphan GPCRs are likely to be receptors for heretofore unidentified signaling molecules, including new peptide hormones and neuropeptides. In 1998, during searching for endogenous ligands for various orphan GPCRs, our group identified orexin A and B as endogenous ligands for two related orphan GPCRs.[1] Because these peptides were localized in the LHA, and intracerebroventricular (ICV) injection of them in rats or mice acutely increased food consumption, they were named orexin A and B after the Greek word *orexis*, meaning appetite. Orexin A and B are produced by cleavage of a common precursor polypeptide, prepro-orexin. The primary structure of orexin A predicted from the cDNA sequences is completely conserved among several mammalian species (human, rat, mouse, cow, sheep, dog, and pig). On the other hand, rat orexin B is a 28-amino

doi: 10.1111/j.1749-6632.2010.05513.x

A

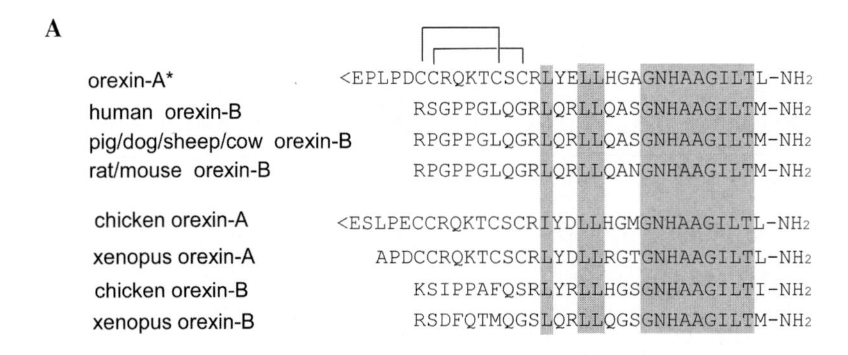

B

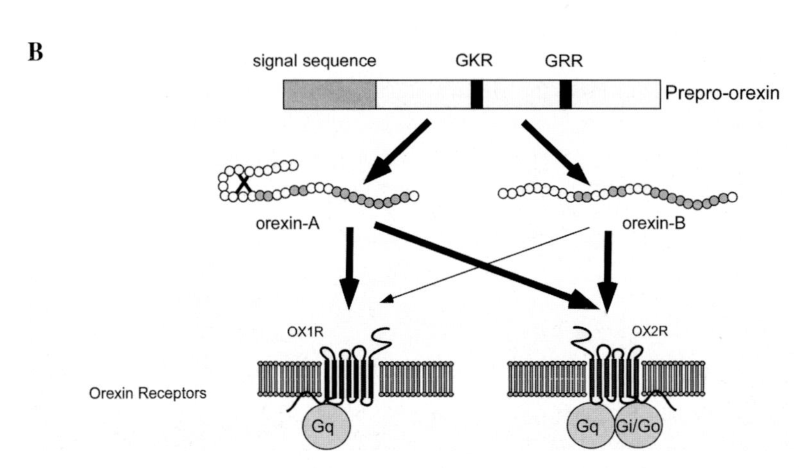

Figure 1. An overview of the orexin system. (A) Structure of mature orexin A and orexin B peptides. The topology of the two-intrachain bonds in orexin A is indicated above the sequence. *Shadows* indicate amino acid identity. Mammalian orexin A sequences thus far identified (human, rat, mouse, pig, dog, sheep, and cow) are all identical, whereas the sequences of orexin B show some differences among species. (B) Orexins and their receptors. Orexin A and orexin B are derived from a common precursor peptide, prepro-orexin. The actions of orexins are mediated via two G protein–coupled receptors named orexin-1 (OX_1R) and orexin-2 (OX_2R) receptors. OX_1R is selective for orexin A, whereas OX_2R is a nonselective receptor for both orexin A and orexin B. OX_1R is coupled exclusively to the G_q subclass of heterotrimeric G proteins, whereas OX_2R couples to $G_{i/o}$ and /or G_q.

acid, C-terminally amidated linear peptide of 2937 Da, which is 46% (13/28) identical in sequence to orexin A. The C-terminal half of orexin B is very similar to that of orexin A (73%; 11/15), whereas the N-terminal half is variable. Orexin B also has a high degree of sequence similarity among species. Several study revealed that the structures of fish, xenopus, and chicken orexin A and B have also conserved structures as compared with mammalian sequences (Fig. 1).[18–20] Furthermore, several studies have shown that orexins play roles in regulation of food intake and sleep/wake behavior in goldfishes and zebrafishes.[21–24] These observations suggest that structures and functions of orexins are phylognetically well conserved.

As an independent work, de Lecea *et al.* identified mRNA expressed specifically within the hypothalamus and identified a cDNA encoding a polypeptide identical to *prepro-orexin*, and named the putative mature peptides hypocretin-1 and -2 (Hcrt-1 and Hcrt-2).[25] The terms *orexin* and *hypocretin* are used as synonyms in many papers currently.

In mammals, the actions of orexins are mediated by two GPCRs, named orexin 1 (OX_1R) and orexin 2 (OX_2R) receptors (also named as HCRTR1 and HCRTR2). OX_1R has one-order higher affinity for orexin A than orexin B, whereas OX_2R binds orexin A and orexin B with similar affinities. OX_1R couples to the $G_{q/11}$ subclass of heterotrimeric G proteins, whereas OX_2R couples to $G_{q/11}$ or $G_{i/o}$ in a neuronal cell line in culture.[26]

Orexin-expressing neurons (orexin neurons) are distributed within the LHA and posterior hypothalamus (PH).[1,27–29] The number of these neurons has been estimated around 3,000 in rat and 50,000 in human brains. Although their cell bodies are exclusively localized in the hypothalamus, orexin neurons send their axonal projections diffusely throughout the central nervous system (CNS), excluding the cerebellum. Especially dense projections are found in the hypothalamus (such as the arcuate nucleus [ARC] and tuberomammillary nucleus [TMN]), and the brain stem (such as the central gray, locus coeruleus [LC], and raphe nuclei).[27–29] Consistent with the broad projections of orexin neurons, OX_1R and OX_2R mRNAs show wide distributions in the CNS with partly overlapping but distinct patterns.[30] For instance, nuclei such as the LC and ventral tegmental area (VTA) mainly express OX_1R mRNA, whereas those including the TMN, nucleus accumbens (NAc), and septal nuclei mainly express OX_2R mRNA. Both mRNAs are observed in the raphe nuclei, laterodorsal tegmental nucleus (LDT), and pedunculopontine tegmental nucleus (PPT). These distributions of both receptor mRNAs suggest partly overlapping and partly distinct roles of orexin receptors.

Narcolepsy

In 1999, two independent studies utilizing dog forward genetics and mouse reverse genetics clearly showed a causal linkage between disruption of the orexin system and narcolepsy–cataplexy. Subsequently, several studies established that loss of orexin neurons is accompanied by human narcolepsy patients. The symptoms and pathophysiology of the sleep disorder narcolepsy, caused by an orexin deficiency, provide insight into the physiological roles of orexin.[5–8] Therefore, we will briefly mention about this disorder in this section.

Human narcolepsy is a debilitating neurological disorder that affects approximately 1 in 2,000 individuals in the U.S.[31] Onset of the condition is usually during adolescence. A cardinal symptom of the disorder is excessive daytime sleepiness (an insurmountable urge to sleep), which manifests itself primarily when the subject falls asleep at inappropriate times ("sleep attacks"). When normal individuals fall asleep, a certain period of non-rapid-eye-movement (NREM) sleep (60–90 min) usually precedes REM sleep. However, the latency of REM sleep is markedly reduced in narcolepsy patients, and REM sleeps are sometimes observed immediately after wakefulness (sleep-onset REM). The existence of "sleep-onset REM" periods is one of the important diagnostic criteria for narcolepsy. In patients, nocturnal sleep is also disturbed and is often accompanied by hypnagogic hallucinations, vivid dreaming, and sleep paralysis, which usually occur when they fall asleep.

Narcolepsy patients often suffer from attacks called "cataplexy," which is a sudden weakening of bilateral muscle tone (muscle atonia), ranging from jaw dropping and speech slurring to complete bilateral collapse of the postural muscles. These attacks are often triggered by emotional stimuli, such as laughter, excitement, and pleasure. Narcolepsy that is accompanied with cataplexy is sometimes referred as "narcolepsy—cataplexy."

Symptoms of narcolepsy–cataplexy can be divided into two distinct pathological phenomena. One is an inability to maintain a consolidated waking period, characterized by abrupt transitions from wakefulness to NREM sleep (i.e., dysregulation of NREM sleep onset). This phenomenon manifests as excessive daytime sleepiness or a sleep attack. This symptom is treated by psychostimulants, such as methylphenidate, methamphetamine, and modafinil. The other key phenomenon is the pathological intrusion of REM sleep into wakefulness (i.e., dysregulation of REM sleep onset): it is during these periods that patients may experience cataplexy, hypnagogic hallucinations, and sleep paralysis. These symptoms are treated by tricyclic depressants and selective serotonin reuptake inhibitors.

Collectively, this disorder is characterized by the inability to maintain each vigilance state, pathological intrusion of NREM, and/or REM sleep into wakefulness, and frequent transitions between states of sleep and wakefulness. This suggests that orexins have important physiological roles in the maintenance and stabilization of sleep and wakefulness.

Orexin or orexin receptor-2 deficiencies cause narcoleptic phenotype

The first clues toward an involvement of the orexins in narcolepsy came from animal models; mice lacking the *orexin* gene or dogs with null mutations in the *orexin receptor-2* (OX_2R) gene show phenotypes remarkably similar to humans with narcolepsy.[7,8] Orexin knockout mice ($orexin^{-/-}$ mice) exhibit frequent sudden collapses that resemble human cataplexy attacks during the dark phase, when mice spend the most time awake and active.[7] These attacks are thought to be homologous to cataplexy.[32] Quantitative sleep state parameters of $orexin^{-/-}$ mice revealed slightly decreased waking time, increased REM sleep time during dark period, decreased REM sleep latency, and, most importantly, a markedly decreased duration of wake time during the dark phase (i.e., inability to maintain a long awake period). Dogs with null mutations in the *orexin receptor-2* (OX_2R) gene show phenotypes remarkably similar to humans with narcolepsy.[7,8]

The link between orexin dysfunction and narcolepsy, especially when accompanied with cataplexy (narcolepsy–cataplexy), has since been supported by studies with human patients. In contrast to normal control individuals, the vast majority of narcoleptic individuals have low or undetectable levels of orexin A in the cerebrospinal fluid (CSF).[33] (In CSF, orexin B is not detectable even in healthy individuals.) A postmortem study of human narcolepsy subjects showed undetectable levels of orexin peptides in the cortex and pons, and an 80–100% reduction in the number of neurons containing detectable *prepro-orexin* mRNA or orexin-like immunoreactivity in the hypothalamus.[5,6] No mutation has been found either in the prepro-orexin or orexin receptor genes of human narcolepsy–cataplexy patients, except for an unusually severe, early onset case associated with a mutation in the signal peptide of prepro-orexin that impairs peptide trafficking and processing.[5] In this case, abnormal signal sequence cleavage results in accumulation of abnormal orexin peptide in cells leading cell death. A recent finding showing concomitant loss of dynorphin, neuronal activity-regulated pentraxin, and orexin, which colocalize in orexin neurons, further suggests a loss of functional orexin neurons, rather than the selective inhibition of orexin expression, in narcolepsy–cataplexy.[34] Based on these observations, as well as a strong association of human narcolepsy–cataplexy with certain HLA alleles,[35] it has been speculated that narcolepsy–cataplexy may result from selective autoimmune degeneration of orexin neurons. Regardless of the cause of the neuron loss, the orexin signaling deficiency in narcolepsy–cataplexy shows that this neuropeptide system plays an important role in the regulation of sleep and wakefulness, especially in the maintenance of long, consolidated awake periods.

We produced transgenic mice in which orexin neurons are ablated by expression of a N-terminally truncated ataxin-3, which induces postnatal apoptotic death of all orexin neurons by adulthood (orexin neuron-ablated mice).[9] Adult mice show essentially the same phenotype of sleep/wake regulation as $orexin^{-/-}$ mice. Therefore, although orexin neurons produce other neurotransmitters such as glutamate and dynorphin,[36–38] orexin is the most important factor among them for regulation of sleep/wake states by these neurons.

Regarding the contribution of each receptor, $OX_2R^{-/-}$ mice have characteristics of narcolepsy–cataplexy, although their narcoleptic phenotype is less severe than that found in $orexin^{-/-}$ mice.[39,40] They show behavioral arrests that are less frequent and severe than those of $orexin^{-/-}$ mice. $OX_1R^{-/-}$ mice do not have overt behavioral abnormalities. $OX_1R^{-/-}$;$OX_2R^{-/-}$ mice appear to be a phenocopy of orexin knockout mice,[41] implying that these two receptors are sufficient to mediate regulation of sleep/wake by orexins. These observations also suggest that despite the lack of an overt $OX_1R^{-/-}$ phenotype, loss of signaling through both receptor pathways appears to be necessary for the emergence of a complete narcoleptic phenotype, suggesting that both receptors are involved in the regulation of sleep and wakefulness.

Therapeutic potentials of orexin agonists and antagonists

The finding of low orexin A levels in CSF in patients with narcolepsy–cataplexy led to the development of a novel, definitive diagnostic test for this disease.[33] A low orexin A levels in CSF has been used as one of the diagnostic criteria for narcolepsy–cataplexy according to the second edition of the *International Classification of Sleep Disorders*.[42]

Moreover, the discovery of causal link between loss of orexin signaling and human narcolepsy–cataplexy has brought about a possibility of novel therapies for this disease. Currently, excessive sleepiness in narcolepsy is treated using psychostimulants, whereas cataplexy is treated with tricyclic antidepressants. γ-hydroxybutyrate (sodium oxybate) is also used to consolidate nocturnal sleep and reduce cataplexy.[43] Treatment with these compounds are problematic due to limited effectiveness, undesirable side effects such as insomnia or symptom rebounds, and the potential for abuse. Because orexin neuron-ablated mice (*orexin/ataxin-3* mice) have an etiology and course of disease similar to those of human narcolepsy–cataplexy, these mice may represent the most accurate pathophysiological model of narcolepsy–cataplexy available.[9] We demonstrated rescue of the narcoleptic phenotype of these mice by genetic and pharmacological means.[44] Chronic overproduction of orexin peptides from an ectopically expressed transgene prevented development of narcolepsy syndrome in orexin neuron-ablated mice. Acute ICV administration of orexin A also maintained wakefulness, suppressed sleep, and inhibited cataplectic attacks in these mice. Intriguingly, ICV administration of orexin A had stronger arousal effects in orexin neuron-ablated mice than in wild-type controls, suggesting that responsiveness of effecter sites for orexins remains intact, or even more responsible for orexins in narcoleptics. These results also indicate that a spatially targeted secretion of orexin is not necessary to prevent narcoleptic symptoms. Unfortunately, however, constitutive production of orexin peptides from a *prepro-orexin* transgene in mice *per se* caused fragmentation of NREM sleep episodes in the light period; when mice spend the most time asleep (our unpublished observations). These results indicate that orexin neurons should be turned on and switched off to maintain consolidated wakefulness and NREM sleep, respectively. These also suggest that orexin receptor agonists with relatively short half-lives (several hours) would be of potential value for treating human narcolepsy–cataplexy. Such agonists might also be useful in the treatment of other conditions of excessive daytime sleepiness in humans.

Conversely, orexin receptor antagonists might be useful for treatment of insomnia patients or as a sleep inducer. Indeed, Almorexant, an orally available antagonist for OX_1R and OX_2R, has been reported to cause subjective and objective electrophysiological signs of sleep in humans,[45] and now it is under Phase III clinical studies.

Neural mechanisms of sleep/wake regulation by orexins

Wakefulness is maintained by multiple neurotransmitters and neuronal pathways. Included in this system are monoaminergic and cholinergic neurons reside in the brain stem. Monoaminergic neurons, including LC noradrenergic, raphe serotonergic, and TMN histaminergic neurons, project diffusely to the forebrain promoting arousal. Other important wake-inducing signals are cholinergic neurons in the brain stem and project to key forebrain targets such as the thalamus, an area critical to regulating cortical activity.[46] Monoaminergic neurons are firing at rapid rates during wakefulness, whereas they reduce their activities during NREM sleep and almost cease discharge during REM sleep. A subset of PPT/LDT neurons is active during both wakefulness and REM sleep (W-REM on), regulating activity of thalamo-cortical projections to generate EEG desynchronization characteristics of wakefulness and REM sleep. Others are active exclusively during REM sleep (REM-on) and thought to induce REM sleep and REM atonia. Conversely, GABA/galaninergic neurons in the ventrolateral preoptic nucleus (VLPO) of the hypothalamus are active during sleep, especially NREM sleep,[46] and thought to initiate and maintain NREM sleep. VLPO neurons and monoaminergic neurons reciprocally inhibit each other. This reciprocal interaction of wake center and sleep center maintains states of sleep and wakefulness.

As mentioned earlier, orexin neurons send dense projections to the monoaminergic/cholinergic nuclei involved in sleep/wake regulation (Figs. 2 and 3). The distribution of the orexin receptors mRNA is consistent with these projection sites; within the brain, OX_1R is most abundantly expressed in the LC, whereas OX_2R is highly expressed in the TMN,[30] and both mRNAs are detectable in the raphe nuclei, PPT/LDT, and BF.

ICV administration of orexin A in rodents potently reduces REM and NREM sleep, and increases wakefulness.[47] Application of orexin directly into the LC, TMN, LDT, and the lateral preoptic area has effects similar to ICV injection on sleep/wake states.[48–52] *In vitro* slice electrophysiology studies has shown that orexin increases firing rates of monoaminergic neurons in the LC,[53,54] raphe,[55,56] TMN,[57–59] and cholinergic neurons in the BF and

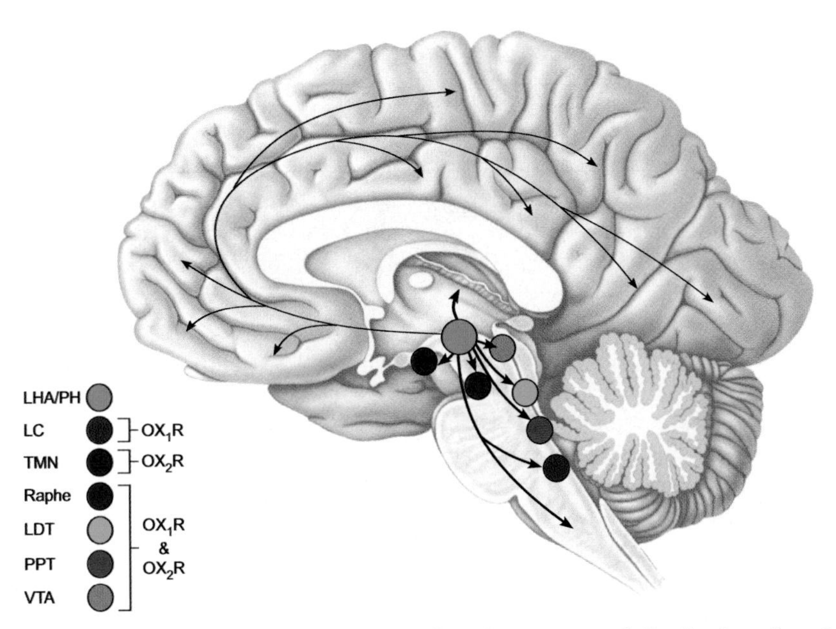

Figure 2. Schematic drawing of main axonal projections of orexin neurons and distribution of orexin receptors in brain. This figure summarizes predicted orexinergic projections in the human brain. Please note that distributions of orexin fibers and receptors (OX_1R, OX_2R) are predicted from the results of studies on rodent brains, because it is on rats or mice that most histological studies on the orexin system have been carried out. *Circles* show regions with strong receptor expression and dense orexinergic projections. Orexin neurons originating in the lateral hypothalamic area (LHA) and posterior hypothalamus (PH) regulate sleep and wakefulness and the maintenance of wakefulness by sending excitatory projections to the entire CNS, excluding the cerebellum, with particularly dense projections to monoaminergic and cholinergic nuclei in the brain stem and hypothalamic regions, including the locus coeruleus (LC, containing noradrenaline), tuberomammillary nucleus (TMN, containing histamine), raphe nuclei (Raphe, containing serotonin), and laterodorsal/pedunclopontine tegmental nuclei (LDT/PPT), containing acetylcholine. Orexin neurons also have links with the reward system through the ventral tegmental area (VTA, containing dopamine) and with the hypothalamic nuclei that stimulate feeding behavior. This figure is adapted with modification from Ref. 80.

LDT,[60,61] but have no effect on the GABAergic neurons in the VLPO.[60] A work using cats showed that orexin A inhibits cholinergic neurons in the PPT *in vivo* through activation of GABAergic interneurons and GABAergic neurons in the substantia nigra pars reticulata.[62] These results indicate that orexin neurons affect the activity of PPT/LDT cholinergic neurons both directly and indirectly to regulate arousal and REM sleep.

More recently, Adamantidis *et al.* succeeded to demonstrate that direct and selective photostimulation of orexin neurons in freely moving mice, in which orexin neurons were genetically targeted to express a photo-activatable cation channel (channelrhodopsin-2), increased the probability of transition to wakefulness from either NREM or REM sleep.[63]

Regulatory mechanisms and input systems of orexin neurons

Considering symptoms of narcolepsy–cataplexy, orexin neurons are expected to be active during wakefulness and to be silent during sleep. In fact, transgenic mice with constitutive activation of orexinergic tone (*CAG/orexin* mice), in which orexin is expressed in a diffuse, ectopic pattern in the brain in unregulated fashion,[44] exhibited abnormal sleep and wakefulness patterns, including fragmented NREM sleep in the light period and incomplete REM sleep atonia with abnormal myoclonic activity during REM sleep (our unpublished results). These suggest that orexin neurons need to be switched off to maintain consolidated NREM sleep and the muscle atonia that accompanies REM sleep.

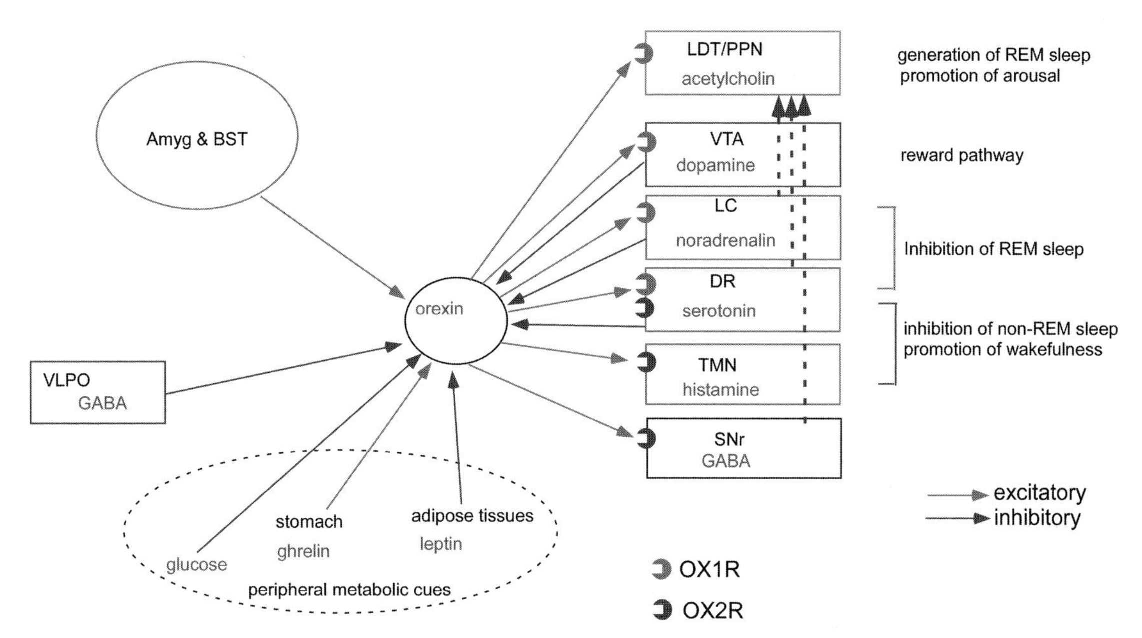

Figure 3. Interactions between orexin neurons with other brain regions. Orexin neurons in the lateral hypothalamic area (LHA) and posterior hypothalamus (PH) are anatomically well placed to provide a link between the limbic system, systems involved in energy homeostasis and monoaminergic and cholinergic neurons in the brain stem. Solid arrows show excitatory projections and broken lines inhibitory ones. Wake-active regions, sleep-active regions and REM-active regions are shown by *red*, *blue*, and *green boxes*, respectively. Orexin neurons promote wakefulness through the monoaminergic nuclei that are wake-active. Stimulation of dopaminergic centers by orexins can modulate reward systems (*purple*). Peripheral metabolic signals such as leptin, ghrelin, and glucose influence orexin neuronal activity to coordinate arousal and energy homeostasis. Input from the limbic system (amygdala and bed nucleus of the stria terminalis [BST]) might regulate the activity of orexin neurons upon emotional stimuli to evoke emotional arousal or fear-related responses. VLPO, ventrolateral preoptic area; DR, dorsal raphe; GABA, γ-aminobutyric acid; LC, locus coeruleus; LDT, laterodorsal tegmental nucleus; PPT, pedunculopontine tegmental nucleus; SNr, substantia nigra pars reticulata; TMN, tuberomammillary nucleus. This figure is adapted with modification from Ref. 80.

Fos expression in orexin neurons in rats is higher during the dark phase (active period) than during light phase (rest period).[64] Consistently, orexin levels in CSF peak during the dark period and decrease during the light period.[65] *In vivo* extracellular recordings further confirmed activity patterns of orexin neurons across sleep/wake cycles with high temporal resolutions.[66–68] Essentially, orexin neurons fired most actively during active waking, decreased discharge during quiet waking, were virtually silent during NREM sleep, and almost silent but exhibited occasional firing during REM sleep.

How these regulations could be achieved? Orexin neurons receive projections from nuclei involved in sleep/wake regulation. GABAergic neurons in the preoptic area, including the VLPO, densely innervate orexin neurons (Fig. 3).[14,15] Orexin neurons are strongly inhibited by both $GABA_A$ receptor agonist, muscimol, and $GABA_B$ receptor agonist, baclofen.[11,69,70] Orexin neurons are also innervated by BF cholinergic neurons.[14] Carbachol, a cholinergic agonist, activates a subset of orexin neurons.[14,71] Thus, orexin neurons are likely to be inhibited by sleep-promoting neurons and activated by wake-promoting BF neurons: these regulations of orexin neurons are in consistent with their proposed function to stabilize wakefulness.

In contrast, wake-active serotonergic neurons and noradrenergic neurons in the brain stem send inhibitory projection to orexin neurons.[14,72,73] Serotonergic and noradrenergic inputs hyperpolarize orexin neurons through activation of G-protein-regulated inwardly rectifying K^+ (GIRK or Kir3) channels mediated by 5-$HT1_A$ receptors and

A

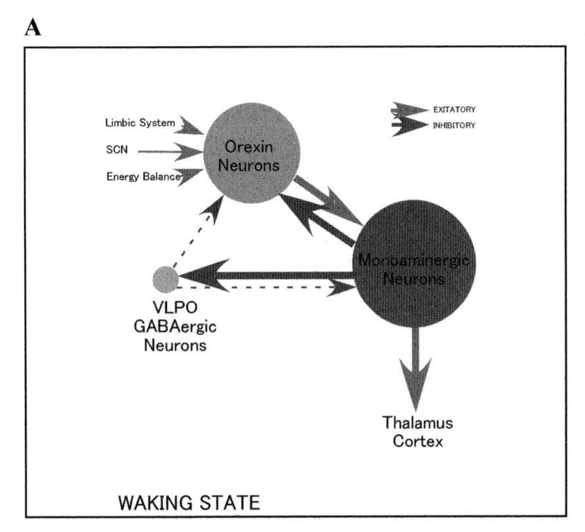

B

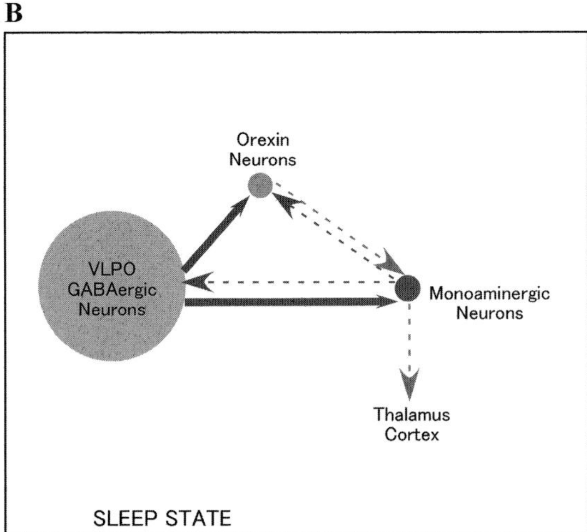

C

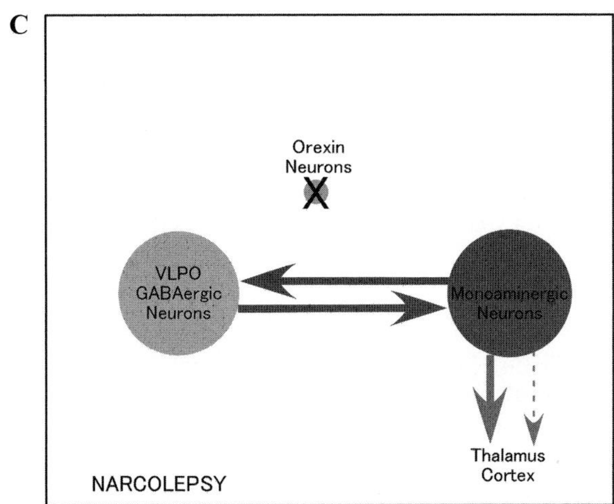

Figure 4. Mechanisms by which the orexin system stabilizes sleep/wake states. The figures represent functional interactions between orexin neurons, monoaminergic wake-active centers, and the ventrolateral preoptic area (VLPO) sleep-active center. Solid arrows show excitatory input, and broken lines inhibitory input. The thickness of arrows and lines represents the relative strength of excitatory and inhibitory input, respectively. Circle sizes represent relative activities of each region. (A) Awake state. Orexin neurons send excitatory influences to monoaminergic neurons, which send inhibitory feedback projections to orexin neurons. This feedback mechanism might maintain the activity of monoaminergic neurons. A slight decrease in input to the monoaminergic neurons results in decreased inhibitory influence to orexin neurons. Orexin neurons, therefore, are disinhibited and increase excitatory influence to monoaminergic cells to maintain their activity. These monoaminergic cells send excitatory projections to the thalamus and cerebral cortex, and send inhibitory projections to the VLPO sleep center. These mechanisms maintain wakefulness states. (B) Sleep state. VLPO sleep-active neurons are activated and send inhibitory projections to monoaminergic neurons and orexin neurons to maintain sleep. (C) Narcolepsy. If orexin neurons are removed, monoaminergic neurons and VLPO neurons set up a mutually inhibitory circuit, which can cause unwanted and abrupt transitions between the states. Activity in one of the competing sides shuts down inhibitory inputs from the other side, and therefore disinhibits its own action. So, when either side begins to overcome the other, the switch abruptly turns into the alternative state. This figure is adapted with modification from Ref. 80.

α_2-adrenoceptors, respectively.[72,73] These negative feedback mechanisms may also be important for fine adjustment of orexin neuronal activity to stabilize wakefulness. Notably, histamine has little effect on orexin neurons.[74]

Local interneurons may also play important roles in the regulation of orexin neurons. In slice preparations, orexin A and orexin B were reported to depolarize orexin neurons.[69] This effect appears to be medicated by orexin-mediated excitation of local glutamatergic neurons that regulate orexin neuronal activity, in part by presynaptic facilitation of glutamate release. On the other hand, GABAergic input from local interneurons to orexin neurons is also important for organization of orexin neuronal activity.[75] Genetic disruption of $GABA_B$ input resulted in marked sleep/wake abnormality.[75]

Horvath and Gao reported an unusual synaptic organization on orexin neurons in which excitatory synaptic currents and asymmetric synapses exert control on the perikarya of these long-projective neurons with minimal inhibitory input, which is sharp contrast to the fact that neuronal cell bodies in the central nervous system are either dominated by inhibitory inputs (long-projection neurons), or have an approximate ratio of excitatory to inhibitory inputs of 1:1.[76] This unique input organization of orexin neurons may be a necessary element for the maintenance of a low threshold for arousal and alertness. On the other hand, this circuitry, along with abundant input from the limbic system, may also be an underlying cause of insomnia. Consistent with these results, an electrophysiological study with slice preparations indicated that orexin neurons are tonically activated by glutamatergic neurons,[69] whereas basal tone of GABAergic input seems to be low, because GABA antagonists do not influence basal activity of orexin neurons. However, in slice from mice with selective deletion of $GABA_B$ receptor in orexin neurons (oxGKO mice), orexin neurons responded with depolarization to bicuculline application. This indicates that $GABA_A$ receptors are tonically activated in orexin neurons of these mice. As a consequence, orexin neurons in these mice showed decreased responsiveness to both excitatory and inhibitory inputs as compared with wild-type mice.[75] With the observation showing that oxGKO mice had highly fragmented sleep/wake state, these observations suggest that proper local GABAergic regulation and normal synaptic organization of orexin neurons are highly important for sleep/wake regulation.

Orexin neurons as a stabilizer of sleep/wake states

How do the orexins physiologically regulate sleep and wakefulness, and why does a lack of orexin signaling result in narcolepsy, a disorder characterized by instability of wakefulness? As mentioned earlier, orexin neurons, monoaminergic/cholinergic centers in the brain stem, and the sleep-active VLPO neurons constitute a triangular interaction; orexin neurons send excitatory projections to monoaminergic neurons, and these monoaminergic neurons send inhibitory projections back to orexin neurons. VLPO sleep-active neurons send inhibitory projections to both monoaminergic neurons and orexin neurons. Monoaminergic neurons send inhibitory projections to VLPO sleep-active neurons. This triangular organization seems to be highly important for stability of vigilance states (Fig. 4). VLPO sleep-active neurons and monoaminergic neurons inhibit each other. This reciprocal inhibition is well suited to avoid intermediate states. Orexin neurons can stabilize waking state by enhancing activity of monoaminergic neurons during wakefulness, avoiding state instability caused by small perturbations. At the same time, negative feedback input to orexin neurons from serotonergic and noradrenergic neurons work to maintain the activities of orexin and monoaminergic neurons within appropriate ranges. During sleep, orexin neurons are turned off together with monoaminergic neurons by VLPO GABA/galaninergic neurons.

In narcoleptics, the orexin-target neurons, such as monoaminergic neurons in the brain stem, might undergo changes in their synaptic organization through plasticity or synaptic scaling mechanisms due to chronic deficiency of orexins. These changes might help the monoaminergic cells to be activated without orexins. Simultaneously, this mechanism may set up a "flip-flop" circuit between sleep-active VLPO neurons and wake-active monoaminergic neurons;[46] these two components reciprocally inhibit each other. In this type of mutually inhibitory circuits, when activity on either side begins to overcome the other, the system will flip into one of two possible extremes. A small perturbation on the activity of one side can easily cause abrupt switching between two states, resulting in frequent state transitions, fundamental problems seen in narcolepsy.

The fact described earlier that orexin neurons are much more active during active waking than during quiet waking *in vivo* clearly suggest roles of orexin go beyond mere global arousal. Recently, Deadwyler *et al.* reported that systemic and nasal delivery of orexin A reduced the effects of sleep deprivation on cognitive performance in nonhuman primates; interestingly, orexin A did not produce facilitative effects if the animals were not sleep deprived,[77] although it is not clear whether orexin A penetrates the blood–brain barrier, or orexin A can influence vigilance states by acting on peripheral receptors. In slice preparations from rat prefrontal cortex, orexin was reported to induce calcium transients in single spines postsynaptic to identified thalamocortical boutons.[78,79] By this cellular mechanism, orexinergic projections to the prefrontal cortex may play a role in prefrontal or "executive" aspects of alertness and attention.

Roles of orexin in other functions

Orexins are thought to be also involved in many other functions, including feeding behavior, autonomic regulation, endocrine functions, nociception, reward system, and emotion. All these systems should be well coordinated. Sleep and wakefulness are regulated to occur at appropriate times that are in accordance with our internal and external environments. Avoiding danger and finding food, which are life-essential activities that are regulated by emotion, reward, and energy balance, require vigilance and thus, by definition, wakefulness. The orexin system regulates sleep and wakefulness through interactions with systems that regulate emotion, reward, and energy homeostasis.

Reciprocal connections between the orexin system and multiple neuronal systems indicate that orexin neurons provide crucial links between multiple brain functions, such as energy homeostasis, reward system, emotion, and arousal. Future studies with full use of mouse molecular genetics, such as selective deletions of genes for particular receptor or signaling molecule in orexin neurons, would lead to further understanding of integrative physiology orchestrated by the orexin system.

From clinical perspective, discovery of the linkage between the orexin system and human narcolepsy–cataplexy led to the development of a novel diagnosis of this disease and to the expectation for development of novel drugs for treatment of narcolepsy–cataplexy. Moreover, future studies may let us understand why orexin neurons degenerate in narcolepsy–cataplexy patients, which would lead to more fundamental therapies for narcolepsy–cataplexy.

Acknowledgments

This study was supported in part by a grant-in-aid for scientific research (S, A, B) and the 21st Century COE Program from the Ministry of Education, Culture, Sports, Science, and Technology (MEXT) of Japan.

Conflicts of interest

The authors declare no conflict of interest.

References

1. Sakurai, T. *et al.* 1998. Orexins and orexin receptors: a family of hypothalamic neuropeptides and G protein-coupled receptors that regulate feeding behavior. *Cell* **92:** 573–585.
2. Haynes, A.C. *et al.* 2002. Anorectic, thermogenic and anti-obesity activity of a selective orexin-1 receptor antagonist in ob/ob mice. *Regul. Pept.* **104:** 153–159.
3. Haynes, A.C. *et al.* 2000. A selective orexin-1 receptor antagonist reduces food consumption in male and female rats. *Regul. Pept.* **96:** 45–51.
4. Edwards, C.M. *et al.* 1999. The effect of the orexins on food intake: comparison with neuropeptide Y, melanin-concentrating hormone and galanin. *J. Endocrinol.* **160:** R7–R12.
5. Peyron, C. *et al.* 2000. A mutation in a case of early onset narcolepsy and a generalized absence of hypocretin peptides in human narcoleptic brains. *Nat. Med.* **9:** 991–997.
6. Thannickal, T.C. *et al.* 2000. Reduced number of hypocretin neurons in human narcolepsy. *Neuron* **27:** 469–474.
7. Chemelli, R. M. *et al.* 1999. Narcolepsy in orexin knockout mice: molecular genetics of sleep regulation. *Cell* **98:** 437–451.
8. Lin, L. *et al.* 1999. The sleep disorder canine narcolepsy is caused by a mutation in the hypocretin (orexin) receptor 2 gene. *Cell* **98:** 365–376.

9. Hara, J., *et al.* 2001. Genetic ablation of orexin neurons in mice results in narcolepsy, hypophagia, and obesity. *Neuron* **30:** 345–354.
10. Boutrel, B. *et al.* 2005. Role for hypocretin in mediating stress-induced reinstatement of cocaine-seeking behavior. *Proc. Natl. Acad. Sci. USA* **102:** 19168–19173.
11. Yamanaka, A. *et al.* 2003. Hypothalamic orexin neurons regulate arousal according to energy balance in mice. *Neuron* **38:** 701–713.
12. Akiyama, M. *et al.* 2004. Reduced food anticipatory activity in genetically orexin (hypocretin) neuron-ablated mice. *Eur. J. Neurosci.* **20:** 3054–3062.
13. Mieda, M. *et al.* 2004. Orexin neurons function in an efferent pathway of a food-entrainable circadian oscillator in eliciting food-anticipatory activity and wakefulness. *J. Neurosci.* **24:** 10493–10501.
14. Sakurai, T. *et al.* 2005. Input of orexin/hypocretin neurons revealed by a genetically encoded tracer in mice. *Neuron* **46:** 297–308.
15. Yoshida, K. *et al.* 2006. Afferents to the orexin neurons of the rat brain. *J. Comp. Neurol.* **494:** 845–861.
16. Harris, G.C., M. Wimmer & G. Aston-Jones. 2005. A role for lateral hypothalamic orexin neurons in reward seeking. *Nature* **437:** 556–559.
17. Narita, M. *et al.* 2006. Direct involvement of orexinergic systems in the activation of the mesolimbic dopamine pathway and related behaviors induced by morphine. *J. Neurosci.* **26:** 398–405.
18. Sakurai, T. 2005. Reverse pharmacology of orexin: from an orphan GPCR to integrative physiology. *Regul. Pept.* **126:** 3–10.
19. Alvarez, C.E. & J. G. Sutcliffe. 2002. Hypocretin is an early member of the incretin gene family. *Neurosci. Lett.* **324:** 169–172.
20. Shibahara, M. *et al.* 1999. Structure, tissue distribution, and pharmacological characterization of Xenopus orexins. *Peptides* **20:** 1169–1176.
21. Appelbaum, L. *et al.* 2009. Sleep-wake regulation and hypocretin-melatonin interaction in zebrafish. *Proc. Natl. Acad. Sci. USA* **106:** 21941–21947.
22. Miura, T. *et al.* 2007. Regulation of food intake in the goldfish by interaction between ghrelin and orexin. *Peptides* **28:** 1207–1213.
23. Volkoff, H. *et al.* 2005. Neuropeptides and the control of food intake in fish. *Gen. Comp. Endocrinol.* **142:** 3–19.
24. Yokogawa, T. *et al.* 2007. Characterization of sleep in zebrafish and insomnia in hypocretin receptor mutants. *PLoS Biol.* **5:** e277.
25. de Lecea, L. *et al.* 1998. The hypocretins: hypothalamus-specific peptides with neuroexcitatory activity. *Proc. Natl. Acad. Sci. USA* **95:** 322–327.
26. Zhu, Y. *et al.* 2003. Orexin receptor type-1 couples exclusively to pertussis toxin-insensitive G-proteins, while orexin receptor type-2 couples to both pertussis toxin-sensitive and -insensitive G-proteins. *J. Pharmacol. Sci.* **92:** 259–266.
27. Date, Y. *et al.* 1999. Orexins, orexigenic hypothalamic peptides, interact with autonomic, neuroendocrine and neuroregulatory systems. *Proc. Natl. Acad. Sci. USA* **96:** 748–753.
28. Nambu, T. *et al.* 1999. Distribution of orexin neurons in the adult rat brain. *Brain Res.* **827:** 243–260.
29. Peyron, C. *et al.* 1998. Neurons containing hypocretin (orexin) project to multiple neuronal systems. *J. Neurosci.* **18:** 9996–10015.
30. Marcus, J. N. *et al.* 2001. Differential expression of orexin receptors 1 and 2 in the rat brain. *J. Comp. Neurol.* **435:** 6–25.
31. Mignot, E. 1998. Genetic and familial aspects of narcolepsy. *Neurology* **50:** S16–S22.
32. Scammell, T.E. *et al.* 2009. A consensus definition of cataplexy in mouse models of narcolepsy. *Sleep* **32:** 111–116.
33. Mignot, E. *et al.* 2002. The role of cerebrospinal fluid hypocretin measurement in the diagnosis of narcolepsy and other hypersomnias. *Arch. Neurol.* **59:** 1553–1562.
34. Crocker, A. *et al.* 2005. Concomitant loss of dynorphin, NARP, and orexin in narcolepsy. *Neurology* **65:** 1184–1188.
35. Kadotani, H., J. Faraco & E. Mignot. 1998. Genetic studies in the sleep disorder narcolepsy. *Genome. Res.* **8:** 427–434.
36. Abrahamson, E. E., R. K. Leak & R. Y. T. Moore. 2001. The suprachiasmatic nucleus projects to posterior hypothalamic arousal systems. *Neuroreport* **12:** 435–440.
37. Chou, T.C. *et al.* 2001. Orexin (hypocretin) neurons contain dynorphin. *J. Neurosci.* **21:** RC168.
38. Rosin, D.L. *et al.* 2003. Hypothalamic orexin (hypocretin) neurons express vesicular glutamate transporters VGLUT1 or VGLUT2. *J. Comp. Neurol.* **465:** 593–603.
39. Willie, J.T. *et al.* 2003. Distinct narcolepsy syndromes in Orexin receptor-2 and Orexin null mice: molecular genetic dissection of non-REM and REM sleep regulatory processes. *Neuron* **38:** 715–730.
40. Hondo, M. *et al.* 2010. Histamine-1 receptor is not required as a downstream effector of orexin-2 receptor in maintenance of basal sleep/wake states. *Acta Physiol.* **198:** 287–294.

41. Willie, J.T. *et al.* 2001. To eat or to sleep? Orexin in the regulation of feeding and wakefulness. *Annu. Rev. Neurosci.* **24:** 429–458.
42. American Academy of Sleep Medicine. 2005. *The International Classification of Sleep Disorders: Diagnostic and Coding Manual.* Author, Rochester.
43. Zeitzer, J.M., S. Nishino & E. Mignot. 2006. The neurobiology of hypocretins (orexins), narcolepsy and related therapeutic interventions. *Trends Pharmacol. Sci.* **27:** 368–374.
44. Mieda, M. *et al.* 2004. Orexin peptides prevent cataplexy and improve wakefulness in an orexin neuron-ablated model of narcolepsy in mice. *Proc. Natl. Acad. Sci. USA* **101:** 4649–4654.
45. Brisbare-Roch, C. *et al.* 2007. Promotion of sleep by targeting the orexin system in rats, dogs and humans. *Nat. Med.* **13:** 150–155.
46. Saper, C.B., T.C. Chou & T.E. Scammell. 2001. The sleep switch: hypothalamic control of sleep and wakefulness. *Trends Neurosci.* **24:** 726–731.
47. Hagan, J.J. *et al.* 1999. Orexin A activates locus coeruleus cell firing and increases arousal in the rat. *Proc. Natl. Acad. Sci. USA* **96:** 10911–10916.
48. Bourgin, P. *et al.* 2000. Hypocretin-1 modulates rapid eye movement sleep through activation of locus coeruleus neurons. *J. Neurosci.* **20:** 7760–7765.
49. Huang, Z.L. *et al.* 2001. Arousal effect of orexin A depends on activation of the histaminergic system. *Proc. Natl. Acad. Sci. USA* **98:** 9965–9970.
50. Methippara, M.M. *et al.* 2001. Effects of lateral preoptic area application of orexin-A on sleep-wakefulness. *Neuroreport* **11:** 3423–3426.
51. Xi, M.C., F.R. Morales & M.H. Chase. 2001. Effects on sleep and wakefulness of the injection of hypocretin-1 (orexin-A) into the laterodorsal tegmental nucleus of the cat. *Brain Res.* **901:** 259–264.
52. Eggermann, E. *et al.* 2001. Orexins/hypocretins excite basal forebrain cholinergic neurones. *Neuroscience* **108:** 177–181.
53. Horvath, T.L. *et al.* 1999. Hypocretin (orexin) activation and synaptic innervation of the locus coeruleus noradrenergic system. *J. Comp. Neurol.* **415:** 145–159.
54. Van Den Pol, A.N. *et al.* 2002. Hypocretin (orexin) enhances neuron activity and cell synchrony in developing mouse GFP-expressing locus coeruleus. *J. Physiol.* **541:** 169–185.
55. Brown, R.E. *et al.* 2001. Orexin A excites serotonergic neurons in the dorsal raphe nucleus of the rat. *Neuropharmacology* **40:** 457–459.
56. Liu, R.J., A.N. Van Den Pol & G.K. Aghajanian. 2002. Hypocretins (orexins) regulate serotonin neurons in the dorsal raphe nucleus by excitatory direct and inhibitory indirect actions. *J. Neurosci.* **22:** 9453–9464.
57. Eriksson, K.S. *et al.* 2001. Orexin/hypocretin excites the histaminergic neurons of the tuberomammillary nucleus. *J. Neurosci.* **21:** 9273–9279.
58. Yamanaka, A. *et al.* 2002. Orexins activate histaminergic neurons via the orexin 2 receptor. *Biochem. Biophys. Res. Commun.* **290:** 1237–1245.
59. Bayer, L. *et al.* 2001. Orexins (hypocretins) directly excite tuberomammillary neurons. *Eur. J. Neurosci.* **14:** 1571–1575.
60. Eggermann, E. *et al.* 2001. Orexins/hypocretins excite basal forebrain cholinergic neurones. *Neuroscience* **108:** 177–181.
61. Burlet, S., C.J. Tyler & C.S. Leonard. 2002. Direct and indirect excitation of laterodorsal tegmental neurons by hypocretin/orexin peptides: implication for wakefulness and Narcolepsy. *J. Neurosci.* **22:** 2862–2872.
62. Takakusaki, K. *et al.* 2005. Orexinergic projections to the midbrain mediate alternation of emotional behavioral states from locomotion to cataplexy. *J. Physiol.* **568**(Pt 3): 1003–1020.
63. Adamantidis, A.R. *et al.* 2007. Neural substrates of awakening probed with optogenetic control of hypocretin neurons. *Nature* **450:** 420–424.
64. Estabrooke, I.V. *et al.* 2001. Fos expression in orexin neurons varies with behavioral state. *J. Neurosci.* **21:** 1656–1662.
65. Yoshida, Y. *et al.* 2001. Fluctuation of extracellular hypocretin-1 (orexin A) levels in the rat in relation to the light-dark cycle and sleep-wake activities. *Eur. J. Neurosci.* **14:** 1075–1081.
66. Mileykovskiy, B.Y., L.I. Kiyashchenko & J.M. Siegel. 2005. Behavioral correlates of activity in identified hypocretin/orexin neurons. *Neuron* **46:** 787–798.
67. Takahashi, K., J.S. Lin & K. Sakai. 2008. Neuronal activity of orexin and non-orexin waking-active neurons during wake-sleep states in the mouse. *Neuroscience* **153:** 860–870.
68. Lee, M.G., O.K. Hassani & B.E. Jones. 2005. Discharge of identified orexin/hypocretin neurons across the sleep-waking cycle. *J. Neurosci.* **25:** 6716–6720.
69. Li, Y. *et al.* 2002. Hypocretin/Orexin excites hypocretin neurons via a local glutamate neuron-A potential mechanism for orchestrating the hypothalamic arousal system. *Neuron* **36:** 1169–1181.
70. Xie, X. *et al.* 2006. GABA(B) receptor-mediated modulation of hypocretin/orexin neurones in mouse hypothalamus. *J. Physiol.* **574:** 399–414.

71. Ohno, K., M. Hondo & T. Sakurai. 2008. Cholinergic regulation of orexin/hypocretin neurons through M(3) muscarinic receptor in mice. *J. Pharmacol. Sci.* **106:** 485–491.
72. Muraki, Y. *et al.* 2004. Serotonergic regulation of the orexin/hypocretin neurons through the 5-HT1A receptor. *J. Neurosci.* **24:** 7159–7166.
73. Yamanaka, A. *et al.* 2006. Orexin neurons are directly and indirectly regulated by catecholamines in a complex manner. *J. Neurophysiol.* **96:** 284–298.
74. Yamanaka, A. *et al.* 2003. Regulation of orexin neurons by the monoaminergic and cholinergic systems. *Biochem. Biophys. Res. Commun.* **303:** 120–129.
75. Matsuki, T. *et al.* 2009. Selective loss of GABA(B) receptors in orexin-producing neurons results in disrupted sleep/wakefulness architecture. *Proc. Natl. Acad. Sci. USA* **106:** 4459–4464.
76. Horvath, T.L. & X.B. Gao. 2005. Input organization and plasticity of hypocretin neurons: possible clues to obesity's association with insomnia. *Cell Metab.* **1:** 279–286.
77. Deadwyler, S.A. *et al.* 2007. Systemic and nasal delivery of orexin-A (Hypocretin-1) reduces the effects of sleep deprivation on cognitive performance in nonhuman primates. *J. Neurosci.* **27:** 14239–14247.
78. Lambe, E.K. & G.K. Aghajanian. 2003. Hypocretin (orexin) induces calcium transients in single spines postsynaptic to identified thalamocortical boutons in prefrontal slice. *Neuron* **40:** 139–150.
79. Lambe, E.K. *et al.* 2005. Hypocretin and nicotine excite the same thalamocortical synapses in prefrontal cortex: correlation with improved attention in rat. *J. Neurosci.* **25:** 5225–5229.
80. Sakurai, T. 2007. The neural circuit of orexin (hypocretin): maintaining sleep and wakefulness. *Nat. Rev. Neurosci.* **8:** 171–181.

Ann. N.Y. Acad. Sci. ISSN 0077-8923

ANNALS OF THE NEW YORK ACADEMY OF SCIENCES
Issue: *Phylogenetic Aspects of Neuropeptides*

Neuropeptide W: a key player in the homeostatic regulation of feeding and energy metabolism?

Fumiko Takenoya,[1,2] Haruaki Kageyama,[1] Kanako Shiba,[1] Yukari Date,[3] Masamitsu Nakazato,[4] and Seiji Shioda[1]

[1]Department of Anatomy, Showa University School of Medicine, Shinagawa-ku, Tokyo, Japan. [2]Department of Physical Education, Hoshi University School of Pharmacy and Pharmaceutical Science, Shinagawa-ku, Tokyo, Japan. [3]Frontier Science Research Center, University of Miyazaki, Kiyotake, Miyazaki, Japan. [4]Division of Neurology, Respirology, Endocrinology, and Metabolism, Department of Internal Medicine, Faculty of Medicine, University of Miyazaki, Kiyotake, Miyazaki, Japan

Address for correspondence: Seiji Shioda, Ph.D., Department of Anatomy Showa University School of Medicine 1-5-8 Hatanodai, Shinagawa-ku Tokyo 142-8555, Japan. shioda@med.showa-u.ac.jp

Neuropeptide W (NPW), recently isolated from porcine hypothalamus, has been identified as the endogenous ligand for both NPBWR1 (GPR7) and NPBWR2 (GPR8), which belong to the orphan G protein–coupled receptor family. NPW is thought to play an important role in the regulation of feeding and drinking behavior, and to be related to the stress response. NPW-containing neurons are localized in several regions of the brain, including the hypothalamus, hippocampus, limbic system, midbrain, and brain stem. Accumulated evidence suggests that hypothalamic neuropeptides, such as neuropeptide Y (NPY), orexin, melanin-concentrating hormone (MCH), and proopiomelanocortin (POMC), are involved in the regulation of feeding behavior and energy homeostasis via neuronal circuits in the hypothalamus. NPW also forms part of the feeding-regulating neuronal circuitry in conjunction with other feeding-regulating peptide-containing neurons within the hypothalamus. We summarize our current understanding of the distribution of NPW and of the neuronal interactions between NPW and the different feeding-regulating peptide-containing neurons. This review also discusses evidence for the dichotomous actions of NPW on energy balance and the potential mechanisms involved.

Keywords: neuropeptide W; hypothalamus; immunohistochemistry; rat

Introduction

G protein–coupled receptors (GPCRs) form the largest protein family of the transmembrane receptors. They detect molecules outside the cell and activate intracellular signal transduction pathways in response to these extracellular messengers.[1] It is estimated that the GPCR family, the largest family of mammalian proteins with approximately 1,000 members, may comprise more than 1% of the human genome and that these receptors are the molecular target for approximately 30% of currently marketed drugs.[2] Moreover, GPCR activation mediates a variety of intracellular responses leading to the regulation of numerous physiological functions, with many ligands used to achieve the physiological control of feeding behavior and energy balance.[3]

Obesity is characterized by a chronic imbalance between energy expenditure and energy intake.[3] Food intake depends upon various influences from the central nervous system (CNS) as well as from the body's energy stores, which secrete stimulating and inhibiting signals such as ghrelin and leptin, respectively.[1,3] Although the mechanisms underlying obesity are far from being fully understood, recently it became clear that obesity is, in part, centrally regulated and that several neuropeptide ligands of GPCRs play an important role in regulating feeding and energy metabolism.

Neuropeptide W (NPW) is known to act as the ligand for two orphan GPCRs, NPB/W receptor 1 (NPBWR1) and NPBWR2 (formerly known as GPR7 and GPR8, respectively).[4–7] NPW is named after the tryptophan residues (single-letter code W) appearing at both its N- and C-terminals in its two

doi: 10.1111/j.1749-6632.2010.05642.x

 © 2010 New York Academy of Sciences.

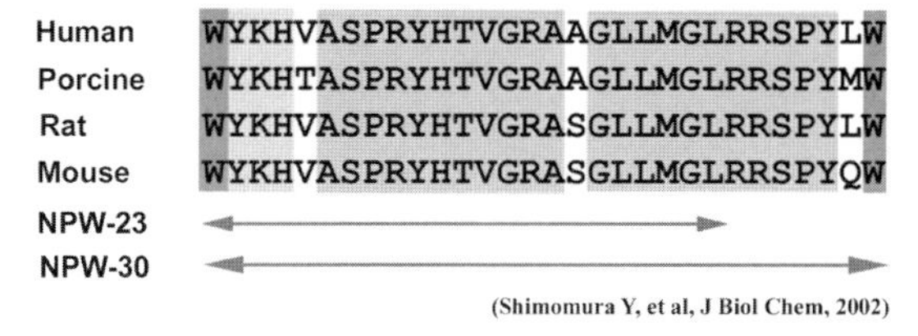

Figure 1. Sequence comparison of NPW-23 and NPW-30. Human, porcine, rat and mouse NPW-23 and NPW-30 sequences are aligned. The gray shaded area indicates a common amino acid sequence of both NPW-23 and NPW-30.

mature forms: NPW-30 (the 30 amino-acid form) and NPW-23 (consisting of 23 amino acids) (Fig. 1). These two peptides are derived from a common precursor peptide, pro-NPW, by proteolytic processing of two pairs of arginine residues at positions 24 and 25, and 31 and 32. Both peptides bind to NPBWR1 and NPBWR2 with similar affinity.[7] Intracerebroventricular (i.c.v.) injections of NPW stimulate food intake in the light phase[8] but suppress it in the dark phase while promoting energy expenditure such as thermogenesis. An anorectic effect was observed following both acute and long-term administration of NPW.[8] These findings suggest that NPW plays important roles in feeding regulation and energy homeostasis, and has both anorectic and orexigenic effects.[8]

Similar to NPW, neuropeptide B (NPB) is also an endogenous ligand for NPBWR1 and NPBWR2. NPB is a 29 amino acid residue with a brominated N-terminal tryptophan.[4] In humans, a shorter 23 amino acid form (NPB-23) is generated by proteolytic processing. NPW-23 and NPB-23 share a 61% amino acid sequence homology, whereas NPW-30 and NPB-29 share a 66% amino acid sequence homology.[4,6] I.c.v. injections of NPB enhance locomotor activity, reduce responsiveness to pain in the formalin test, and induce a biphasic feeding effect at low doses (early, mild orexigenic action, followed by anorexia), whereas higher doses cause anorexia at all time points.[9] NPB may as such be involved in feeding, weight regulation and pain response through either direct or indirect actions in the CNS.

Here we review the findings from a number of studies that advance our understanding of the regulation of feeding behavior and energy metabolism, and that have been shown to involve NPW neuronal circuits in the brain. These findings include localization and regulatory studies, which may help clarify the precise roles of NPW in the regulation of feeding behavior and energy homeostasis.

Receptors for NPW: NPBWR1 and NPBWR2

In 1995, O'Dowd *et al.*,[10] using oligonucleotides based on the opioid and the somatostatin receptors, amplified genomic DNA, and isolated fragments of two novel GPCR genes. These genes, named GPR7 (NPBWR1) and GPR8 (NPBWR2), share 70% identity with each other, and significant similarities with the transmembrane regions of the opioid and somatostatin receptors. NPBWR1 has been found in humans and rodents, whereas the NPBWR2 gene has not been detected in rat or mouse.[1] However, the gene of NPBWR2 has been discovered in other mammalian species such as the human and the rabbit.[1] In humans, experiments using northern blot analysis have shown that NPBWR1 mRNA is expressed primarily in the cerebellum and frontal cortex, whereas NPBWR2 is located mainly in the frontal cortex. Studies using RT-PCR, have also revealed the presence of NPBWR2 mRNA at high levels in the human CNS.[1] High levels of NPBWR1 mRNA were found in the hippocampus and amygdala,[6] whereas in the rat, strong NPBWR1 mRNA expression was detected in the hypothalamus and amygdala.[4] Similarly, *in situ* hybridization studies have revealed that NPBWR1 mRNA was present in the rat hypothalamus, including the arcuate nucleus (ARC), ventromedial nucleus (VMH), paraventricular nucleus (PVN), dorsomedial nucleus (DMH), suprachiasmatic nucleus (SCh), and supraoptic nucleus (SON),[11] which are well known as centers for the regulation of feeding and energy homeostasis.

Moreover, Singh *et al.*,[12] using $^{[125]}$I-NPW receptor autoradiography, detected radiolabeled NPW in the dorsal vagal complex (DVC), periaqueductal gray (PAG), superficial gray layer of the superior colliculus (SuG), medial amygdaloid nuclei (MeA), subfornical organ (SFO), bed nucleus of the stria terminalis (BNST), medial preoptic area (MPA) and SCh in rat brain.[12] In particular, expression of NPBW1 was found predominantly in the amygdala and hypothalamus,[5,10,12] On the basis of the distribution of NPBWR1 expression, it is possible that species differences between humans and rodents

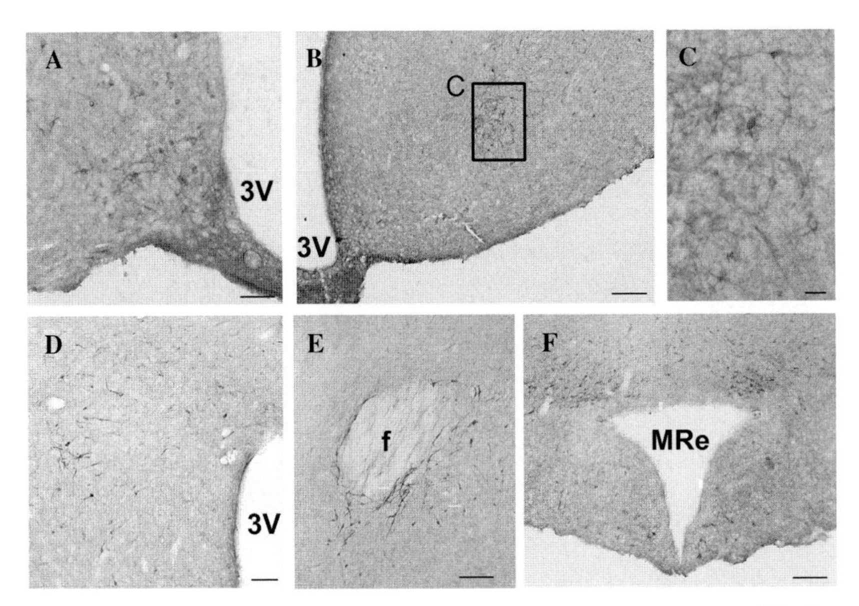

Figure 2. Distribution of NPW-producing neurons in the hypothalamus. NPW-positive cell bodies are located in the ARC, VMH, PVN, LH, and PMD. Scale bar = 20 μm.

exist in relation to the physiological functions of NPW. These findings suggest that NPBWR1 receives feeding-related signals from the feeding-regulating center of the hypothalamus, and that these receptors are involved the regulation of feeding behavior. To this extent, NPBW1 knockout mice exhibit hyperphagia and develop adult-onset obesity.[13]

Distribution and localization of NPW in the brain

Quantitative RT-PCR showed that NPW mRNA was highly expressed in the substantia nigra and spinal cord, and moderately expressed in the hippocampus, amygdala, hypothalamus, corpus callosum, cerebellum, and dorsal root ganglia in the human CNS.[6] More precisely, we demonstrated using RT-PCR analysis, that NPW mRNA is expressed in the rat hypothalamic PVN, lateral hypothalamus (LH), ARC, and VMH.[14] However, Kitamura *et al.*[15] have reported that the distributions of both NPW-like immunoreactivity (NPW-LI) and NPW mRNA-expressing cell bodies were exclusively found in the periaqueductal gray (PAG), ventral tegmental area (VTA), Edinger-Westphal nucleus (EW), and dorsal raphe nucleus (DR), but not in the ARC, PVN, DMN, LH, POA, SCh, or VMH of rats using *in situ* hybridization and immunohistochemical methods. Dun *et al.*[16] demonstrated that NPW-immunoreactive neuronal cell bodies were present within various hypothalamic nuclei such as the PVN, SON, accessory neurosecretory nuclei, dorsal, and lateral hypothalamic areas, perifornical nucleus, ARC, and anterior and posterior pituitary. Their findings are consistent with the Yanagisawa group's results[5] in that NPW mRNA was expressed in the VTA and DR of mouse, suggesting that the distribution of NPW is not species-dependent. Very recently, we reported that NPW-LI cell bodies are observed in many regions of the rat brain, including the islands of Calleja (ICj), ARC, PVN, LH, and VMH (Fig. 2), the lateral division of the bed nucleus of the stria terminalis (BNST) and amygdala.[14] In addition, NPW-LI cell bodies were detected in the midbrain, premammillary nucleus, dorsal part (PMD), PAG, EW, lateral parabrachial nucleus (LPB), amygdala, medial parabrachial nucleus (MPB), prepositus nucleus (Pr), and the raphe nucleus in rat brain.[14] Moreover, we were the first to identify the presence of NPW-LI cell bodies in the PVN, VMH and amygdala at the electron microscopic level.[14] These NPW-LI neurons were shown to make synaptic contact with unknown axon terminals in the PVN and ARC. These findings differ from those of other groups with regard to the distribution of NPW-LI cell bodies in the rat brain. Although the notion of NPW-expressing neurons in the hypothalamus remains controversial, one plausible explanation for this discrepancy is due to different antibody sources, rather than to species-specific differences. Differences in findings by *in situ* hybridization and immunohistochemistry may be due to the very low expression levels of NPW mRNA in the PVN, ARC, LH, and VMH, thus making it difficult to detect NPW mRNA without signal amplification such as the tyramide signal amplification (TSA) system.[17] Further studies using novel approaches and technologies are required to address the issue of the precise location of NPW-containing neurons in the hypothalamus and other brain regions.

NPW-LI nerve fibers have been identified in the SON, retrochiasmatic nucleus, dorsal and lateral hypothalamic areas, median eminence, amygdala, and posterior pituitary in rat brain.[16] NPW-LI nerve fibers are observed in the septum, BNST, dorsomedial and posterior hypothalamus, CeA (central amygdaloid nucleus), CA1 field of the hippocampus, ARC, VMH, interpeduncular nucleus, ventral tegmental area, inferior colliculus, lateral parabrachial nucleus, facial nucleus, and hypoglossal nucleus.[15] Moreover, we have reported that NPW-LI nerve fibers are present in the lateral septal nucleus (LS), the ventral part of the lateral division of the bed nucleus of the stria terminalis (BSTLV), amygdala, LH, LPB, and the lateral PAG (LPAG).[14] Given the abundant distribution of NPW-LI nerve fibers in the midbrain and limbic system such as CeA and BNST, these findings suggest that NPW may play a role in the regulation of fear and anxiety as well as in feeding behavior.[15]

Regulation of feeding and energy metabolism by NPW

Several neural nuclei that are associated with energy homeostasis are present within the hypothalamus. It is well known that the LH is a feeding center, the VMH is a satiety center, and that the ARC is an integrated center for feeding regulation. To date there have been no neuroanatomical studies carried out to examine the neural colocalization of NPW with other feeding-regulating peptides. Typical orexigenic peptides are neuropeptide Y (NPY), orexin, and melanin-concentrating hormone (MCH). Our results showed very close neuronal interactions between NPW-containing nerve fibers and orexin- or MCH-containing neuronal cell bodies and nerve fibers in the LH.[18] (Fig. 3) Moreover, Levine *et al.*[19] demonstrated, using dual immunostaining, that *c*-Fos was induced in orexin-containing neurons in the perifornical region of the LH after i.c.v. infusion of NPW. In particular, *c*-Fos immunoreactivity was increased in the perifornical region of the LH after NPW infusion. Orexin is recognized as a peptide that promotes food intake as well as playing a role in the control of sleep and wakefulness.[20] These findings suggest that the effect of NPW on food intake and wakefulness may be mediated via activation of the orexin pathway. Interestingly, we identified NPW-LI cell bodies in the VMH, which is known to be a center for satiety.[14] Leptin acts on the VMH where its receptors are abundantly expressed, and reduces food intake via the signal transducer and activator of transcription-3 (STAT3) and suppressors of cytokine signaling (SOCS-3).[21,22] Changes to the synaptic activity of projections from the VMH to proopiomelanocortin (POMC)-containing neurons in the ARC are known to inhibit feeding behavior[23,24] and to regulate glucose levels and energy homeostasis.[25] Very recently, Date *et al.*[26] demonstrated that leptin treatment promotes phosphorylation of STAT3, then induces SOCS-3 in the NPW-containing neurons isolated from the VMH. It is possible that leptin mediates satiety signals via NPW neurons in the VMH. It was demonstrated, using a double immunostaining method, that NPW-containing neurons project toward the POMC-containing neurons in the ARC. Furthermore, treatment of NPW was shown to inhibit the firing activation of NPY neurons and to decrease the frequency of spontaneous inhibitory postsynaptic current (sIPSCs) in POMC neurons in the ARC using loose-patch extracellular recording of these neurons (identified by promoter-driven green fluorescent protein expression).[26] These findings strongly suggest that NPW inhibits NPY and activates the POMC neuronal pathway, thus inducing the suppression of food intake.

Based on the distribution of NPW-LI neurons, we speculate there are other functions of NPW in the brainstem because NPW-LI fibers abundantly project into the NTS (nucleus tractus solitarii) and parabrachial nucleus. Neuroanatomical studies have shown descending projections from the LH to the NTS and parabrachial nucleus,[27–30] which contain the second- and third-order gustatory neurons, respectively. Therefore, it is possible that NPW is related to the taste response. Further

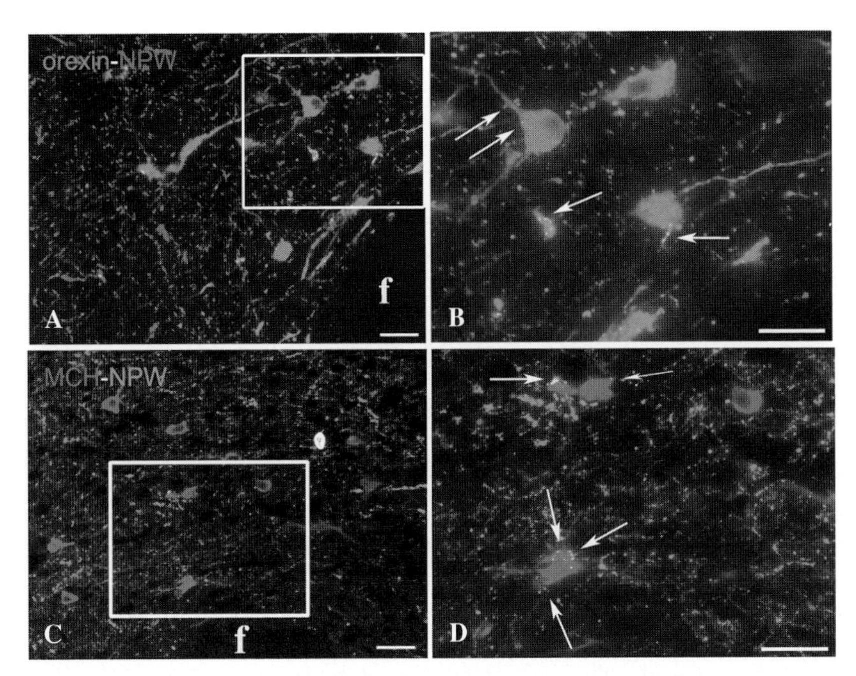

Figure 3. Dual immunofluorescence photomicrographs of the neural interaction between NPW and orexin- and MSH (melanin-concentrating hormone)-immunoreactive neurons in the LH. (A) NPW-immunoreactive fibers (*red*) are close apposition (*arrows*) with orexin-immunoreactive neurons (*green*) in the LH. A magnified image of area in A is shown in B. Scale bar = 20 μm. (B) NPW-immunoreactive fibers (*green*) are closely apposed (*arrows*) to MCH-like immunoreactive neurons (*red*) in the LH. (C) A magnified image of area in C is shown in D. Scale bar = 20 μm.

studies are required to analyze precisely the neuronal networks underpinning each of these physiological functions.

Mondal *et al.*[8] could not detect enhanced locomotor activity when NPW was administered to rats, but did observe increased O_2 consumption and increased CO_2 production, while also demonstrating an increase in body temperature compared to rats administered vehicle alone. This latter effect, independent of locomotor activity, may result from protein and triacylglycerol catabolism induced by corticosterone. Interestingly, another study by the same group found that the levels of NPW isolated from rat stomach antral cells decreased in fasted animals and increased in animals that had been re-fed.[31] In other experiments, the administration of NPW has been shown to increase arterial blood pressure, heart rate and plasma catecholamine levels in rats.[32] Baker *et al.*[33] also demonstrated that administration of NPW-23 elevates prolactin, corticosterone, and growth hormone levels. Moreover, Taylor *et al.*[34] reported that i.c.v. infusion of NPW activates the hypothalamic–pituitary–adrenal (HPA) axis, as demonstrated by plasma corticosterone levels in conscious rats. NPW expression was decreased in tissue after glucocorticoid treatment or hyperthyroidism. Conversely, hypothyroidism induced a marked increase in the expression of NPW in the rat stomach. Overall, these data indicate that NPW released from the stomach may regulate nutritional and hormonal status.[35] The presence of NPW mRNA has been confirmed in the pituitary gland and adrenal gland by RT-PCR analysis.[17] Furthermore, NPW-LI was found to be colocalized in cells that express dopamine beta-hydroxylase, but not phenylethanolamine-N-methyltransferase. NPW is expressed in noradrenalin-containing cells in the adrenal medulla and may play an important role in modifying the endocrine function of the adrenal gland.[17]

NPW significantly increases insulin release from isolated pancreatic islet cells in the presence of high concentrations of glucose. Fura-2 microfluorometry experiments with rat single β cells demonstrated an NPW concentration-dependent increase in cytosolic Ca^{2+} concentration ($[Ca^{2+}]_i$)

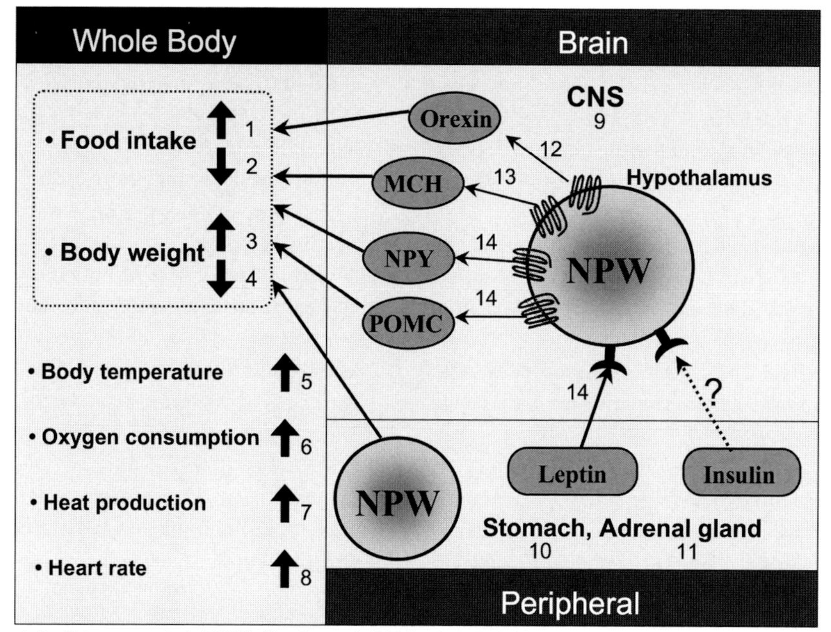

Figure 4. Summary of NPW actions on feeding and energy metabolism in the brain and peripheral organs.

at high glucose levels, which was abolished under external Ca^{2+}-free conditions and in the presence of an L-type Ca^{2+} channel blocker.[36] Double immunohistochemical analysis showed that NPW-immunoreactivity was found in pancreatic islets and colocalized with insulin-containing β cells, but not glucagon-containing α cells and somatostatin-containing δ cells.[36] These results suggest that NPW may serve as a local modulator of glucose-induced insulin release in rat islets.

Conclusion

In this review, we highlighted NPW distribution and its role in feeding regulation (Fig. 4). Understanding the role of NPW in the CNS will prove helpful for the discovery of therapeutic drugs and for designing therapies that overcome obesity and other life-style-related disorders.

Acknowledgments

This study was partially supported by grants from the Ministry of Education, Culture, Sports, Science and Technology of Japan, Grant-in-Aid for Scientific Research (C) (to FT (20500634) and HK (21590222)). The study was also supported in part by a grant from the Ministry of Education, Culture, Sports, Science and Technology of Japan, Grant-in-Aid for Exploratory Research 21659059 (to SS) and by a High-Technology Research Center project grant from the Ministry of Education, Culture, Sports, Science and Technology of Japan (to SS).

Conflicts of interest

The authors declare no conflicts of interest.

References

1. Lee, D.K. *et al.* 2003. Continued discovery of ligands for G protein-coupled receptors. *Life Sci.* **74:** 293–297.
2. Maguire, J.J. *et al.* 2005. Regulation of vascular reactivity by established and emerging GPCRs. *Trends Pharmacol. Sci.* **26:** 448–454.
3. Xu, Y.L. *et al.* 2004. Orphan G protein-coupled receptors and obesity. *Eur. J. Pharmacol.* **500:** 243–253.
4. Fujii, R. *et al.* 2002. Identification of a neuropeptide modified with bromine as an endogenous ligand for GPR7. *J. Biol. Chem.* **277:** 34010–34016.

5. Tanaka, H. *et al.* 2003. Characterization of a family of endogenous neuropeptide ligands for the G protein-coupled receptors GPR7 and GPR8. *Proc. Natl. Acad. Sci. USA* **100:** 6251–6256.
6. Brezillon, S. *et al.* 2003. Identification of natural ligands for the orphan G protein-coupled receptors GPR7 and GPR8. *J. Biol. Chem.* **278:** 776–783.
7. Shimomura, Y. *et al.* 2002. Identification of neuropeptide W as the endogenous ligand for orphan G-protein-coupled receptors GPR7 and GPR8. *J. Biol. Chem.* **277:** 35826–35832.
8. Mondal, M.S. *et al.* 2003. A role for neuropeptide W in the regulation of feeding behavior. *Endocrinology* **144:** 4729–4733.
9. Yamamoto, T. *et al.* 2005. Anti-hyperalgesic effects of intrathecally administered neuropeptide W-23, and neuropeptide B, in tests of inflammatory pain in rats. *Brain Res.* **1045:** 97–106.
10. O'Dowd, B.F. *et al.* 1995. The cloning and chromosomal mapping of two novel human opioid-somatostatin-like receptor genes, GPR7 and GPR8, expressed in discrete areas of the brain. *Genomics* **28:** 84–91.
11. Lee, D.K. *et al.* 1999. Two related G protein-coupled receptors: the distribution of GPR7 in rat brain and the absence of GPR8 in rodents. *Brain Res. Mol. Brain Res.* **71:** 96–103.
12. Singh, G. *et al.* 2004. Identification and cellular localisation of NPW1 (GPR7) receptors for the novel neuropeptide W-23 by [125I]-NPW radioligand binding and immunocytochemistry. *Brain Res.* **1017:** 222–226.
13. Ishii, M. *et al.* 2003. Targeted disruption of GPR7, the endogenous receptor for neuropeptides B and W, leads to metabolic defects and adult-onset obesity. *Proc. Natl. Acad. Sci. USA* **100:** 10540–10545.
14. Takenoya, F. *et al.* 2010. Distribution of neuropeptide W in the rat brain. *Neuropeptides* **44:** 99–106.
15. Kitamura, Y. *et al.* 2006. Distribution of neuropeptide W immunoreactivity and mRNA in adult rat brain. *Brain Res.* **1093:** 123–134.
16. Dun, S.L. *et al.* 2003. Neuropeptide W-immunoreactivity in the hypothalamus and pituitary of the rat. *Neurosci. Lett.* **349:** 71–74.
17. Seki, M. *et al.* 2008. Neuropeptide W is expressed in the noradrenalin-containing cells in the rat adrenal medulla. *Regul. Pept.* **145:** 147–152.
18. Takenoya, F. *et al.* 2008. Neuronal interactions between neuropeptide W- and orexin- or melanin-concentrating hormone-containing neurons in the rat hypothalamus. *Regul. Pept.* **145:** 159–164.
19. Levine, A.S. *et al.* 2005. Injection of neuropeptide W into paraventricular nucleus of hypothalamus increases food intake. *Am. J. Physiol. Regul. Integr. Comp. Physiol.* **288:** R1727–1732.
20. Sakurai, T. *et al.* 1998. Orexins and orexin receptors: a family of hypothalamic neuropeptides and G protein-coupled receptors that regulate feeding behavior. *Cell.* **92:** 573–585.
21. Bjorbaek, C. *et al.* 1998. Identification of SOCS-3 as a potential mediator of central leptin resistance. *Mol. Cell.* **1:** 619–625.
22. Rosenblum, C.I. *et al.* 1996. Functional STAT 1 and 3 signaling by the leptin receptor (OB-R); reduced expression of the rat fatty leptin receptor in transfected cells. *Endocrinology.* **137:** 5178–5181.
23. Koyama, K. *et al.* 1998. Resistance to adenovirally induced hyperleptinemia in rats. Comparison of ventromedial hypothalamic lesions and mutated leptin receptors. *J. Clin. Invest.* **102:** 728–733.
24. Elmquist, J.K. *et al.* 1997. Leptin activates neurons in ventrobasal hypothalamus and brainstem. *Endocrinology* **138:** 839–842.
25. Schwartz, M.W. *et al.* 2000. Central nervous system control of food intake. *Nature* **404:** 661–671.
26. Date, Y. *et al.* 2010. Neuropeptide W: an anorectic peptide regulated by leptin and metabolic state. *Endocrinology* **151:** 2200–2210.
27. Hosoya, Y. *et al.* 1981. Brainstem projections from the lateral hypothalamic area in the rat, as studied with autoradiography. *Neurosci. Lett.* **24:** 111–116.
28. Roberts, W.W. 1980. [14C]Deoxyglucose mapping of first-order projections activated by stimulation of lateral hypothalamic sites eliciting gnawing, eating, and drinking in rats. *J. Comp. Neurol.* **194:** 617–638.
29. Saper, C.B. *et al.* 1976. Direct hypothalamo-autonomic connections. *Brain Res.* **117:** 305–312.
30. Matsuo, R. *et al.* 1984. Lateral hypothalamic modulation of oral sensory afferent activity in nucleus tractus solitarius neurons of rats. *J. Neurosci.* **4:** 1201–1207.
31. Mondal, M.S. *et al.* 2006. Neuropeptide W is present in antral G cells of rat, mouse, and human stomach. *J. Endocrinol.* **188:** 49–57.
32. Yu, N. *et al.* 2007. Cardiovascular actions of central neuropeptide W in conscious rats. *Regul. Pept.* **138:** 82–86.
33. Baker, J.R. *et al.* 2003. Neuropeptide W acts in brain to control prolactin, corticosterone, and growth hormone release. *Endocrinology* **144:** 2816–2821.

34. Taylor, M.M. *et al.* 2005. Actions of neuropeptide W in paraventricular hypothalamus: implications for the control of stress hormone secretion. *Am. J. Physiol. Regul. Integr. Comp. Physiol.* **288:** R270–R275.
35. Caminos, J.E. *et al.* 2008. Expression of neuropeptide W in rat stomach mucosa: regulation by nutritional status, glucocorticoids and thyroid hormones. *Regul. Pept.* **146:** 106–111.
36. Dezaki, K. *et al.* 2008. Neuropeptide W in the rat pancreas: potentiation of glucose-induced insulin release and Ca^{2+} influx through L-type Ca^{2+} channels in beta-cells and localization in islets. *Regul. Pept.* **145:** 153–158.